Novel
SCIENCE

Adelene Buckland

Novel
SCIENCE

FICTION *and the* INVENTION *of*
NINETEENTH-CENTURY GEOLOGY

The University of Chicago Press
Chicago & London

Adelene Buckland is a lecturer in nineteenth-century literature at King's College London. She is coeditor of *A Return to the Common Reader: Essays in Honour of Richard Altick*.

The University of Chicago Press, Chicago 60637
The University of Chicago Press, Ltd., London
© 2013 by The University of Chicago
All rights reserved. Published 2013.
Printed in the United States of America

22 21 20 19 18 17 16 15 14 13   1 2 3 4 5

ISBN-13: 978-0-226-07968-4 (cloth)
ISBN-13: 978-0-226-92363-5 (e-book)
ISBN-10: 0-226-07968-6 (cloth)
ISBN-10: 0-226-92363-0 (e-book)

Library of Congress Cataloging-in-Publication Data

Buckland, Adelene, author.
  Novel science : fiction and the invention of nineteenth-century geology / Adelene Buckland.
      pages cm
  Includes bibliographical references and index.
  ISBN 978-0-226-07968-4 (cloth : alkaline paper)
  ISBN 978-0-226-92363-5 (e-book)
  1. Geology in literature. 2. Geology—Great Britain—History—19th century. 3. Literature and science—England—History—19th century. 4. English literature—19th century—History and criticism.   I. Title.
PR468.G38B83 2013
823'.80936—dc23   2012044802

♾ This paper meets the requirements of ANSI / NISO Z39.48-1992 (Permanence of Paper).

for **ISOBEL EDITH**

CONTENTS

INTRODUCTION. Formations * 1

PART ONE
Stories in Science
{
ONE. Fictions of a Former World * 31
TWO. The Story Undone * 56
THREE. Lyell's Mock Epic * 95
FOUR. Maps and Legends * 131
}

PART TWO
Science in Stories
{
FIVE. Kingsley's Cataclysmic Method * 179
SIX. Eliot's Whispering Stones * 221
SEVEN. Dickens and the Geological City * 247
}

CONCLUSION. Losing the Plot * 274

{
ACKNOWLEDGMENTS * 277
APPENDIX. "Lines on Staffa," by Charles Lyell * 281
NOTES * 283
BIBLIOGRAPHY * 327
INDEX * 365
}

Formations { **INTRODUCTION**

In nineteenth-century Britain the study of the ancient earth gripped the public imagination. Digging and cutting into the ground to create canals, mines, railways, and sewers, men and women discovered and named those creatures we now call "dinosaurs," fossilized lizards of often gargantuan proportions whose broken skeletons were quickly reassembled for museums and shows, engraved on the pages of books and periodicals, and enthused about in prose and verse.[1] The bones of these terrible monsters were found in soft layers of rock that a new band of men, the "geologists," would argue had been set down on the beds of ancient seas and lakes and raised into modern cliffs, mountains, and valleys. Deep beneath these soft, fossil-rich layers were other rocks stretching far into the earth's crust, in which the remains of tinier and simpler creatures like the trilobites yielded evidence of a dim and distant primeval world (fig. I.1). Deeper still, though sometimes also protruding violently out across the surface of the land, were huge masses of hard rock, crystallized granites and basalts in which fossils were so rare it was uncertain whether life had existed in them at all. Though studies of the workings and structure of the earth were nothing new, it was in the nineteenth century that the strata of rocks and fossils became widely understood as chapters in a history of the earth spanning unimaginable millions of years (see fig. I.2), and it was in the nineteenth century that the study of the earth became central to the economic and cultural life of the nation (fig. I.3).

Novel Science contributes to our growing understanding of this "golden age" of geology, and it focuses on the specifically literary dimension of the new science. It is well known that during this period geologists invented and

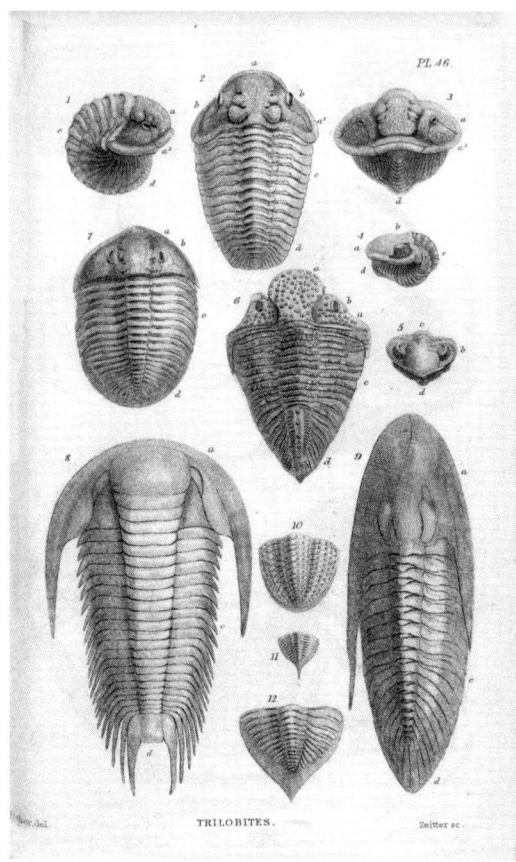

Fig. I.1. "Trilobites," plate 46. In William Buckland, *On Geology and Mineralogy* (1836), 396. Reproduced by permission of the syndics of Cambridge University Library.

elaborated new visual and material forms by which to give shape and meaning to the millions of specimens they were collecting from across the globe. Two- and three-dimensional maps, geological columns, sketches, and a variety of new kinds of museum, exhibition, and private collection made it possible for geologists to comprehend in new ways the structure of the earth over millions of years of its history.[2] But it is less well established that, at the same time, geologists elaborated new *literary* forms with which they would explain, interpret, order, describe, argue about, and bring into existence a science whose claims and insights were both complex and new. This book sets out to prove that the literary activities of geologists did not merely represent ancient worlds already discovered or understood, but were integral to the methods and practices of the new geology, just as important in generating new ways of understand-

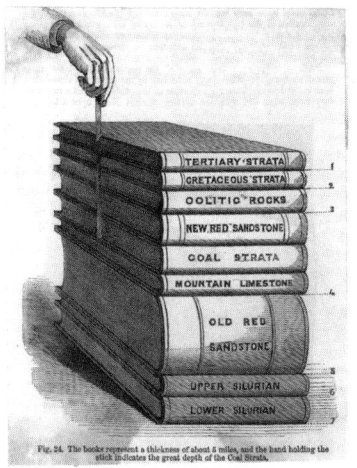
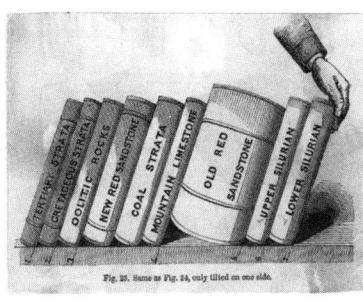

Fig. I.2. Stratification diagrams. Each book represents a volume of the earth's history laid out in stratigraphic form. Reproduced by permission of the Mary Evans Picture Library.

Fig. I.3. Interior of the Museum of Practical Geology. The museum was designed to offer public instruction in the economic and industrial utility of geology, and was arranged almost like a stratigraphical column, the earliest rocks and fossils exhibited on the lowest balconies, ascending to the newest in the higher eschelons. Even the portal was built of polished English stones. See Yanni, *Nature's Museums* (1999), 51–61; Forgan, "Bricks and Bones" (1999), 194–200. Reproduced by permission of the Mary Evans Picture Library.

ing the geological past as mapmaking, collecting, fieldwork, observation, or museum-building.³ This was not uncontroversial: indeed, *Novel Science* will argue that many kinds of writing were often viewed with suspicion by geologists, who feared polished prose and seductive stories as anathema to good scientific work. Nonetheless, always embroiled in literary debate, geology was *written* into existence in the nineteenth century as much as it was found, discovered, collected, mapped, or modeled. *Novel Science* tells the story of how that happened.

The Story of the Rocks

The word "geology" entered into usage in the late eighteenth century. In its first coinage, it was used to refer to "theories of the earth," or generalized "systems" explaining the entire workings of the earth from its earliest origins to its putative end. Such all-encompassing systems, almost cosmological in scope, offered explanations of all the features of the earth according to a simple, comprehensive historical pattern.⁴ By the early nineteenth century, however, the term's meaning had shifted and was premised on the rejection of such theories of the earth. As defined by the famous French anatomist Georges Cuvier, "geology" was a descriptive, empirical science based on the firsthand study of specific local regions and on the gathering of facts and observations rather than on the cosmological speculations of high-level theory.⁵ It was in this sense that the word was adopted by the Geological Society of London, founded in 1807. At this Society the new science of geology would avoid theorizing on the cosmogony and cosmology of the earth and would instead undertake descriptive and empirical activities through a commitment to outdoor fieldwork.⁶ Geology absorbed a wide range of sciences of the earth. Like mineralogy, it involved the collection, identification, and classification of specimens of rock. Cabinets and museums would be filled with its treasures. Like physical geography, the new science would provide accurate and painstaking descriptions of the natural world, both in texts and images, and one of its principal objectives would be to create maps of the physical world. Also like physical geographers, the new "geologists" placed great emphasis on fieldwork, on traversing vast swaths of terrain in order to study rocks *in situ* (figs. I.4 and I.5). Furthermore, geology was partly derived from "geognosy," a science identified with the mining school at Freiberg. Geognosy was concerned less with mapping the surface of the landscape than with producing "columns" and "sections"—vertical slices of rock as it descended into the depths of the earth or was exposed

Fig. I.4. "Geologists hunt crystals at Mont Blanc in the Alps" (1870), emphasizing the geologists' physical skill and daring. Reproduced by permission of the Mary Evans Picture Library.

at a cliff or quarry face. Geology would be unique in combining the maps of physical geographers with the columns and sections of geognosy to produce a three-dimensional vision of the land and its structure. Finally, geology would borrow from earth physics, the attempt to explain the events that had produced these physical features via causal mechanisms such as fire, volcanic activity, earthquakes, and the manifold actions of water on the land. Earth physics was much more limited in scope than those cosmological theories of the earth, for it attempted to explain individual and local features rather than, through a single law, the overall character of the earth from its beginnings to the ends of time.[7]

The problems inherent in the writing of geology were clear from the outset. In *Geological Inquiries*, their founding manifesto, the members of the Geological Society of London set out their research program. Maps and columns would be of paramount importance, and geologists were to limit their

Fig. I.5. "Quarry strata exposed." Here the strata are clearly and idealistically exposed in neat layers. Reproduced by permission of the Mary Evans Picture Library.

speculations on the beginning or ending of the world, on patterns of geological change, and on wholesale causal mechanisms by which such change could be thought to have occurred through time. "Geology," they wrote, "relates to the knowledge of the system of our earth, of the arrangements of its solid, fluid, and aeriform parts, their mutual agencies, and the laws of their changes," but crucially this overarching "system" and its "laws" could only be understood once a vast mass of facts and observations had been "collected from a number of particular and minute instances."[8] That time was not yet nigh. For the moment, it was sufficient that "the Miner, the Quarrier, the Surveyor, the Engineer, the Collier, the Iron Master, and even the Traveller" would collect facts and observations through direct encounters with nature, and send them to the "Philosophers" of the Geological Society. These men, with sufficient time, money, leisure, and education on their hands, would compile those facts and observations and begin to explain their significance.[9] The "inquiries" of the title were, therefore, a series of technical questions, laid out in the pamphlet, on the structure and locations of particular geological features. Of valleys, for instance, geological observers were to ask such questions as, "Is the bottom or floor even or rugged? —nearly level or much inclined? If inclined, whether regularly or interruptedly, and in what direction?"[10] Furnished with detailed descriptions of this kind, the gentlemen of the Geological Society would compile "mineralogical maps," take charge of the development of a sound geological nomenclature, make available this knowledge for "public improvement and utility," and ensure that any "theoretical opinions" they advanced were strictly subordinate to and "compared with the appearances of nature."[11]

Historians have made it clear that we should not take these statements at face value.[12] In the first place, it is clear that by the 1830s the strongly empirical rhetoric of the *Inquiries* was relaxed, and that the Society had broadened its research program as "a younger generation of geologists argued for a more liberal methodology, and put it into practice."[13] By this time the emphasis on outdoor fieldwork, travel, mapmaking, and direct observation of "the appearances of nature" had evolved into a much more theoretical concern to produce a universal "stratigraphic column," an abstract visual illustration of all the earth's rocks in the order in which they were thought to have been originally deposed.[14] Still the focus was on the structure rather than on the story of earth history. As one historian has put it, "Where Enlightenment cosmogonists had argued to a stalemate about the entire world, Victorian geologists fruitfully focused attention on some crucial strata in a single part of Britain."[15]

This was no easy task. As many nineteenth-century geological textbooks made clear, a variety of local disturbances may have knocked individual exposures of rocks out of their original sequence. As figure I.6 shows, the action of pressure on soft, semimolten layers of rock as they formed beneath the earth's waters could "fold" the rocks into waves, producing synclinal depressions (or troughs) or anticlinal formations sloping downward from a crest. If only a single exposure of these rocks could be seen, it would be almost impossible to tell in which order the rocks had been deposited. Moving from east to west along a trough, the traveler might first encounter it from the outside, meeting the lowest and oldest layer of rocks first and progressing through a series to the newest rocks, positioned on the inside. If he continued his journey in the same direction, and if he was lucky enough that the other half of the trough was exposed above the surface of the land, the traveler would then encounter the newest rocks as he journeyed past the inside curve and moved toward the oldest layer on the outside. If, however, only one side of the trough was exposed or accessible, it would be very difficult for him to decide whether the rocks went from oldest to newest as he moved from east to west, or from newest to oldest.

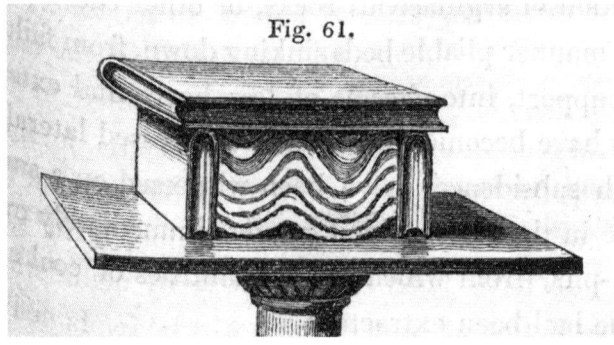

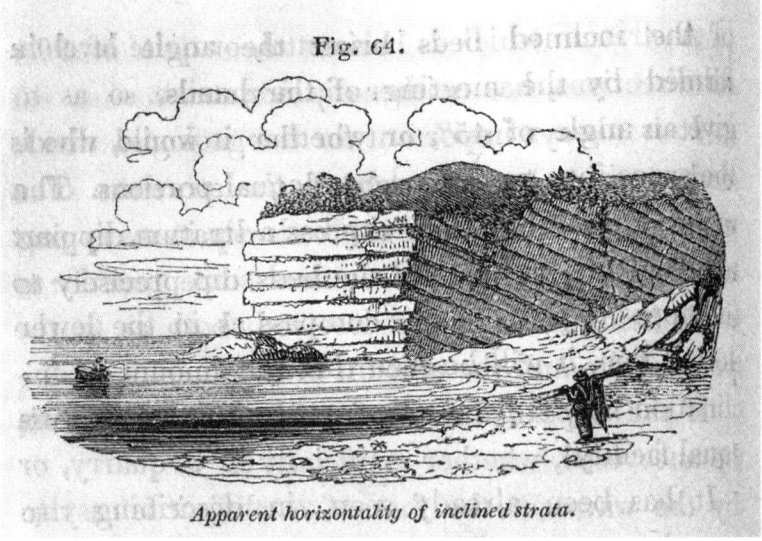

Fig. I.6. From Charles Lyell, *Elements of Geology* (1838), 103. Reproduced by permission of the syndics of Cambridge University Library.

Fig. I.7. From Charles Lyell, *Elements of Geology* (1838), 108. Reproduced by permission of the syndics of Cambridge University Library.

In addition, strata in individual locations might also be "broken, and their planes interrupted," creating a displacement sometimes of "one hundred or even two hundred yards" between rocks that had once been continuous.[16] Such faults created rift valleys or inland cliffs, and often inverted the strata. Plate 1 reveals a series of faults in different locations across Britain. In the topmost image, the rocks colored blue once ran continuously but have been forced apart, leaving the inland cliff of Moughton Scar looming over Ribblesdale in Yorkshire. In the image directly below it, the vertical lines indicate faults, and it is clearly possible to see that the beds have slipped out of alignment. Along with folds and faults the particular orientation of exposed rocks might disguise their true features from the geological observer. As figure I.7 shows, the "dip" of the strata (or their inclination from the horizontal) could be concealed if the exposed layers were viewed from the line of their "strike," orthogonal to the direction of the dip. Here the fisherman on the shoreline can see the dip of the rocks because he views them from the side, while his friend in the boat is misled by their apparent horizontality as they are exposed in the cliff face.[17] Metamorphosis, too, could transform previously stratified rocks by heat or pressure, leading the unskilled geologist to assign them to the wrong positions in a sequence. In the face of all this perplexing detail the evidence of fossils became of increasing importance in determining rock sequence, though this evidence could also often be precarious or misleading.[18] In order to create general views of a geographical area, the evidence of a wide range of quarries, cliff faces, faults, mine shafts, exposed rock, topographical information, and fossil evidence would need to be correlated across as much of the area as possible. Even then, it was more than likely that the geologist would need to correlate the evidence from that region with that of other regions—across Europe or the world—that exhibited similar sequences of rocks and fossils—in order to ascertain their true identity. Since nobody could hope to cover every inch of ground in all these areas, and would never be able to see everything that lay beneath the surface of the land, stratigraphy was an inherently tricky and theoretical business.

Members of the Geological Society have been described as "pioneers of this kind of stratigraphical fieldwork," ushering in a heroic age for geology in which almost all the major subdivisions of the stratigraphic column were thrashed out.[19] While the empirical focus of *Geological Inquiries* was not abandoned, it was certainly relaxed in support of this more theoretical research, and this went hand in hand with a deepening of the Society's gentlemanly deportment. Leisured, erudite, and well-acquainted, the leading geologists of the Society are often said to have had no tangible interests in the economic or practical dimensions of their science.[20] "A cohort of young men," borne on the "current of idealistic earnestness" and with "a commitment to enthusiasm and activity," they have been described as having given themselves up to a "passionate Romantic engagement" with nature, inspiring them to make "hair-raising traverses of the Alps" and undertake "headlong horse-riding across Britain."[21]

Disdaining the merely practical applications of geology in favor of a philosophical commitment to truth for its own sake, geologists of this period have been said to have been thrilled with "*Wanderlust*," dazzled by "Nature in her wild and unspoilt forms," and drawn primarily to the study of the oldest, least fossiliferous rocks, such as granite and basalt, which tended to exist in the earth's most dramatic and awe-inspiring locations.[22] Nonetheless, this view has recently become open to debate. Though the dominance of a gentlemanly elite is indisputable, it is now clear that despite the geologists' disdainful rhetoric for practical geology, and despite the triumphal processions of gentlemanly individuals in early histories of the science, mining academies and geological surveys were of critical significance to geology even in England.[23] Furthermore, the stratigraphic activities of this band of "heroic" gentlemen were often motivated by such economic considerations as the search for a base limit for coal in the strata.[24] And it is possible to overstate the case that the gentlemen dominated the science to the exclusion of all others. In fact, the practical observation of the natural world by a broad range of practitioners, from miners and quarrymen to middle-class ladies on seaside vacations, continued to be of significance both in generating specimens and insights for leading geologists and for contributing to the science's popularity and prestige.[25] For, as *Geological Inquiries* had made clear, if the disaggregated "appearances of nature" were to be the building blocks of the new science, it was imperative that as many different observations of the natural world were gathered together as possible. The authors of the pamphlet had called the British public of all classes to examine that world for themselves, and that appeal never left the science. During the Napoleonic Wars, travel abroad was difficult, and geology gave added excitement to travel within the British Isles.[26] Geological fieldwork fitted well, too, with other gentlemanly outdoor pursuits, such as hunting and field sports, emulated by many middle-class Victorians.[27] Later, the Victorian cults of masculinity gave the middle classes additional incentive to risk their lives by dangling off precipices taking measurements or chipping out fossils the country over.[28] And as figure I.8 suggests, miners, quarrymen, navvies, and laborers all worked with the rocks and possessed detailed local knowledge, and they could try to turn a profit by selling that knowledge, or selling fossils to middlemen and geologists. One historian has claimed that "geology's most popular aspect," indeed, was "the collecting of fossils and minerals."[29] Through the practices of observation, collecting, and fieldwork, "nineteenth-century geology, often perceived as the sport of gentlemen," was, in fact, "reliant on all classes."[30]

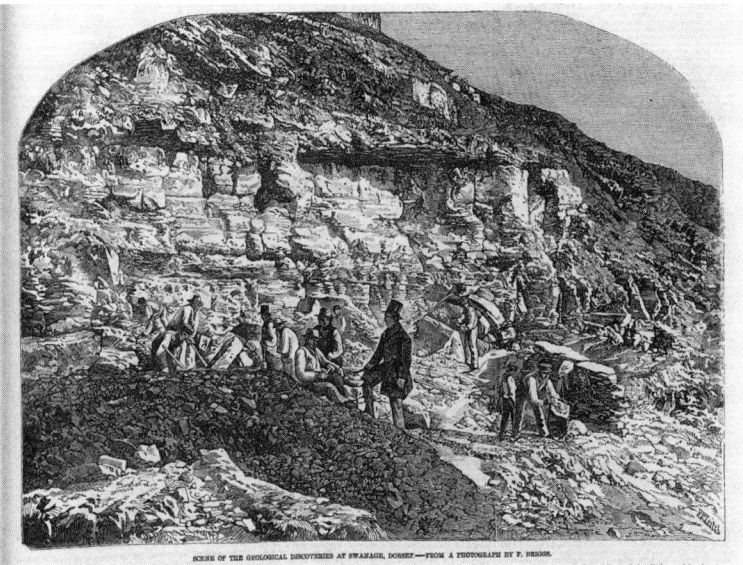

Fig. I.8. "Geological discoveries at Swanage." *Illustrated London News*, issue 895 (December 26, 1857), 637. Note the geologist in the foreground, presiding over the diggers. Reproduced by permission of the syndics of Cambridge University Library.

All this meant that, despite its gentlemanly comportment, geology garnered an increasingly wide cultural cachet through the period. As was emphasized by David Page, the author of a series of successful geological textbooks published at midcentury, more and more people could afford the equipment requisite for venturing out as a field geologist. All that was needed, he wrote, was

> a hammer to detach specimens, and a bag or basket to carry them in—a pocket magnifying-lens to detect minuter structures—a compass and clinometer to determine the strike and dip of strata—a sketch-book to note unusual phenomena—an observing eye and a pair of willing limbs.[31]

By the time Page was writing, in the 1850s, almost every county in England could boast its own natural history society, many of which centered on the acquisition of fossils, rocks, minerals, and other natural-historical specimens, organized railway field excursions, and housed museums and collections of varying degrees of importance. These were "emblems" of the "cultural erudition" of the "emerging bourgeoisie," to be considered "alongside art galleries, libraries and gardens" as "manifestations of civic pride, evidence of the sophistication of one town in contrast to its neighbours, the capital and the wider empire."[32] The collections created for such local museums were, moreover, often of central importance to the gentlemen who visited them from the auspicious Geological Society of London.[33]

By midcentury newspapers and periodicals had occasion to poke fun not only at the gentlemanly "geologists" of the Geological Society, as they donned ragged clothing to clamber up cliff faces and descend deep into mine shafts, but also at a much broader group of enthusiasts and practitioners who cluttered up their wives' linen drawers with specimens of rocks and fossils or forgot themselves for days on end in the description of a broken bit of strata. An indicative example of this lampooning of the practical geologist comes in Charles Dickens's novel *Bleak House*, first published in 1852–1853:

> "People objected to Professor Dingo when we were staying in the north of Devon after our marriage," said Mrs. Badger, "that he disfigured some of the houses and other buildings by chipping off fragments of those edifices with his little geological hammer. But the professor replied that he knew of no building save the Temple of Science. The principle is the same, I think?"
>
> "Precisely the same," said Mr. Badger. "Finely expressed! The professor made the same remark, Miss Summerson, in his last illness, when (his mind wandering) he insisted on keeping his little hammer under the pillow and chipping at the countenances of the attendants. The ruling passion!"[34]

In *Bleak House* the example is a telling one. Mrs. Badger lectures the heroine, Esther Summerson, on the importance of young men attaining a steady vocation. The warning is crucial, for in failing to find his vocation and stick to it the young protagonist, Richard Carstone, is sucked into the interminable machinations of the Chancery suit that the novel viciously satirizes. Like that legal case, Richard simply wastes away until he dies the tragic death that follows a life spent in futile pursuits. Professor Dingo has not met that fate, for he has been steadfast to the last to his geological purpose, chipping away with his geological hammer at the faces and mantelpieces of his servants and friends, unable to let go of his passion even when his mind has gone. The practice of geology was obsessive to the point of imbecility, the story implies, but it was also comfortingly concrete.

Dingo was not the only one, moreover. Nineteenth-century periodicals are filled with stories like these about the geological "ruling passion!" In a short series in 1853 *Punch* described the "hardly thoughtful" Fred who abandoned his wife while they were on honeymoon "only to pick up pebbles," she notes in disgruntled fashion, "and knock and chip at the rocks with that hammer which he always carries about with him, and which . . . he seems now and then to think more of than *his own wife*."[35] Dickens's periodical *Household Words* (1850–1859) published the story of a "stone-mad" husband who converted his wife's linen drawers into a geological cabinet after stumbling across his son's copy of Page's textbook, and it published another story about a man turning his back on marriage altogether in devilish pursuit of the rich geological deposits buried beneath the soil.[36] For these men too, geology was an all-consuming obsession. Nonetheless, even as it poked fun at geological enthusiasts, *Household Words* also included geological instructions for its readers, offering them the timetables of steamers leaving London for the Isle of Sheppey, for instance, along with the names of the best fossil collectors to visit on that island.[37] Or again, in 1880, the otherwise uxorious geologist-hero of Isabella Banks's novel *Wooers and Winners* is so frequently to be found out on geological fieldwork, or buried in his study classifying rocks and fossils, that he is entirely unobservant of the needs of his children and wards. "Absent, indeed!" the narrator notes drily. "But why should a man, pondering the occult secrets of creation, be expected to note the actions or development of young people, even though one should be his own? The fossilised past had a more intelligent voice for him than had the human present."[38] Doing geology, we might be tempted to conclude, was at least as popular as reading or writing about it, and literary parodies of geological practices abounded. But *Novel Science* is less concerned with representations of geology or geologists in literary texts than it is with the fact that one of the principal practices of the geologist was, itself, literary. Doing geology, I want to argue, was not anathema to reading or writing about past worlds from one's armchair. Doing geology meant writing it too.

Form and Practice

Focus on the practice of geology, rather than on its representations in written texts of the period, may seem counterintuitive in a book about the literary cultures of the science.[39] Indeed, the "literary projection" of geology has been the focus of much of the most exciting recent research in the field.[40] Geologists, historians have shown, used verbal descriptions, poetry, and stories to attract public support for their often startling researches and to give shape to their imaginings of a long-dead past for which they had only fragmentary evidence.⁴¹ Narrative was a powerful tool for geological writers, who generated literary "sensations" around the telling of geological stories and borrowed devices and patterns from novels, plays, history, and poetry to give order and meaning to the complex worlds they encountered beneath the soil.⁴² Nineteenth-century geologists developed other representational strategies too, creating new conventions for illustration, model-making, painting, and museum display, and those burgeoning visual and material cultures were always interlocked with vibrant textual imaginings of the geological world.⁴³ Guidebooks, descriptions, lectures, museum labels, fictional vignettes on the giant fossil saurians, and poetic quotations scattered through geological books animated a newly discovered past for increasingly broad readerships. During the first half of the century, Ralph O'Connor has shown, these textual accounts of earth history grew in confidence, stature, and public appeal, moving out of the relatively private arenas of the dinner club, notebook, or university lecture where they began and into the public domain of exhibitions and writings intended for general readers.⁴⁴ Showmanlike, epic, and apocalyptic rhetoric abounded in geology, and there was no clear boundary between those who sought to use texts to appeal to wider audiences and the practitioners of the science.⁴⁵ Alfred Lord Tennyson and Charles Dickens put geological ideas and images to diverse literary uses.⁴⁶ In many ways, as O'Connor has put it, "scientific writing *was* literature" in the nineteenth century, read and consumed by the many rather than the few in a culture in which the specialized disciplines we live with today had not yet crystallized.⁴⁷ "Literature," for one, was not yet a category that excluded nonfiction, and geologists, coming into contact with ancient worlds, needed imagination to recreate a past neither they nor any human forebear could directly observe. Literature, in this broad sense, had a vital role to play in creating geology as a modern science, both by attracting new readers and by shaping the artistic and literary conventions by which former worlds could be understood.

In particular, "romance" and "epic" have come to be seen as vital genres that shaped nineteenth-century geological writing. It is well established that geologists frequently deployed imagery from the chivalric medieval romances. William Buckland, for instance, first Reader of Mineralogy at the University of Oxford, was "dubbed Sir Ammon Knight," a pun on the fossil ammonites he had spent long hours in the field researching, and geologists imagined themselves as "brethren of the hammer, knights errant, a spiritual fraternity in search of a stratigraphical grail."⁴⁸ In this configuration, geology was heroic, and like their knightly forebears geologists crusaded in pursuit of truth in far-flung lands. Geology was often portrayed as a romantic science, with geologists traveling through time into a forgotten and sometimes violent past, contemplating temporal infinitude, and subjugating creatures resembling the ancient dragons. Here "romance" meant something more than knightly tales of love and heroism. It also betokened wildness, immensity, and wonder, transporting geologists and their readers into a tempestuous world of "specific geological features . . . sharply distinguished from developed landscape," marked by "the broken or dislocated character of the landforms" of many landscapes in Britain.⁴⁹ Geologists, moreover, littered their texts with quotations from Virgil, Homer, Dante, and Milton, transforming the underground they explored into an underworld of epic magnitude, "dovetailing geohistory with sacred history" and subjugating such devilish monsters as the mylodon or the iguanodon.⁵⁰ The association of geology with epic poetry gave the new science a perceived compatibility with classical literature and the Bible, those staples of the gentlemanly education taught at Oxford and Cambridge.⁵¹

If science *was* literature in the nineteenth century, it is the premise of this book that literature was science too. Writing was not simply a means of imagining or publicizing geology, but rather was a kind of scientific practice. Restructuring a geological textbook, for instance, or writing a series of lectures could, no less than reorganizing the objects on display in a museum, bring new ideas or evidence into sharper focus for geologists who were still searching for appropriate forms for their rapidly developing science. As in a museum, of course, there were many outlandish, untruthful, or misleading forms in which geological evidence could be cast, and there was significant debate about which forms worked best for which purposes, or about the different emphases different forms could give to the evidence.⁵² Nineteenth-century geologists were constantly alert to the importance of choosing the best literary forms with which to write geological science into existence. Powerful narratives of earth history could seduce and tantalize readers, they knew, just as the biblical cosmogony or—heaven forbid—evolutionary cosmologies had caught their admirers, but at what cost? How much detail, geologists asked, might be lost in the telling of too-simple stories for a public too quick to pant for the latest literary sensation? And might not some readers, if the yarn was ripping enough, be seduced into believing almost anything?

Imagining geological history in proportion to the truncated timescale of the biblical narrative, for instance, one of the century's best-known geologists, Charles Lyell, wrote in a famous passage:

> Such a portion of history would immediately assume the air of a romance; the events would seem devoid of credibility, and inconsistent with the present course of human affairs. A crowd of incidents would follow each other in quick succession. Armies and fleets would appear to be assembled only to be destroyed, and cities built merely to fall in ruins. There would be the most violent transitions from foreign or intestine war to periods of profound peace, and the works effected during the years of disorder or tranquility would be alike superhuman in magnitude.[53]

That "air of a romance" linked bad geology with fantastical fictional forms by means of a rhetorical tradition that had been commonplace since at least the Restoration. It was nothing new to describe the scientific works of one's opponents, or the novels with which one's own novel was a competitor, as mere "romance." In this tradition, "Romance comes to stand for a species of deceit that undiscriminatingly includes lying and fictionalising," whose opposite is "'true history.'"[54] Thomas Sprat, founding member of the oldest scientific society in the world, the Royal Society, famously used the term in this sense, calling for the modern man of science to write in the style of "the ethically and socially humble recorder of reality," worrying at how to turn travel journals, for instance, "into ... narrative[s]" without "diluting" their "crucial historicity."[55] Members of the Geological Society like Lyell drew on this long-established rhetoric without hesitation. In the passage quoted here, the "quick succession" of events Lyell describes in a truncated historical narrative, their lack of credibility and "superhuman ... magnitude," is clearly romantic in just such a pejorative sense. Far from conjuring up the quests of medieval or epic heroes to discover and defend the truth, this kind of "romance" is thrilling, improbable, inventive, and artificially plotted. Such romance stripped geological events of their historicity. As such, Lyell's repeated use of the conditional tense—"would seem," "would be," "would appear"—lends a sense of inevitability to his hypothetical set of scenarios: each constructive event, the assembling of armies or the building of cities, predetermines its own destruction or ruination in a seemingly unavoidable pattern. It is as if the model of "violent transition" from one state to another and back again cannot be altered, no matter what the circumstances. Here, the falsified patterns of "romance" determine the way in which the bad geologist sees the world before his eyes. The bad geologist is subservient to the logic of a form.

The definition of "form" I am using here is considerably wider than that usually deployed by historians of science and literary critics when they talk about style, metaphor, and narrative in scientific writing. It is now common practice to consider literary form as one of the "ways in which knowledge tends to become systematized, codified, and legitimated."[56] Narrative, for instance, has been seen as a systematizer of geological knowledge, helping geologists to order the unwieldy past according to certain kinds of plots, such as progression or evolution.[57] And it has been seen to have legitimating powers for geology in this period, capturing new audiences and readerships, aligning geology with culturally authoritative narratives from classical and biblical literatures, and justifying Britain's place at the apex of geological history.[58] But here, while continuing to broaden the range of that discussion and to test it against a wider variety of texts, I also want to consider that "form is never simply a tool of knowledge," an epistemological implement, but is also "an attribute of being, a category of ontology."[59] In other words, things do not exist without forms. In that sense, as Henry S. Turner has suggested, "form" does not simply refer to the literary or linguistic forms deployed in texts but is fundamental to the very existence of concepts, images, ideas, and materials—scientific or otherwise.[60] To take one example, Turner has demonstrated that "what counted as 'form' in the first place for early modern writers" was shaped by the structural and spatial mathematics of geometry, "one of the oldest and most enduring ways of thinking about the problem of form." As these notions "began to compete with rhetorical notions of form that were primarily linguistic, stylistic, and qualitative" in the seventeenth century, new forms began to be generated between a surprisingly wide range of groups, from critics, playwrights, and mathematicians to surveyors, carpenters, and costumers.[61] Geologists in the nineteenth century were engaged in a similar set of debates. Studying the structures and forms of the natural world with unprecedented success, and grappling for the material, visual, and verbal forms by which to instantiate them, their repudiation of "romance" was a symptom of this struggle to formalize the geological past, and to analyze the forms of the land with which they were confronted. The organizing of the geological landforms and the pasts they embodied through verbal means was a fundamental practice by which the new science came to be.

The point is perhaps clearer when we remember that the principal activity of geologists engaged in the stratigraphic project was to recognize that the earth *had* a specific form and to ascertain what that form might be. The earth could be traveled over and seen by means of a variety of other forms, moreover (the Grand Tour, the railway routes, through letters and instruments), and the hunt was on for the literary, visual, and material forms that would translate as much of the form of the natural world into comprehensibility as possible. It was in a combination of each of these forms that "geology" could be said to exist. This approach, moreover, allows us to challenge the argument that the many claims made by geologists to have superseded "an age of romance" in their science were largely disingenuous. As O'Connor puts it, for instance, the "characteristic trope of setting 'truth' against 'imagination' is a blunt polemical instrument and one of the oldest tricks in the rhetorical armoury of Western literature for discrediting an opponent. It should not be taken as evidence that science and imagination, or science and speculation, were necessarily seen as mutually exclusive in all circumstances by the person wielding this rhetorical instrument."[62] I agree with this last point, and O'Connor powerfully reveals that by the 1820s geologists had reconstructed the primeval world by drawing heavily on imaginative and speculative forms. O'Connor's work also reveals that vigilance about the deceptive powers of storytelling was an essential part of geological practice, and he explores the ways in which geology appropriated "romance," as something potentially dangerous, as the imaginative negotiation of a *tension* between fact and fiction in a quest to make the science appealing within a culture where antifiction sentiment ran high.[63] But it is also true that the practices of geology (fieldwork, travel, observation, and collecting) were as important to its popular appeal as the fossil monsters on which O'Connor dwells, and there the "anti-romance" rhetoric could be accorded more significance. It may be an well-worn rhetorical trick but it often signaled, as it does for Lyell, a suspicion of self-determining fictional forms as they encroached upon the mind and vision of the geological observer, and it was a trick that was useful at this moment precisely because the form of the earth and of earth history was still unknown and, without an adequate form, unknowable. Geology was only as good as the forms that constituted it.

As Lyell's example reminds us, problems in literary form were often used as a conceptual tool for thinking through the problems of geological form. Geologists were keen to experiment with the ways in which different forms of writing could help them see the "truth" better, and to delimit those kinds of writing—like cosmological "theories of the earth"—that they felt encouraged excessive speculation. This was because novels, too, becoming dominant at the same time geology was crystallizing as a discipline, sought to claim intellectual or cultural authority by parodying and rejecting such forms as "romance" and the "epic."[64] From *Don Quixote* on, novelists had claimed that their works were literally "true" by asserting that they were definitely not "romance," and the convention of denigrating romantic fictions persisted into the nineteenth century as novelists in a self-consciously "realist" tradition claimed their works to be authoritative and truthful (if not actually true), and to have serious things to say about the contemporary and historical worlds they described.[65] Literary critics have been wont to take this rhetoric more seriously than historians of geology. As Michael McKeon puts it, for instance, the repudiation of romance "registers an epistemological crisis, a major cultural transition in attitudes toward how to tell the truth in narrative."[66]

Part 1 of *Novel Science*—"Stories in Science"—argues that geologists at the Geological Society, struggling to throw off the shackles of the biblical cosmogony and of all-encompassing cosmological theories of the earth, were key protagonists in "a major cultural transition in attitudes toward how to tell the truth in narrative" that spanned novel writing, literary criticism, poetry, and science. In chapters 1 and 2 I focus on the geologists' often rabid devouring of the fictions of Walter Scott—the man who did most to reinvent the novel as a credible and authoritative literary form. Scott's ironic, often episodic, loosely plotted narratives, self-consciously distanced from the romantic and epic traditions they had inherited, chimed with the geologists' need to narrate geological events without falling victim to the too-seductive machinations of fiction. In chapter 3 I explore the work of perhaps the greatest geological writer of his age, Charles Lyell, as he struggled to articulate a new "uniformitarian" methodology for his science by turning to Byronic mock-epic as a mode for shaking off the shackles of too-rigid and traditional stories for discerning earth history. Associating themselves with such fashionable literary figures as Scott and Byron, exploiting their narrative devices and techniques for scientific ends, the gentlemen geologists invented and reformed their science, its values, and its methods, ironically distancing themselves from the pleasures of plot. Indeed, this turn from structure to story was realized in the most fundamental form by which the geological past was formalized: the geological map. Focusing in my fourth chapter on geological, travel, linguistic, journalistic, and fictional texts about Yorkshire, including Emily Brontë's 1847 novel *Wuthering Heights*, I chart the ways in which maps and texts together made *place* the load-bearer of historical understanding, immersing readers in the structures of a particular location rather than its story. This is not by any means to say that location and story are mutually exclusive, but rather to register a nineteenth-century effort to give physical, or structural, form to the past in ways that rendered its narratives—important as they were—only implicit or unarticulated.

The problem with plots—that they seemed self-determining, like that denigrated kind of "romance," subordinating all of the complex and contradictory evidence to a single pattern or design—was all the more acute because geology was inherently a historical science. It lent itself to, and borrowed from, historical patterns for describing the past. But Lyell's example hints at the distinction that was lurking beneath geological discourse of the period. Geologists were happy enough to reveal, in order, a loosely chronological series of events that had happened during the course of earth history. What they were wary about was *plot*, a single explanatory mechanism, by which all events could be understood and related, such as evolution or progressive development. The difference is subtle, but it is central. In narratology, the difference is "between *story* (the totality of the narrated events), *narrative* ('the discourse, oral or written, that narrates them') and plot."[67] While a "story," or a set of events, can be related in simple or minimal narrative form, as in a diary, in historical annals, or in a chronicle, where events follow one from another without much (if any) explicit causal linkage, "plot" provides the basis of just such a causal relation. In doing so it connects, interprets, and gives meaning to the events of the story. As E. M. Forster famously put it in *Aspects of the Novel*, "The king died, and then the queen died" is a story, but "The king died, and then the queen died of grief" is a *plot*.[68] As that example implies, "plot" offers a fundamental, underlying, and explanatory structure on which the events of a story may be hung. In doing so, it makes sense of those events. It imposes interpretation upon them.

Quite obviously, this distinction is a loose one. As the narratologist Seymour Chatman points out, we might assume the connection of the king's and queen's deaths simply *because* of their being told one after the other. That is, we might infer causal links between sequentially related events even where none are made explicit in the narrative.[69] Furthermore, "Texts can have widely differing degrees of plot connectivity: some are tightly and linearly plotted ... others make use of 'mosaic plots' ... whose causal coherence is not immediately obvious; others again are loosely plotted, episodic, accident-driven, and possibly avoid plotting altogether."[70] As part 1 of *Novel Science* will make clear, nineteenth-century geologists, wary of too much plot, exploited all of these ambivalences, suggesting that there may well be a "plot" but that geologists did not yet have enough information to reveal it, leaving plots implicit or unarticulated, or threading narratives together in episodic or deliberately disjointed fashion. A key example here is the "pageant of earth history," which forms O'Connor's principal evidence for his case that narrative was an important component of the representation of geology in Victorian literature. O'Connor suggests that pageants offered readers theatrical "sequences of scenes" from the deep past, and that this meant that "the grand narratives of eighteenth-century cosmology returned in a new guise to bring the deep past before the eyes and imaginations of a wider readership." But there is a key distinction between the "pageant" or the "sequence of scenes" offered in many popular geological texts of the period and "grand narratives." O'Connor's examples make it clear that such writings often resisted precisely the plotting that was so essential to cosmology and instead presented "earth history as a discontinuous succession of scenes rather than as a gradually evolving narrative," dividing scenes "by a concatenation of brief clauses" marked off from one another by semicolons, by "sudden jolt[s]" between scenes, and with "the theatrical agency of mist" functioning as a rising or falling curtain between scenes from different geological epochs. O'Connor argues that such blank spaces, inserted between the descriptions of different periods in geological history, highlighted "the ignorance of the scientific spectator" of the mechanisms and causes that might link those scenes.[71] But they also, I think, allowed the geologist to construct a chronological sequence or narrative of earth history without *requiring* an underlying plot to explain it.[72] This is not to say that an underlying plot was not hinted at, suggested, or left hanging on a tantalizing thread for the reader, or the viewer, to supply. But the form of the pageant meant that plots could be left hanging in precisely this way—hinted at, but rarely fully or explicitly articulated.

The notion of "form" I am using here, moreover, allows me to situate writing within a broad spectrum of other formalizing activities intrinsic to geology, and specifically to link it with the practical, spatial, and material cultures of geological science I have begun to outline above. Formal analysis can help us understand the museum collections that needed to be first discovered and then given order (an order that could reflect personal, local, commercial, or national interests, for example), the journeys geologists took over a series of routes made possible by formalized networks of roads, railways, and sea passages, and the maps and sections that gave abstract form to rocks whose physical forms were too complex to be understood at a single glance. All of these, equally, are forms, which together go some way to constituting "geology" as it was in the nineteenth century. As will be argued in part 2 of this book, men like Charles Dickens and the comparative anatomist Richard Owen, for instance, in restructuring the earth as a commercial spectacle, transformed themselves into commercially successful giants on the urban stage. Collecting and displaying rocks and fossils in arrangements rooted in their knowledge of particular local landscapes, novelists and fossil collectors like Charles Kingsley and George Eliot gave shape to worlds gone by in ways that gave them cultural prestige, a voice for telling the truth. Form, we might say, mattered.

This is not to reduce science to text, to a play of endless representations without any real purchase on the truth. Nor is it to suggest that geologists simply told stories about the past, none of which were more or less true than the others, or more or less true than the fictions about that past printed in the pages of novels and magazines. Nor is it to reduce literature to the laws of scientific explanation, as if the rich and complicated literary texts that have challenged readers over centuries were nothing but data sets for the scientific analysis of the human imagination.[73] It is certainly not the case that geologists had free play with their literary imaginations to conjure up the geological past in any manner they wished, or that they simply fit the evidence to a preconceived narrative pattern. In fact, the opposite is the case. Geologists, determined to render the world in all its empirical detail, were constantly alert to the problems and dangers inherent in writing about that world. They were not naive to the threat of fictionalization that lurked in the process of "emplotment." A legacy of the claims made by deconstructionists that science is merely text, and of attempts to deal with subsequent criticisms of the literary study of science, has been that critics have found it difficult to see that nineteenth-century men and women of science—in this case geologists—were themselves always concerned to find appropriate stylistic and formal modes for apprehending, collecting, and organizing their data truthfully and accurately. They anticipated the skepticism of the literary that we saw on the "science" side of the science wars in the 1990s, even as they embraced the explanatory and organizational powers of narrative and its ability to convert new readers and thinkers to the science. It is only right, I contend, that we pay attention to the historically specific concerns of these geologists, and that we see that the writing of science is neither a secondhand representation of "facts" already ascertained, nor, in admitting that it may make a real contribution to scientific knowledge or understanding, a way of telling whatever stories we might like, regardless of the evidence. As many nineteenth-century geologists knew, writing was a useful—albeit often a dangerous—scientific tool.

The Problem with "Literature and Science"

In part 2 of *Novel Science*—"Science in Stories"—I am concerned to overcome one particularly damaging legacy of these debates about literature and science. Namely, I take issue with the notions either that "science" furnishes facts and ideas that "literature" reflects, reproduces, or contests or, as in more recent "literature and science" studies, that there is a more active, "two-way" relationship that nonetheless preserves science and literature as discrete, demarcated fields. In particular, this kind of criticism has tended to see the Victorian novel as an appropriator of geological "plots" in a way I consider to be fundamentally at odds with the widespread cultural interrogation of the powers and limitations of plot as a mode of accurate worldly apprehension.

Long after the discoveries of geology had been absorbed into the cultural mainstream, its aesthetic and moral connotations continued to be liable to suspicion. Opponents of scientific naturalism in the 1860s and 1870s, for instance, frequently associated evolutionary theory and the worlds it described with the paganism and sensuality of the "fleshly" poetry of Algernon Charles Swinburne and Dante Gabriel Rossetti, or with the aestheticism of the fictions of Walter Pater, hinting at the formal and sexual looseness of both.[74] As Gowan Dawson has recently shown, the serial parts of novels published in monthly instalments, or on the pages of weekly, fortnightly and monthly magazines, were often compared unfavorably to the disjointed spectacle of the fossil bodies assembled from fragments of bone by comparative anatomists.[75] Critics of the novel frequently compared the laws and characters of fiction to physiological laws and phenomena, and it was commonplace to describe the effects of reading fiction in physiological terms.[76] As new literary, scientific, and cultural forms emerged through the course of the nineteenth century, the epistemological and moral status of those forms was always under intense scrutiny. Those deconstructionist critics who once labeled both empiricism in science and realism in literature as "naive" could not have had it more wrong.

Nonetheless, as Dawson has pointed out, those critics who have been keen to make the counterclaim that there was an easy, coterminous relationship between science and literature in the nineteenth century may also have been a little too "sanguine."[77] As should, by now, be clear, the forms and styles of fictional, imaginative, and scientific writing were under complex stress throughout the period, with each form keen to reject and dissociate itself from others as part of a process of self-definition. In the 1980s influential critics such as George Levine and Gillian Beer described evolutionary science as a new "source" of "authoritative organisation" for novelists seeking to construct fictional worlds after the model of the "real" world.[78] For these critics there was a genuine connection between the hidden and fecund connections among events, creatures, and people described in the novels of such writers as Eliot, Dickens, and Thomas Hardy, and the entangled, complicated, and lateral interrelations among species in Darwin's *Origin of Species* (1859).[79] Subsequently, criticism interrogating the multifarious relationships between the forms of "literature" and the forms of "science" in the nineteenth century has become a booming area of research. To take just a few examples, critics have charted the descriptive and observational practices of natural history and those of the realist novel, the suspense-making patterns of Victorian fictions and the development of hypothesis as a scientific method, and the fictional narration of pathology and sickness and their narration in contemporary medical and psychological discourse.[80] In each of these cases and many more besides literature and science have been considered in active and multivalent relation, sharing "one culture" rather than divided into two, and each has been considered equally capable of directing and contesting the patterns, interests, conclusions, and attitudes of the other.

Geology has had something of a raw deal within this fertile critical field. In the current literature, geology has tended to be cast as a mere handmaiden of evolutionary biology, and the same kinds of relationships asserted by Beer and Levine between Darwinian biology and realist literature have been considered to have existed between geological texts and novels in the same period.[81] According to Beer and Levine, novelists including George Eliot, Charles Dickens, and Thomas Hardy shaped their fictional worlds by absorbing or contesting the patterns of evolutionary development.[82] Subsequent critics have adopted this model and applied it to nineteenth-century geology, identifying plots of "uniformitarianism" (Lyell's gradualistic model of change, to which Darwin was indebted) and a violent "catastrophism," and associating them with different forms of fiction.[83] In general, "uniformitarianism" has been allied with realism. *Principles of Geology* is perhaps the most famous scientific text of the nineteenth century after *Origin of Species*, and Darwin acknowledged his debt to Lyell's vision of ordinary, everyday geological causes working over an immense span of time to produce vast changes in the appearance of the globe.[84] Critics have suggested that this view had much in common with the incremental patterns of change and attention to minute causes held to be important to literary "realism" in the century.[85] By contrast, in the same works "catastrophism," usually (mis)identified as a belief that geological change was predicated on wholesale global disasters, has been allied with some opposite—most often with "romance" as a byword for nonrealism.[86]

But under historical scrutiny, arguments that may work for evolutionary science do not work nearly so well for geology. In the first place, it is no longer by any means clear that those arguments are unequivocally true for evolutionary science. Historians of science have revealed a vast body of pre-Darwinian scientific literature, research, and public discussion, and a rejection of "natural selection" and embracing of diverse kinds of non-Darwinian evolutionary theory in the later nineteenth century, bringing into question just what kinds of patterns and texts may have been the object of discussion and the basis for fictional narratives in the nineteenth century. At the same time literary scholars have begun to unearth the ways in which the "literature" discussed by many critics in the 1980s staked out its coexistence with scientific ideas by excluding certain *kinds* of literature (such as the productions of literary aestheticism) from its definitions.[87] If in geology there was a commonplace rhetoric of "romance" by which other geological theories and works, or earlier phases of the study of the earth, could be dismissed, then as critics we have found ourselves casting "geology" in this earlier, romance mode in relation to evolutionary biology. To some extent we have simply reproduced the rhetorical arguments of our subjects.

Furthermore, the uniformitarian/catastrophist debate misreads the history of geology in two ways. First, as will be explored more fully in chapter 3, there were never two schools of geologists under these appellations. Charles Lyell did not label his theory "uniformitarianism" (he did not, in fact, label it a "theory" at all, since he, like his colleagues, was suspicious, if not of theorizing in general, then at least of its liability to excess). His research was praised by geologists at the Geological Society for its empirical prowess, and with some of his concepts many were in general agreement. Second, the arguments made against Lyell by his colleagues were not, as has sometimes been assumed by critics, made in order to defend the biblical account of creation, and thus of a young earth, or because of a preference for inexplicably violent and "romantic" forces, as against Lyell's "realist" view. Differences between geologists in this period were subtle and complex, determined much less by religious denomination (as is sometimes asserted) than by forms of publication, institutional affiliations, networks of friendship and collaboration, access to travel, and regional differences, and to the exhibitions, museums, lectures, and written works these men produced and consumed.[88] If as literary critics we are fully to get to grips with nineteenth-century geological science, we must learn more from historians of geology in order to unpack the full range of this complexity, and we must complement our concern with "narrative" form with a concomitant interest in the other forms (the lecture, the museum collection, the dinner party, the gentlemen's-club debates, the Mechanics' Institute, and the map, for example) that constituted the sciences of which we speak.[89] Too much of what has been written on Lyell has called him the mere purveyor of a simple plot of earth history readily transposable into nineteenth-century fictions. But plots were problems for geologists, not proofs, and Lyell's theory was self-reflexive about the place of story in his science in ways that literary critics have not yet begun to unpack. I seek to begin that process here.

What that entails, of course, is considering the complex and multivalent forms—visual, material, cultural, and literary—that make knowledge possible, and accepting that writing is a fundamental means by which human beings can apprehend the world around them. As such, there can be no meaningful distinction between science and literature, since writing can be (though it is not restricted to being) a mode of scientific practice, a means by which a science becomes a science. Over the course of the nineteenth century, a "realist" form emerged across literary culture (in which I include scientific writing) that focused on the humdrum, the everyday, the minute, and the particular, and that positively reveled in the raggedy forms such a focus might create. It also absorbed techniques and structures from the very forms it proposed to reject. The rejection of heavily plotted forms was never entirely stable. Furthermore, this was specifically located, for geology, in practice—in the difficulties of travel and fieldwork, in the threats to the life of the geologist as he scrambled up a denuding cliff face, in the paucity of geological evidence, in the ruins of the rocks as they stood, very often, next to the ruins of ancient monuments. For novelists and geologists alike, there was an intellectual satisfaction to be had in the *breakdown* of plot, in pauses in its development in which reader and writer were asked to stand back and scrutinize the events related with a dispassionate eye, in the irretrievable piece of missing evidence, in the inability to keep focus on the story because the details kept cluttering up the pace, and in dry disquisition or ostentatiously long description. In my three chapters on nineteenth-century novelists, I see this geological breakdown at play. Charles Kingsley was both a geologist and a novelist, though he continually professed himself to be much better suited to the former role and disgusted with himself in the latter. In his social-problem novels *Yeast* (1848), *Alton Locke* (1850), and *Two Years Ago* (1857), geologist-protagonists profess their disregard for geological theorizing and test their mettle by immersing themselves in concrete geological detail. As they do so, the plot structures of the novels they inhabit invariably and often self-consciously unravel. In the last of those novels, indeed, the structure of the text is based less around a coherent plot than around stratigraphic correlations between rocks in Devon, the Eifel region, and North Wales. George Eliot, always alive to the problems and possibilities of metaphor, of language, and of plot structures, draws consistent attention in *The Mill on the Floss* and *Daniel Deronda* to the pleasures and power of geological fragmentation as a site of meaning. And Dickens, in *Dombey and Son, Bleak House*, and *Our Mutual Friend*, stages urban chaos and fragmentation as a geological spectacle. For these novelists, geology did not offer an authoritative plot pattern on which to hang a fiction, as many critics have suggested. It offered instead a useful form for narrative breakdown.[90] Tell stories they might, but the gaps and blanks of the stratigraphic column could usefully suggest dark and disturbing ruptures in the continuous flow of the plot. Geology, this book will argue, offered the Victorians a language for the breakdown that might secure narrative authenticity. If there is one thing I would like my readers to consider as they close the pages of *Novel Science*, it is that for realist novelists, questing to give the novel a new cultural authority, and for geologists, seeking to do the same for their science, narrative had a much more ambiguous and much more complex role to play than we, with our own narratives of the relations between "science" and "literature," have hitherto been accustomed to believe.[91]

PART ONE *Stories in Science*

Fictions of a Former World { **CHAPTER ONE**

On 14 June 1837, less than a week before Victoria's accession to the throne, the young anatomist Richard Owen attended his first meeting of the prestigious Geological Society of London. Entering Somerset House on the Strand, the neoclassical palace in which the Society's apartments were housed, Owen had reason to feel excited. Geology was the most controversial and the most celebrated of all the new sciences, and its Society combined these qualities in equal measure. The only major scientific society in Britain to permit discussion after the reading of papers, it was well-known for the liveliness of its debates. And on the night in question, passions were running high over the classification of the oldest fossil-bearing rocks. "We had a grand battle at the Geological Society last night," wrote one member the next morning, "in which I bore the brunt on our side; but, though well banged, I was not beaten."[1] Watching the most eminent men of science of the day battle, bang, and beat one another as they revolutionized understandings of the structure and history of the earth, Owen must have felt he had ringside seats to the best show in town.

And yet, in later years it was not the "grand battles" of Society meetings that he would recall. Rather, his memory settled on events that occurred much later in the evening. "After supper," his wife Caroline wrote in her journal, at the London rooms of Lord Cole, Owen witnessed a coterie of leading geologists "play 'high jinks,' as immortalised by Sir Walter Scott in 'Rob Roy.'"[2] Caroline muddled her Scott novels: "high jinks" is played by a group of lawyers in the Edinburgh of the 1780s in *Guy Mannering*, not *Rob Roy*. But the rules of the game as Scott had set them out were adhered to precisely. Following the throwing of dice, "those upon whom the lot fell were obliged to assume and maintain, for a time, a certain fictitious character, or to repeat a certain number of fescennine [obscene] verses in a particular order."[3] In *Guy Mannering* we have a description of one such "fictitious character" impersonated by the players of the game: "a monarch, in an elbow-chair, placed on the dining-table, his scratch wig on one side, his head crowned with a bottle-slider, his eye leering with an expression betwixt fun and the effects of wine."[4] The "fescennine verses" there included "such crambo scraps of verse as these: 'Where is Gerunto now? and what's become of him / Gerunto's drowned because he could not swim, etc. etc.' "[5] Though we can fairly assume that the verse got a little bawdier than that within the boozy all-male confines of Cole's rooms, Scott's crapulous monarch enthroned on a dining table was faithfully impersonated by the geologists. In this case the prominent natural philosopher George Stokes apparently "took the chair as king, and was excellent as the arbitrary monarch."[6] In both Scott's 1815 description and its imitation here over twenty years later, falling out of character or failure to recall the verses in question "incurred forfeits": drinking or paying a fine.[7] And just as the Edinburgh lawyers are forced to drink or cough up if they slip into the language of the law, for their geological imitators "all kind of scientific discourse was prohibited on pain of forfeit."[8]

What significance can we attach to the playing of such a game, by such men, in such circumstances? To what extent is the literary culture of nineteenth-century men of science—the books they read, the jokes and stories they told, the characters they admired or impersonated—worth considering as a category through which they created new knowledge? Should our understanding of scientific exchange in the nineteenth century be confined solely to spaces, such as Somerset House, deliberately designated "scientific"? In this instance it seems obvious that the geologists themselves would have answered with a resounding yes. The rules of high jinks marked a boundary between science and saturnalia "on pain of forfeit, and geological expressions" were banned "on pain of fighting the champion (Lord Cole's brother) with hammers."[9] Indeed, Scott's recording of the game encoded this division between professional life and fiction; the drunken monarch of *Guy Mannering*, Mr. Pleydell, is on weekdays "a lively, sharp-looking gentleman, with a professional shrewdness in his eye, and . . . a professional formality in his manners." But on Saturdays he casts off his lawyer's wig and coat to carouse with his colleagues until the early hours of the morning.[10] The conditions of the game itself, with its element of performance and its ban on professional discourse, signal its separation from ordinary working life.

And yet, both for Mr. Pleydell and for the geologists, high jinks had its serious side. Guy Mannering is initially bewildered to stumble upon such a screwy Pleydell outside office hours, but his timing actually turns out to have been quite fortuitous. As Pleydell admits, it is his policy to "always speak truth of a Saturday night." " 'And sometimes through the week I should think,' said Mannering." "Why yes!" replies the honest Pleydell, "as far as my vocation will permit."[11] The irony is obvious. It is only in play, only in fiction, only when he is drunk on a Saturday night, that Pleydell the lawyer can really tell the truth. Even more importantly, for the geologists, high jinks was an odd game to play in 1837, given that, in Scott's account, "the custom of mixing wine and revelry with serious business" was, by the 1780s, already "considered as old-fashioned," and had been "forgotten" entirely by the time he penned the novel in 1815. In imitating this extinct Scottish drinking game, we can therefore presume that the London geologists were precisely imitating the mode of social interaction that Scott revived and represented for them: a rabble-rousing, all-male, collegiate identity in which business and revelry were nominally separated but nonetheless remained closely linked.[12] They were, after all, playing only with other geologists. In Owen's account, the very same members of the Society who had "banged" and "beaten" one another in debate just a few hours beforehand were to be found celebrating together at Cole's rooms, paying their fines into the same coffers and knocking back the same tipple. Suddenly the prohibition on shop talk acquires an added charge. For if the Geological Society was unique in allowing the odd "battle" to be fought at its meetings, it would rarely allow scientific disagreement to develop into all-out warfare. To do so would undermine geology's reputation as both a gentlemanly and an empirical science, built on indisputable fact. In banning discussion of science after meetings and replacing it with activities promoting friendship and fellowship, and with the performance of fictional identities far removed from the real-life clashes performed in the debating chambers, "high jinks" helped ensure that the gentlemanly culture on which geological science depended remained intact.

Earthly Forms

In their reenactment of the eighteenth-century world of *Guy Mannering*, the London geologists claimed affinity with the prestigious intellectual enterprise of the Scottish Enlightenment. But in doing so, they deliberately overlooked perhaps the most famous episode in the history of geology, which had taken place in Edinburgh at the very same time that Scott had written

Guy Mannering. For between the convivial Edinburgh of the 1780s and its mirthful reenactment in the 1830s, there had occurred a famously indecorous geological controversy, which had convulsed the Edinburgh literati and threatened to turn earth science into a public joke. Indeed, it was specifically in order to overcome this ungentlemanly episode that the Geological Society of London had first been formed. Articulating the power, prestige, and gentlemanly respectability of geology at a time when science was still marred by "social insecurity," as Richard Yeo has put it, the gentlemen of the Geological Society committed themselves to rescuing a science from the tatters in which it lay on Edinburgh's newly built boulevards.[13]

It is hardly surprising that such geological pugnacity should have emerged in the Scottish capital. Edinburgh is perhaps the most geologically dramatic of any city in Europe: its buildings nestle in the crevices and shadows of the Salisbury Crags, Calton Hill, and Castle Rock, which looms over Prince's Street. And the geological dimensions of the landscape were inseparable from the city's intellectual culture, as they were often used as evidence in its theoretical battles over the formation of the earth. In *Guy Mannering,* the protagonist is introduced to such Enlightenment luminaries as Adam Smith, Henry Home, and David Hume, but also to James Hutton, the author of a controversial *Theory of the Earth*, which would come to dominate the geological fray that emerged at the turn of the nineteenth century. Hutton's "theory," like other Enlightenment cosmologies, considered the earth as a complex but ordered whole whose past, present, and future could be fully explained by reference to eternal law.[14] Hutton's world had no beginning or end. Continents were perpetually wasted to the depths of the ocean floor, where they were melted by the earth's internal fires and fused into new rock masses. Powered by heat, these new rocks were then thrust above sea level, forming new continents, which, in turn, were eroded back whence they came. The cycle was designed to ensure the ongoing habitability of the earth: the decay of the rocks produced life-sustaining soils, and their renovation ensured that the supply never ran out (fig. 1.1). The earth's eternal habitability in turn ensured the preservation of mankind and the ongoing development of his reasoning powers, of which Hutton's text itself was evidence.[15]

Hutton's theory that all rocks were created by melting and fusion in the bowels of the earth had relied upon some controversial interpretations of the rock record. The most problematic of these was his interpretation of granite. In the eighteenth century, "following well-established precedent," the famous German mineralogist Abraham Gottlob Werner had argued that the earth had

Fig. 1.1. "Unconformity of basalt on the river Jed, south of Edinburgh." In James Hutton, *Theory of the Earth* (1795). Reproduced by permission of the syndics of Cambridge University Library.

begun as a vast primordial ocean constituted by a thick mix of minerals and water.[16] Most layers of rock had been deposited, one atop the other, out of this fluid, and could be classified with reference to the order of their deposition. The harder, relatively insoluble "Primary" or "primitive" rocks, such as granite and marble, were often found at the bottom of the rock pile, out of all predictable sequence, and rarely contained fossils. These were usually interpreted as the oldest rocks. They were followed by a series of softer, often fossiliferous "Secondary" or "Flötz" rocks, such as limestone, which lay above them in more obviously sequenced layers. This classification had been very successful: Werner had built upon largely consensual evidence from mineral chemistry to coin the concept of "rock formations," each viewed as a historical entity, formed at a unique historical moment, rather than as a static "natural kind."[17] This concept gave geologists a powerful new research program by which they could map those formations in space, interpreting rocks in broadly historical sequence. In this scheme, Primary rocks were generally considered to be older than the Secondaries, and granite was the oldest rock of all. Granite was the beginning. In Hutton's cyclical scheme, of course, beginnings could not exist.

FICTIONS OF A FORMER WORLD   35

In the Highlands, in Galloway, and on the Isle of Arran off the west coast of Scotland, Hutton found evidence to suggest that granite had not been deposited by water as the first in a long sequence of depositions. It had been squirted upward in molten form through preexisting rock formations, as all other rocks before it had been, and as all newly formed rocks would be in the future. Granite, he speculated, was not the oldest rock in the world. In Hutton's system, there was no such thing as the oldest rock at all.[18]

Hutton first published his theory in 1788, in an eighty-page paper which attempted to reproduce, in its literary form, the cyclical structures of the system he had described. After proselytizing that the entire globe was a machine or organism, Hutton considered, in isolation, several of the processes that made up his system. He gave chemical proofs that heat and pressure could consolidate loose sands and gravels into molten substances, and he used cabinet specimens to show that these molten substances could penetrate already formed rocks from below (fig. 1.2). Though he concluded by explaining the cyclical system of "decay and renovation" that sustained the earth, he never quite joined the dots between his arguments about specific processes within the cycle, he barely discussed the "decay" that constituted one-half of his system, and he never fully returned to his opening sense of the global comprehensiveness that his theory claimed. What's more, when he published a longer version of the theory in 1795, Hutton did not provide any additional proofs that the earth was in perpetual decay, but merely strengthened his emphasis on the ferocious, active power of fire, unbalancing the text even further. Adding hundreds of pages of appendices in defense of his principles, Hutton rendered a system premised on the elegant design of the earth in prolix and ungainly prose. The form of the text and the form of the world it described were out of kilter. By the turn of the century, Hutton was dead and the theory was fading from memory.[19]

But in 1802 John Playfair, professor of natural history at the University of Edinburgh, rewrote Hutton's theory. Aged fifty-four, Playfair was a distinguished mathematician, a fashionable bachelor known for championing female intellectual equality, and a lively essayist on the verge of a seventeen-year career reviewing for the prestigious *Edinburgh Review*. Three years later he would fight to ensure that university posts in Edinburgh could remain secular, and he used both his gregarious personality and his journalistic powers to bemoan the state of mathematics at Cambridge and to promote the study of continental algebra. This broad intellectual span was integral to Playfair's rewriting of Hutton's theory in a new, more secular guise as *Illustrations of the Huttonian Theory of the Earth*, and his widely admired literary talents enabled

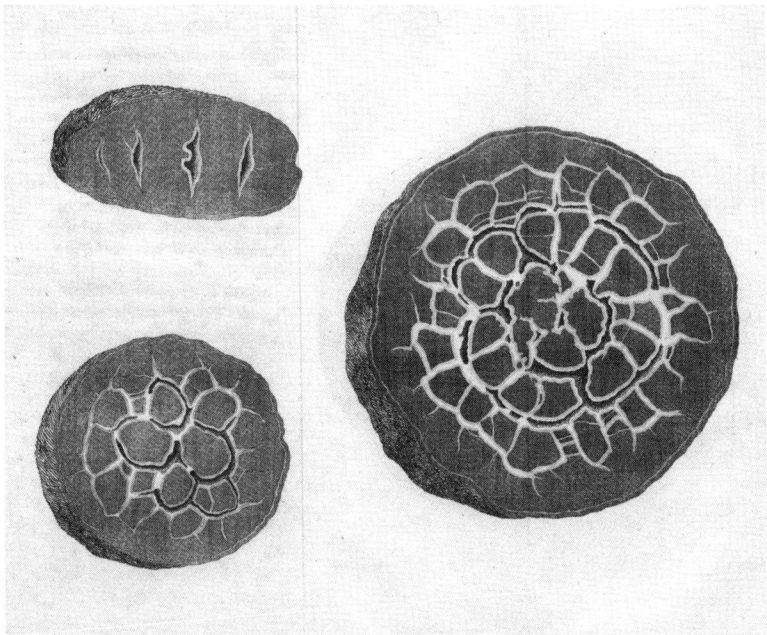

Fig. 1.2. "Cabinet specimens," showing molten rock thrusting up beneath settled strata, producing rocks and soils for farming on the surface, which tumble to the base of the image and are melted once more. In James Hutton, *Theory of the Earth* (1795). Reproduced by permission of the syndics of Cambridge University Library.

him to integrate argument and proof more fully than his predecessor had managed. Playfair removed Hutton's deistic metaphysics from the text, "offering the new century a bowdlerised version that reduced it to a purely mechanical system of physical processes interacting in dynamic equilibrium," aligning it with the prestigious mechanical system of Newton's physics.[20] The early chapters of Playfair's volume established evidence for the widespread decay of rocks, which Hutton had glossed over, and synthesized Wernerian interpretations of the strata with a more fully fleshed-out elaboration of Hutton's cycle. Proving through prose the "originality, grandeur, and simplicity" of Hutton's theory, Playfair made it newly elegant and readable.[21]

In Playfair's reformulation, Hutton's theory had not met "the general approbation" not because it lacked evidence or proof, but because the human imagination was naturally resistant to conceiving of the world in such vast terms. "The greatness of the objects which it sets before us, alarms the imagination," he wrote:

the powers which it supposes to be lodged in the subterraneous regions; a heat which has subdued the most refractory rocks, and has melted beds of marble and quartz; and expansive force, which has folded up, or broken the strata, and raised whole continents from the bottom of the sea; these are things with which, however certainly they may be proved, the mind cannot soon be familiarised.[22]

"The greatness of the objects which it sets before us, alarms the imagination." The first part of Playfair's sentence sets the pattern for the rest, dividing a complex subject, the "greatness" of the theory's matter and forces, from its verb and object, "alarms" and "imagination." In this case the phrase is interrupted by a modifying clause, the action of Hutton's theory which "sets" that greatness "before us." The effect is to attribute agency to Hutton himself, if not for creating the "greatness of the objects of the earth," then for bringing those subterraneous "powers" into existence for the human imagination. The earth is forceful and formidable, partly because Hutton has made it so. The inversion into the passive voice at the end of the sentence further accentuates the sense of Hutton's imaginative audacity: the heat and rocks and forces of his vision are "things with which . . . the mind cannot soon be familiarised." Syntactically at least, disbelief in Hutton's theory is a failure not of the theory itself but of the imagination.

As the sentence swells, a litany of clauses massed between semicolons, Playfair presents his reader with a grand vision of the earth that moves beyond the level of the individual imagination. The modifying clauses are dropped, and the emphasis on Hutton's reimagining of the earth is dissolved in its sublime, self-perpetuating "heat" and "expansive force." A new world comes into being, free even from the creative mind of the philosopher, and Playfair overwhelms us with its power. The abstract nouns with which Playfair begins each new clause pile atop one another, each performing grander and more extended actions than the last: while the "powers" are passively "lodged" in the earth's bowels, the "heat" has actively "subdued" *and* "melted" rocks, and the "expansive force" has violently "folded up," "broken," and "raised" strata and continents. Playfair lingers long on the bewildering amplitude of Hutton's earth before bringing the sentence into order. Raising Hutton's theory into grandeur by transforming particular rocks and continents into the objects of vast and abstract "powers" and "forces," accumulating those forces beyond the bounds of syntactic coherence before bringing them into grammatical relation, Playfair enacts both the capaciousness of Hutton's theory of the earth and its power as a single, general law that makes comprehensive sense of seemingly disparate phenomena.

Such dramatic prose enables Playfair to make provocative claims about scientific method. "The change and movement also," he continues,

> which this theory ascribes to all that the senses declare to be most unalterable, raise up against it the same prejudices which formerly opposed the belief in the true system of the world; and it affords a curious proof, how little such prejudices are subject to vary, that as Aristarchus, an ancient follower of that system, was charged with impiety for moving the everlasting VESTA from her place, so Dr Hutton, nearly on the same ground, has been subjected to the very same accusation.[23]

There is more than a hint of the controversial Humean skepticism of the senses here. The senses misleadingly "declare" the world "to be most unalterable" when in fact, it is the subject of perpetual "change and movement." Hutton's system moves beyond commonsense experience of the world, Playfair proclaims, but only insofar as did Aristarchus's "true system of the world": that the earth, as Copernicus later proved, moved around the sun. Superimposing this more expansive historical narrative on the limited evidence of the senses, Playfair reminds his readers, if they were not already overwhelmed by the sensualism of his prose, that the call to override the senses by performing prodigious feats of the imagination has been a feature of scientific development since the days of the Greeks. Moreover, in a neat rhetorical move, he implies that the unconvinced public no longer includes the reader, who is invited to share in Playfair's reminiscence on the story of Aristarchus, which he does not deign to tell in full. His reader, presumably educated, knows Greek literature, and the history of science, well enough to fill in the details. We are not the readers who need to be convinced, for we already know the "curious proof" offered by Aristarchus.

Though historians agree that, even under Playfair's pen, Hutton's theory attracted very few scientific followers even in Edinburgh, suddenly it acquired the force to make people angry. By 1808 the Wernerian Robert Jameson was so exasperated with Playfair's dominance at the Royal Society of Edinburgh that he defected and set up a rival club, the Wernerian Natural History Society, for those who continued to believe that the earth's strata had been deposed in sequence from a thick aqueous solution, and were not being perpetually formed by fire in the bowels of the earth. Edinburgh's savant community was split in two. By 1812 Scott himself had become involved. Having commissioned a play for the Theatre Royal Edinburgh from his friend George Steuart Mackenzie, Scott was shocked when, at its premiere, laughter erupted inexplicably and uncontrollably from the stalls. The play was supposed to be a serious melodramatic tragedy. Nobody was supposed to find it funny. On further investigation, Scott realized that geology was to blame. Angered by Mackenzie's belligerently Huttonian *Travels in the Island of Iceland*, a group of Wernerians had attended the opening night with the intention of exacting their revenge on Mackenzie by laughing his play off the stage. "The most mortifying part of the business," Scott wrote, "was that at length even those who went as the author's friends caught the infection and laughd most heartily."[24] Six years later, when *Blackwood's Edinburgh Magazine* mercilessly parodied every leading figure of the Edinburgh literati, to the general outrage of the city, Scott reserved his most vociferous defenses for Playfair, whom *Blackwood's* had accused of atheism. By now Scott was up to date with the Huttonian controversy, and he was offended by the "severe and personal" nature of the affront. "Although Playfair has never been suspected of orthodoxy," he wrote, "yet I know not that he has upon any occasion made any attack on religion."[25]

Scott knew Playfair because they ate dinner and drank expensive wines together at the Friday Club they had together founded in 1803, along with such literary luminaries as Dugald Stewart, James Hall, Francis Jeffrey, the famous editor of the *Edinburgh Review*, and the publication's cofounder, the acerbic wit Sydney Smith. The club styled itself as the most exclusive literary gathering in Edinburgh. It had no committee, no rules, and no votes on new admissions. Instead, existing members could informally suggest an eminent friend for invitation to dinner. If he was deemed suitable, and if he proved his suitability as both sociable clubgoer and intellectual spark, he could tentatively be admitted into the fold. His "qualifications" would most importantly include "a taste for literature," with "literature" referring not solely to imaginative writing but also to the humanistic scholarship of the Edinburgh Enlightenment, including history, science, economics, and sociology. He would need "agreeable manners, and above all, perfect safety," the absolute conviction that theological debate did not constitute atheism or heresy.[26] And he would need to be rich. The club incurred deliberately high dinner bills to ensure it remained "more comfortable, and more select," than other literary societies.[27] "In twenty-four years three of our members, and only three have complained of this," one member later recalled, "which has generally made us more extravagant when they were present."[28] Starting out as a weekly supper club, the men drank "a very pleasant but somewhat dangerous beverage, of

Rum, Sugar, Lemons, Marmalade, Calves foot jelly, and hot water."[29] Later, when it became a monthly dining club, claret and extravagant French wines were the libations of choice. Though the more dignified vinosity of the Friday Club seems not to have produced antics of the order of "high jinks," still, at a particularly drunken meeting, Playfair was said to have attempted (and failed) to steal an image of Galen's head from above an apothecary's shop.[30] As at the Geological Society of London, liquor and literature were the principal ingredients in a specifically gentlemanly concoction.

It has been claimed that the Friday Club was formed in order to defend Huttonian principles.[31] Certainly Jameson and his fellow members of the Wernerian Society never belonged, and some of the club's members were proponents of the Huttonian theory. But most were not involved in the dispute at all. Writing to Joanna Baillie about Mackenzie's play in 1812, Scott, for instance, seemed not even to have heard of that "set of chemists called Wernerians, who it seems differ in their opinion concerning the cosmogony of the world from [another] sect of philosophers the Huttonians."[32] This is a peculiar statement indeed from a man who was apparently a founding member of an avowedly Huttonian literary society. And yet Scott defended Playfair more vehemently than any other figure in the course of the *Blackwood*'s scandal. He did so, we might conclude, on broader grounds, as a defense of the gentlemanly sociability embodied by the Friday Club rather than on conviction of Huttonian principles. Committed to gentlemanly intellectual discussion in surroundings of "perfect safety," in which party politics and theoretical disputes were "avoided as much as possible," Scott could hardly sanction an accusation of atheism leveled at his personal friend in scurrilously public print.

The Friday Club was not a Huttonian society, then, but it did constitute a powerful literary coterie of which Playfair was a leading member. And in this sense, it gave him power. John Gibson Lockhart, one of the *Blackwood*'s authors and later Scott's son-in-law, was canny enough to see that he needed to make literary reparations to the aging philosopher. In his semifictionalized satire of Edinburgh society, *Peter's Letters to His Kinsfolk* (1819), Lockhart rewrote Playfair not as an atheist but as an uxorious septuagenarian "specimen of buoyant spirit and muscular strength," taking part with "easy hilarity and good humour" in "a trial of strength in leaping" in the gardens of Jeffrey's villa.[33] Jeffrey himself commissioned a prodigious sixty articles from Playfair for the *Edinburgh Review* between 1802 and 1819, and claimed that he was the greatest writer of his generation—better, indeed, than Edmund Burke, Samuel Johnson, or Edward Gibbon.[34] Science and literature were not separate beasts in this collegiate Edinburgh culture. Science was *part of* literature, and Playfair was better placed than his opponents to give it resounding voice. It was this that constituted his greatest threat to the Wernerians.

Most importantly, Playfair was not simply the pied piper of Hutton's cosmology, attracting followers with mellifluous prose in one hand and a bottle of claret in the other. As his essays for the *Edinburgh Review* show, the literary dexterity required to write good scientific prose was of serious and central significance to his philosophic practice. As a mathematician, Playfair was part of a British educational system in which Euclidean geometry was used to train young men in the appropriate habits of thought for gentlemanly life.[35] The canonical texts of Euclidean geometry operated via a synthetic mode of reasoning, meaning that they began with general truths and deduced from them particular facts or solutions. Such a method presented the world as a fixed system, as befitted a sound liberal education for the educated classes. Young men, moving through a series of logically organized problems, were expected slowly to deduce the logical arrangement of the universe. As they did so, they would acquire the mental strength and organization to think accurately about the world. In short, they would become gentlemen. For Playfair, not only should the organization of the world and of the mind reflect one another, but the written texts used to teach this mathematics should also exhibit logical arrangement. "Fully aware of the value of the truths he had to unfold," Euclid had "never sought to enhance that value by any artificial means either of concealment or display" on the level of his prose.[36] Only writers with faulty reasoning to hide would resort to ornamental or showy style; bad mathematics would inevitably lead to esoteric phraseology and poorly constructed prose. Form, we might say, mattered.

The world, the textbook that described it, and the mind of the gentleman: each would reflect the other in perfect harmony and balance. A poorly written textbook could throw the whole scheme off center. "The works in that science," Playfair wrote of geometry,

> by the adaptation of their parts, may exemplify *quantum series, juncturaque pollet*; but can hardly illustrate any other of the rules of the critic or the orator. They admit no expression purely ornamental, and reject everything that can withdraw the attention from the main object. To metaphor and variety of expression they are peculiarly averse, and the geometer must never forget, that the transparency of a medium may be injured by the flowers scattered on its surface, no less than by the mud diffused through its mass.[37]

"Quantum series, juncturaque pollet" is a quotation from Horace's *Ars Poetica*, in which Horace claims he "would so execute a fiction taken from a well-known story, that anybody might entertain hopes of doing the same thing; but, on trial, should sweat and labour in vain. *Such power has a just arrangement and connection of the parts: such grace may be added to subjects merely common*" (my italics).[38] The same story told differently, Horace says, would not have the same effect. The devil is not in the detail, but in the form, in the precise mode of arrangement by which "subjects merely common" were related. Good geometry, in Playfair's conception, shares that much with Horace's fiction, and his rewriting of Hutton depended for its scientific persuasiveness on its elegant construction.

This fictionalizing aspect of literary form in scientific arrangements would become critical to the Geological Society's rejection of cosmogony and cosmology in its first few decades. Playfair's interpretation of Horace's phrase is telling, for it also alludes to Francis Bacon's seminal text *The Advancement of Learning*, which had been published in a new edition in 1808. Here Bacon criticized Euclid's synthetic, step-by-step description of his methods precisely *because* they added grace to mathematical propositions by their persuasive arrangement. As such, they were "more fit to win consent or belief" than to stimulate further inquiry of their propositions, "carry[ing] a kind of demonstration in orb or circle, one part illuminating another," satisfying readers more by form than by truth. *Tantum series, juncturaque pollet, Tantum de medio sumptis accedit honoris*, Bacon wrote, "a man shall make a great show of an art, which, if it were disjointed, would come to little."[39] But for Playfair, geometry was licensed to make these "adaptations" of its parts to form a plausible whole, and to engage in pseudo-rhetorical descriptions of its methods, not in the fear that fiction might intrude into mathematics, but because these literary techniques revealed the beautiful symmetry of mathematics itself. Indeed, Bacon's suspicion of the power of literary form to "carry a kind of demonstration in orb or circle, one part illuminating another," represented the highest point of philosophical completeness for Playfair. It might be a description of the theories of Newton or Hutton, each part revealed in harmonious grammatical relation to the whole.

That is not to say that the problems of fiction did not trouble Playfair. In his essays for the *Edinburgh Review* he promoted the superiority of French mathematics, controversial enough in the decades of the Napoleonic Wars, but also controversial for its *analytical* method and use of imaginary symbols to denote unknown quantities and of negative numbers that have no existence in physical reality.⁴⁰ In Britain, such a mode of reasoning carried more than a touch of metaphysical and moral ambiguity: negative numbers were seen as "mysterious and fantastick," and reasoning with numbers that did not exist was derided as "discoursing, since it deserves not to be called reasoning."⁴¹ It was, rather, a play on words, a set of signs without referents. Playfair, too, worried about the unknown principle by which "symbols alone," to which "no idea is attached," could be mechanically manipulated to produce solutions and "quantities that really exist," but he justified the analytic method on the grounds of its usefulness.⁴² Algebra, he wrote, constituted "the most general and most perfect language that has yet been employed in science," the least prolix language in existence, "invented expressly for the purpose of assisting the mind in the management of thought," serving "to make the individual think with more accuracy or extension."⁴³ It was equally well suited to a gentlemanly liberal education as Euclidean geometry.

If only, Playfair lamented, Werner had learned the lessons of the continental mathematicians. As he put it, "the history of Nature should be written" in plain descriptive prose, free of any "inference which may be drawn from it" and unencumbered by theory. "The fact however is," he sighed,

> that language is but ill provided with terms adapted to this sort of description. The mind passes so rapidly from external characters to that which it conceives them to signify, that it hardly rests at all upon the former, and is not always at the pains to give them names. This method of proceeding is very well adapted to the ordinary business of life, where we are, in general, more concerned about the things that are signified, than about the signs which point them out. In philosophy, however, it is often quite otherwise. When the thing signified is not obvious, and is not to be found out but by the comparison of many instances, and an extensive induction from particulars, there is no surer barrier against the discovery of truth than those imperfections of language which force us to a premature interpretation of the signs, before a due comparison of instances can be made.⁴⁴

In using the term "formation" to denote not "the act of forming, but ... the thing formed," Werner had constantly inferred in his description of the rocks "the fabulous origin of the strata, and involves the notion of rocks formed by simultaneous deposition from water" in what should simply have been mere description.⁴⁵ For Playfair, the mathematician's ability to manipulate "signs," and to make "comparison of many instances" before attributing meaning to them, was, in Werner's description, bypassed. Werner's attempt at the "vast power of Generalization" achieved by Hutton was constantly undermined by its increasing inconsistency "with known facts," as granite and serpentine and porphyry were discovered in places "inconsistent" with the theory of their deposition and in "situations which they were not originally meant to occupy."[46] The proof of Werner's poor arrangement of the earth was in the pudding. Unlike Euclid's writings, Werner's prose contained everything "mysterious, obscure," and "paradoxical," was filled with "imperfections of language," technical vocabulary "uncouth in its sound, inartificial in its formation, and ... studiously inaccurate," and was veiled in "theoretical obscurity."[47] If Werner had succumbed to the disease of inaccuracy, we might conclude, it was the dark "veil" of his prose that had enabled the diagnosis.[48]

Only by a perfect harmony between synthetic and analytic methods could the charge of fiction be avoided. Playfair wrote in the *Illustrations* that while some men had understood "that all theory is mere fiction, and that the only alternative a philosopher has, is to devote himself to the study of facts unconnected by theory, or of theory unsupported by facts," this is patently untrue:

> The philosopher who has ascended to his theory by a regular generalization of facts, and who descends from it again by drawing such palpable conclusions as may be compared with experience, furnishes the infallible means of distinguishing between *perfect science* and *ingenious fiction*. Of a geological theory that has stood this double test of the analytic and synthetic methods, Dr. Hutton has furnished us with an excellent instance, in his explanation of granite. [Playfair's emphasis.][49]

Hutton had ventured out into the field to make original observations on granite, which suggested to him that "it had been melted, and injected with fluid, among the stratified rocks already formed." From observation he had proceeded to theory, for which he searched for more evidences in Arran and Glen Tilt as confirmation. Theory unsupported by facts was "ingenious fiction," but theory generated from inductive evidence and confirmed by new proofs was "perfect science." It was just that it had been left to Playfair to prove it so.

Hutton's prose, Playfair claimed, creating a space for his own rescue of the theory, had also been obscure and inaccurate, his theory "proposed too briefly, and with too little detail of facts," supposing "in the reader too great a knowledge of the thing described," and embodying "peculiar notions of arrangement." But these faults derived from Hutton's "care" "to render" his theory "strictly logical," and they were false indicators that the theory itself might be untrue.⁵⁰ Rewriting Hutton's theory, Playfair needed to reestablish the perfect fit between world and text that constituted good science. He would prove that Hutton's theory made grammatical sense. In its stylistic sophistication, then, his *Illustrations* transformed Hutton's theory into the comprehensive synthesis it had initially claimed to be. It was not just that he wrote more elegantly than Hutton. It was that, from within a powerful literary coterie in Edinburgh, and on the pages of the most influential literary review of the day, Playfair's elegant prose had created for Hutton's theory the kind of persuasive philosophical completeness of which Horace would have been proud. It was by giving Hutton's earth a new form, in making the kind of persuasive arrangement of Hutton's proofs that Bacon might have considered a "fiction," that Playfair had manage to pack an implausible theory with such plausible punch.⁵¹

Novel Forms

If the charge of fiction hovered around scientific theorizing in Edinburgh at this time, scientific theorizing did not go undiscussed on the pages of the city's fiction. A year after Playfair's death his friend Scott became president of the Royal Society of Edinburgh, one of the most eminent scientific societies in Britain, of which Playfair had been general secretary for fourteen years. Characteristically, Scott seems to have found the whole thing quite funny. On his appointment he joked that "in future" he would have his friends "respect my opinion in the matter of chuckie-stanes, caterpillars, fulminating powder, and all such wonderful works of nature. I feel the spirit coming on me, and never pass an old quarry without the desire to rake it like a cinder-sifter."⁵² "I have been chosen President of the Royal Society," he wrote to a leading Tory politician, "which keeps one feather out of a Whig bonnet."⁵³ Four years later, Scott included the following stanza in "St. Ronan's Well":

And some rin up hill and down dale,
knapping the chucky stanes to pieces with hammers,
like sae many road makers run daft.
They say it is to see how the world was made.⁵⁴

Not for Scott the desire to know the world and its systems better. Rather, he seems to have accepted the presidency as little more than a tactical political move—and, perhaps even more tellingly, for the chance to have a few more good dinners and bottles of rum and claret.

But Scott also saw his lack of scientific interest as a *qualification* for the role of president rather than a potential barrier to his success. Criticizing those men who were "born to be splitter[s] of hairs in argument and gatherer[s] of pebbles in science," he claimed he would become no such pedant:

> Being an anxious vindicator of prerogative in all establishd authorities I am not likely to forsake my claim to that which is thus happily vested in my own person and therefore uphold myself to be a better judge how the World is made than if I had been a looker and capable ex officio.... Meantime I have only thought it necessary to get up for my inaugural oration the well-worn opinion of Mr. Jenkinson in the Vicar of Wakefield upon the Cosmogony of the World.[55]

Ephraim Jenkinson is the amiable conman in Oliver Goldsmith's 1766 novel. He first dupes the young son of Charles Primrose, the eponymous vicar, into selling one of the family horses for "a groce of green spectacles" with worthless copper rims, then dupes the vicar into selling a second horse for nothing at all. As it later transpires, he has also worked his charms on the vicar's next-door neighbor. Most importantly, Goldsmith plays on the literary cliché of the gullible, unworldly hero by making his villain pretend to possess knowledge about "how the World is made." "The world is in its dotage," Jenkinson tells Primrose,

> and yet the cosmogony or creation of the world has puzzled philosophers of all ages. What a medly of opinions have they not broached upon the creation of the world? Sanconiathon, Manetho, Berosus, and Ocellus Lucanus, have all attempted it in vain. The latter has these words, *Anarchon ara kai atelutaion to pan*, which imply that all things have neither beginning nor end. Manetho also, who lived about the time of Nebuchadon-Asser ... formed a conjecture equally absurd; for as we usually say *ek to biblion kubernetes*, which implies that books will never teach the world; so he attempted to investigate.[56]

With a heavy dose of dramatic irony, Goldsmith shows the vicar a fool for being bamboozled by the sound of garbled Greek and a hotchpotch of ancient philosophies.[57] As the text progresses and we hear Jenkinson repeating his speech, always attempting to trick yet another unwilling dupe, he too is made absurd. Each time we hear it, the speech becomes less powerful—indeed, the last time we hear it, Jenkinson is in prison, and his trickeries have got him less than nowhere. But the central structural irony of the novel is that Jenkinson ends up deliberately tricking the vicar not with his cosmogonical speculations and not for his own gain, but for the vicar's own good. With this comic deflation of the power of cosmogony, Goldsmith mocks the virtue of worldliness and ensures his gullible hero comes up trumps. Within this structure, cosmogony becomes as "absurd" as Jenkinson claims. Just as invoking cosmogony is a useful trick, by which he can feign a cleverness he does not possess, cosmogony itself, his speech hints, might be a preposterous philosophical game by which men of science make claims to knowledge of the universe that stretches far beyond what they can possibly understand or see. In the end, the domestic, the prosaic, and the trivial win the day.

Scott was only joking, of course. But jokes are no less revealing than drinking games or stories. The claims circulating around fiction here recall those claims made by Horace and Bacon that the formal arrangement of "common subjects" could make those subjects far more attractive and plausible than they might be on their own. Such formal arrangements might merely be fictions, and fiction, as the *Vicar of Wakefield* showed all too well, could quickly become lies. Under Jeffrey's editorship of the *Edinburgh Review*, this too-persuasive dimension of fictional form was a frequent target of criticism. "The great objection" to novels, Scott and Jeffrey claimed,

> is that they are too entertaining—and are so pleasant in the reading, as to be apt to produce a disrelish for other kinds of reading which may be more necessary, and can in no way be made so agreeable. Neither science, nor authentic history, nor political nor professional instruction, can be conveyed in a pleasant tale; and, therefore, all these things are in danger of appearing dull and uninteresting to the votaries of those more seductive studies.[58]

The "too entertaining" form of the novel coils readers into its rhythms and structures, absorbing them until they lose themselves, lose thought, in the fictional world. Women readers were often depicted in the *Edinburgh Review* as unable to put down their books until their desires to know the plot had been fully satiated, physically exhausting themselves and neglecting their domestic duties. As it is described here, the novel also "seduced" young men, diverting them from their more serious studies of "science," "history," or "professional instruction"—studies given extensive coverage on the pages of the *Edinburgh*. In such cases, the damage could be permanent, for novelistic pleasures could "produce" a *habitual* "disrelish for other kinds of reading." Young men, inveigled by the tantalizing form of the novel, might become incapable of turning back to the more rigorous reading demanded of them as they trained to become lawyers or politicians or men of science. The isolating power of the novel, its pleasures most easily consumed in bedrooms or armchairs, in private spaces and in snatched moments, was directly opposed to the more sociable intellectual pursuits on which masculine civil society was seen to rest—the pursuits of university, club, and learned society. And its engrossing powers directly opposed it to the kinds of detached rational thinking required for the intellectual training young men needed to undergo.[59] Euclidean geometry this was not.

As several critics have noted, the construction of the novel both as feminine *and* as potentially desirable opened up a space for its potential reconfiguration in masculine terms. Enter Walter Scott. Certainly Scott did not abandon compelling plot. Reviewers in the *Edinburgh Review* and the *Quarterly Review* praised Scott's truth to nature and his descriptive powers, and were only partially able to hide their enthusiasm for his wild landscapes and stories.[60] Subsequent attempts to claim Scott as a "realist" novelist have only uneasily contended with the gypsies, pirates, robbers, abductions, prophecy, love affairs, invasion scares, and battles in which his plots revel.[61] And two hundred years after the fact, some critics continue to accuse Scott precisely of having been "too entertaining," his fictions converting Scotland into a "romance" territory that still functions as a locus of European fantasies, holidays, or film sets, unable to achieve political credibility on the international stage.[62] Scott's plots were at least as thrilling as the plots of his Gothic contemporaries. But they were also infused at every turn with the convivial, tongue-in-cheek spirit Scott had represented in "high jinks." Even as he plunged his readers into the historic worlds of the Highland clans or the medieval courts, plumbing folk ballads and courtly "romance" literature for his source materials, Scott took canny joy in jolting his readers from their private engrossments in the fictional world and back into comic acknowledgment of the present. In an early chapter of his first novel, *Waverley* (1814), for instance, Scott jokingly reminds us that his mode of telling the story is that of "a humble English postchaise, drawn on four wheels, and keeping his majesty's highway":

> Those who dislike the vehicle may leave it at the next halt, and wait for the conveyance of Prince Hussein's tapestry, or Malek the Weaver's flying sentry-box. Those who are contented to remain with me will be occasionally exposed to the dulness inseparable from heavy roads, steep hills, sloughs, and other terrestrial retardations; but, with tolerable horses, and a civil driver, (as the advertisements have it) I also engage to get as soon as

possible into a more picturesque and romantic country, if my passengers incline to have some patience with me during my first stages.⁶³

The narrative, as journey, is inconvenient, incommodious, dull, and requires the "patience" of its reader-travelers, who are invited to "leave it at the next halt" if they prefer to travel aboard magic carpets or flying sentry-boxes. Scott warns us, quite appropriately as it turns out, that he will tell us a story, but we will never be allowed to fully immerse ourselves in it, for it will take the most circuitous routes to its logical ends, so that we notice the bumps and jolts, the "heavy roads, steep hills, sloughs, and other terrestrial retardations" from which the plot is constructed more than we will care for the plot itself. Whenever we might forget ourselves in the pleasures of Scott's story, he pulls us back in this way to a disarming recognition of the arbitrariness of the route he has constructed, allowing us the critical distance to stand back and judge the scene, even to turn away from it. He never quite lets us become absorbed in his tales without wrenching us back into an awareness of the practices by which they have been told.⁶⁴

Scott's most distinctive narrative trick was to swing to the opposite extreme, to the equally obsessive but much graver pursuits of the antiquarian collector. As a renowned antiquary of ballads, folk songs, dialects, customs and traditions, Scott not only drew on these collected materials as he constructed his historical fictions. He also claimed antiquarianism as the basis of his fictional form. This is surprising, given that, while antiquaries collected and debated the significance of pots, statues, medals, coins, fossils, ballads, forgotten manuscripts, and a whole host of other historic objects, they were well known for resisting the incorporation of those objects into the grand narratives of historical progress promulgated by the philosophical historians of the Edinburgh Enlightenment.⁶⁵ Caricatured for their monomaniacal obsessions with particular objects or debates, with the endless accumulation of proofs, evidences, and objects, antiquaries were marginalized in Edinburgh as misers and pedants unable to connect isolated facts into overarching plots. Enlightenment philosophers likened antiquarianism to language in its barbaric phase, an ever-proliferating clutter of nouns that remain unconnected by the more sophisticated verbal constructions unique to advanced languages and cultures. Here again is the sense of fit between language, form, and patterns of thought so important to Playfair's geological systematizing. But for Scott, purveyor of some of the most exciting fictional narratives of the early nineteenth century, the discombobulated prose of the antiquary became both the source of an important comic vein in his writing and a crucial strategy by which he could differentiate his work from what had been caricatured as the frivolous excesses of the contemporary novel.

On one level the voice of the antiquarian pedant enabled Scott to construct a defense of the arts of fiction and storytelling. In the prefaces and "dedicatory epistles" framing many of the novels, and the introductions to the 1829–1833 Magnum Opus edition, Scott invented an array of comical antiquarian-critics who repeatedly reminded readers of the imperfections of his fictional practices. Such ironically named figures as "Captain Clutterbuck" or "the Rev. Dr. Dryasdust" chastise Scott for debasing his historical sources by mingling them with fiction, for getting his facts wrong, for "causing history to be neglected, readers being contented with such frothy and superficial knowledge as they acquire from your works, to the effect of inducing them to neglect the severer and more accurate sources of information."[66] Appropriating the complaints made by the quarterly reviews against frothy feminine fictions, and applying them in grave tone to Scott's work, Dryasdust argues that his novels mislead "the young, the indolent, and the giddy, by thrusting into their hands works which ... leave their giddy brains contented with the crude, uncertain, and often false statements which your novels abound with."[67] Scott refutes the charges, claiming that his fictions create, rather than quench, a "thirst for knowledge" in his readers, and accords himself as much poetic license as he requires. Clearly these antiquarian critics enable Scott to preempt and partially control the reception of his novels as gentlemanly texts. But on another level, Scott ironically claims that, like an antiquarian, he is resistant to the machinations and self-determinations of plot itself. In the preface to *Fortunes of Nigel* (1822), for example, the authorial persona claims affinity with "Smollett, Le Sage, and others," who had written histories "of the miscellaneous adventures which befall an individual in the course of life," rather than telling

> the plot of a regular and connected epopoeia, where every step brings us a point nearer to the final catastrophe. These great masters have been satisfied if they amused the reader upon the road; though the conclusion only arrived because the tale must have an end, just as the traveller alights at the inn because it is evening.[68]

Preferring "miscellaneous adventures" to "regular and connected" plots, convenient endings to meaningful ones, the author of *Waverley* is rarely encountered without the claim that he "could never form a plot." This is "a very commodious mode of travelling, for the author at least," retorts Clutterbuck, in this instance: "In short, sir, you are of opinion with Bayes—'What the devil does the plot signify, except to bring in fine things?'"[69] In an anonymous review of his own *Tales of My Landlord* for the *Quarterly Review* Scott wrote (of himself) that

> we must own that his stories are so slightly constructed as to remind us of the showman's thread with which he draws up his pictures and presents them successively to the eye of the spectator. He seems seriously to have proceeded on Mr. Bays's maxim—"What the deuce is a plot good for, but to bring in fine things?" Probability and perspicuity of narrative are sacrificed with the utmost indifference to the desire of producing effect.[70]

Even Scott's fictional antiquarian critics accuse of him of paying scant attention to the plot, of careless fictional constructions, of a lack of verbal connection structuring the clutter of descriptive material that accumulates and multiplies on the pages of his novels. In this, Scott presents himself as more of an antiquarian than even Clutterbuck or Dryasdust, justifying the antiquarian clutter of details as the formal basis on which his fictions rest.

The ironic treatment of the narrative-as-journey trope and the assumption of the antiquarian's verbal paucity come together most explicitly in the opening chapters of *The Antiquary* (1816) as we wait, alongside the novel's young English protagonist, in "impatient" mood for a carriage.[71] An older passenger, the antiquary of the novel's title, berates a woman in the coach office for the carriage's lateness in terms so prolix that we seem to be being deliberately bored before the narrative has even begun. As in the opening chapters of *Waverley*, the author seems to invite us to step out of his carriage and to warn us that tedium will be par for the course. When the carriage finally arrives, momentum is again arrested as the antiquary discovers that his younger companion possesses some knowledge of Roman antiquities and begins to drown him (and us) in "a sea of discussion concerning urns, vases, votive altars, Roman camps, and the rules of castrametation."[72] This long-winded discussion overtakes narrative movement, for "the pleasure of this discourse had such a dulcifying tendency, that although two causes of delay occurred, each of much more serious duration than that which had drawn the antiquary's original "wrath," he now "rather seemed to regard the interruption of his disquisition than the retardation of his journey."[73] Description has become more important than event. Finally, the antiquary deliberately obstructs the onward journey in order to show "his companion a Pict's camp, or Round-about, a subject which he had been elaborately discussing" and which happens to lie close to the route. They consequently miss the boat they were hoping to catch. At this point, we still have absolutely no idea where these two men are going, why they are going there, or what place they will occupy in the ensuing narrative. It is not until the second chapter that we learn either of their names, and in the third and fourth chapters we are confronted with even longer, plotless sermons from the antiquary, Jonathan Oldbuck, on his favorite antiquarian debates, many of which he has clearly misinterpreted according to his own peculiar obsessions. Oldbuck is hardly the giddy young reader seduced by the feminine charms of the novel—he is middle-aged, cranky, misogynistic, and scholarly. The narrative momentum of Scott's novel is grounded on his perverse, antinovelistic pleasure in the fundamental irreconcilability of things and ideas, in the quiddity of objects, in the breakdown of narrative schemes. Indeed, Scott has to remind us in the prefatory advertisement to the novel, that he, like Oldbuck, had "been more solicitous to describe manners minutely than to arrange my case in an artificial and combined narrative, and have but to regret that I felt myself unable to unite these two requisites of a good Novel."[74] Scott and his antiquary take mutual delight in pedantic scholarly debates surrounding things and objects to the exclusion of the relations that might be plotted between them. The only thing we are left doubting here is Scott's "regret" that his novel lacks narrative combination.

Invoking antiquarianism, Scott gave himself license to tell enticing stories, and gave his readers license to forget themselves in those stories, safe in the knowledge that the form of his novels would always break down, draw attention to itself, getting stuck in the quagmires of intellectual disquisition in its most extreme form and the clutter of proliferating description. Fluctuating between these two obsessive extremes—the giddy, youthful, sexual pleasures of novelistic plot and the middle-aged, pedantic bachelorhood of the monomaniacal antiquary—Scott ironized them both, so that the overall effect is for the novels to emerge as a middle ground between two polarized literary forms.

Moreover, as Ferris has noted, Scott's paratextual materials frequently "assume the form of letters or dialogues and invoke the recollection of sites and scenes of conversation and exchange." As such, they define "the act of narration as an extension of the sociability" on which so much of Edinburgh culture was founded.[75] Scott combined the thrills and spills of fictitious narrative— and, we might add, their opposite, the serious and pedantic researches of the grave antiquarian—with the "prose ... of the Edinburgh Enlightenment: a professional, institutional language, administrative and legal and academic," a language that reminded his readers, as Clutterbuck puts it, that his work contained "too many things *Quae maribus sola tribuunter* [which may be allotted to males alone]" to have been written by a woman.[76] Accordingly, in Scott's company, the same Jeffrey who had previously chastised the novel for its "too entertaining" plots was now, with no sense of contradiction, actually pleased that his novel could draw "us irresistibly away from our graver works of politics and science, to expatiate upon that which every body understands and agrees in."[77] The operative terms here are clearly "us," "our," and "every body," referring quite specifically to men whose habitual reading matter is made up of "politics and science," men less likely to lose themselves in the "too entertaining" forms of the novel than in the "graver works of politics and science." They might even, we might add, pull the miserly antiquarian back into this mediating social realm, which "every body understands." The author of *Waverley* was "evidently a person of a very sociable and liberal spirit—with great habits of observation—who has ranged pretty extensively through the varieties of human life and character, and mingled with them all."[78] With their oscillations between the addictive plots of the contemporary novel and an antiquarian rebellion from story, the novels shifted focus from the tale onto its telling, adopting an ironic stance that was of a piece with their vivacious depictions of a "high jinks" brand of male sociability. In this way, they constructed themselves as texts infused with the "sociable and liberal" spirit of the clubs and learned societies on which sociable and intellectual interaction was founded, and gave the novel the literary status so desired by Playfair and the reviewers themselves.

It is in this light that we can consider Scott's claim that his lack of scientific knowledge made him a suitable president of a scientific society. He would be no gatherer of pebbles, but neither would he become engrossed in the grand cosmological and cosmogonical disputes about, as he put it, "how the World is made," disputes that had threatened to destroy the very fabric of the Royal Society just a few years before. Having learned the lesson of Ephraim Jenkinson, Scott would not become so passionate about scientific questions as to forget himself defending them. As his speech to the Royal Society of Edinburgh proposed, with tongue firmly in cheek, viewing empirically observed data through the lens of one theory or other had led to too many rival systems, to a level of generality that had made science absurd. It was this that had provoked such vehement debate in Edinburgh and had threatened to turn science, like George Mackenzie's play, into a public joke. And it was this that the Geological Society, much more seriously than Scott, would publicly proclaim to avoid. Just as Pleydell only tells the truth on Saturdays, in *The Vicar of Wakefield* untruths become the ironic route to the establishment of order and justice. It was this ironic approach to knowledge that Scott celebrated, and it was this approach that the gentlemen of the Geological Society appropriated when they played "high jinks" in the rooms of William Cole. If geology entered its most prestigious phase in the hands of these men, it did so, the next chapter will argue, precisely by rejecting the cosmological thinking of Hutton and Playfair and choosing instead the more quotidian, contingent, and self-ironizing forms with which Scott had bartered a new prestige for the novel.

The Story Undone { **CHAPTER TWO**

Soaring over the eastern end of Prince's Street in Edinburgh is a soot-soaked, two-hundred-foot monument to the memory of Walter Scott. Built in the 1840s, at a cost of over sixteen thousand pounds, the cumbrous Gothic spire stands partly as testament to the imaginative spell Scott cast over a generation of nineteenth-century geologists. Several of the leading gentlemen of the Geological Society had visited Scott in the 1820s at Abbotsford, his mock-ancestral castle in the Scottish Borders, and others had excitedly made his acquaintance in London in the same decade.[1] Charles Darwin saw him in the presidential chair at the Royal Society of Edinburgh, and "read and re-read" Scott's novels "until they could be read no more," in the cozy armchairs of the Athenaeum club and out loud to his wife, Emma.[2] When Scott died in 1832, several eminent geologists, wishing to express their "national gratitude" to the author, became committee members of a fund to purchase Abbotsford and secure it to his children. When this plan failed, they later contributed money to the monument itself.[3]

It is hardly surprising that they should have had so much respect for the inventor of "high jinks." Under Scott's gentlemanly penmanship, the suspect form of the novel, with all its seductive stories, had been given a new respectability. Threatened by the cosmologies of biblical literalists on the one hand, who urged that the earth was only six thousand years old and had been created in six days by God, and by the discredited philosophy of Hutton and Werner on the other, geology too needed to rewrite a story—the story of the earth—in a newly credible form. And the terms of that credibility were very much defined by the values of the heavyweight quarterly periodicals. Many of the gentlemen geologists of the Geological Society wrote essays for the quarterlies and regularly socialized with such men as John Gibson Lockhart, Scott's son-in-law and editor of the *Quarterly Review* from 1825, and Sydney Smith, the famously acerbic wit who had cofounded the *Edinburgh Review* in 1802. Everybody read the quarterlies, which provided literary gossip and hot topics of debate for their parties.[4] When one geologist met another, as when Charles Lyell met Alexander von Humboldt in Paris in 1823, they could use their shared reading of the *Edinburgh* and the *Quarterly* to break the ice. Indeed, Lyell was befriended by the geologist Leonard Horner, the son of Scott's dramatist friend Joanna Baillie and the man who had introduced Darwin to the Royal Society, precisely because he had "got into the 'Quarterly Review' an article on the liberal side of geology," for which Horner was deeply grateful.[5] The literary culture of the Geological Society gentlemen, then, was the literary culture forged by Scott, and John Playfair, in the clubs of Edinburgh a generation before.[6]

Like Playfair, the geologists would use the quarterlies to carve out a culturally authoritative space for their particular brand of science. But it would not be Playfair's cosmological theorizing that they would plead for. In the *Edinburgh Review,* the geologist William Fitton advertised the new "matter-of-fact methods" in geology, and asked geologists and general readers to ignore "well-organised theories which, a short time since, created so much controversy."[7] Geology would establish itself as a venture that "deserves ... the name of science" not by telling tall tales about the system of the earth, but by patiently and humbly unraveling the order and structure of the earth's rocks.[8] The patterns of causation that might make narrative sense of the strata, connecting them in a simple story, were to be rigorously circumscribed. As Fitton described it, "fire and water are still indeed opposed to each other, as in days of yore; but it is now a modest and well-ordered struggle; more frequently resolving itself into a question about particular rocks, than embracing, as formerly, the whole series of mineral formations."[9] Pulling back from imaginative immersion in a totalizing system of the earth, geologists would instead submit themselves to the more difficult labor of "minute processes of observation" on particular rather than general issues.[10]

Avoiding "those more easy and tempting speculations which, till of late, it has been the fashion to dignify with the name of Geology" was necessary if geology was to shake off its controversial reputation, especially important in the socially turbulent 1820s and 1830s.[11] "A modest and well-ordered struggle" was not likely to produce disciples so zealous that they would laugh tragedies off the stage or storm out of the learned societies, or convert readers to atheistic stories of a godless universe. But redefining geology as a discipline of modesty and order also ran the opposite risk. This kind of geology might attract no disciples at all.[12] Recalling the burial of his friend, the botanist Robert Brown, to few mourners on Kensal Green in 1858, the eminent imperial geologist Roderick Murchison later recalled "how different was the day before yesterday when the popular novelist [William Thackeray] was interred in the same place!" to a crowd of thousands. Novelists, he feared, were better loved than men of science. But then again, "a different verdict could hardly be expected from the masses or the fashionable world. Every one knows *Cornhill* and *Punch*, *Pendennis* or *Vanity fair*, or some one of Thackeray's good novels." It was unsurprising, then, that Thackeray "obtained a good share of the public applause which the nation accorded to Walter Scott, whilst the *Princeps Botanicorum* of Europe dies unknown by English scribes."[13] Scott's achievement for the novel, giving rational, ironic, gentlemanly shape to captivating historical romance, had performed the double task of capturing the public imagination *and* securing an authoritative voice for the novelist among the gentlemanly elites—a voice quickly appropriated by Thackeray. In this regard, Murchison might not only have had gratitude on his mind when he helped set up the committee for the Scott memorial. Playing a part in the heritage of the nation's most beloved author, the notoriously expansionist and nationalistic Murchison presumably hoped for a little of Scott's limelight for himself, and for his geological colleagues. As this chapter explores, the geologists' imaginative and cultural identification with Scott offered them a crucial strategy for this process of scientific self-definition and helped give form to their new vision of the earth.

Sinking the Triple-Decker

Murchison may have been a little put out at the ignorance of science among "the masses" and "the fashionable world," but it was his most important collaborator, Adam Sedgwick, Woodwardian Professor of Geology at Cambridge from 1818, and president of the Geological Society in 1830 and 1831, who most acutely felt the tension between the new geological methods and the need to create public appeal for geology. Blunt, bluff, and affable, Sedgwick was a vicar's son from Yorkshire who had matriculated at Trinity College in 1804 as a sizar, paying reduced fees in exchange for servant duties. Finishing fifth wrangler, Sedgwick went on to a prebendal stall at Norwich Cathedral as well as a distinguished geological career, and was widely known for his jocularity on the pulpit as well as in geological disputes. He liked a "blow-out" with his friends, and is fondly remembered in all the lives and letters of the geologists of his generation.[14] In 1855, aging and increasingly irascible, Sedgwick sat in the audience of the British Association for the Advancement of Science, watching Murchison, his one-time friend and colleague, poach swaths of strata from the "Cambrian" system it had been his life's work to define, and insert them into a wholly different section of the rock record, Murchison's own "Silurian" system.[15] Sedgwick stood up to speak. Theatrically "taking off a thick heavy great-coat in which he had been sitting," Sedgwick cleared his throat and looked up. A titter ran through the audience. "Oh," Sedgwick reassured them, breaking into a grin, "I'm not going to fight him!" As one geologist later remembered, "The smile passed at once into a good laugh and general applause through the room."[16]

The story of their friendship, collaboration, and subsequent disagreement is now well known, but it may be fruitfully recapitulated here. In 1831 both Sedgwick and Murchison decided to spend their summer vacations hammering at the "Palaeozoic" (or oldest fossil-bearing rocks). Sedgwick set out from the slaty regions of North Wales and moved southward, as Murchison advanced toward him from Shropshire and the Welsh borderlands in the south. By the end of that summer they had agreed: there were two stratigraphic systems that constituted the early history of the globe: Sedgwick's "Cambrian," in which there were very few fossils, and above it Murchison's "Silurian." But from 1839, with his publication of *Silurian System*, Murchison began to annex the Cambrian strata and incorporate them into his Silurian rocks. By the 1850s the dispute had spilled out into the popular press. Murchison's Siluria won the battle during the two men's lifetimes, despite Sedgwick's counterclaims that there was a stratigraphic break in Murchison's system, such that the "Lower Silurian" could be reinterpreted as Cambrian, and despite the subsequent discovery of a distinctive Cambrian fauna by continental geologists. When the two men died, Sedgwick felt he had been sidelined from the Geological Society, and the imperially minded Murchison had extended his Silurian interpretation over great swaths of rock the world over. He was the self-appointed "King of Siluria."[17]

It is often written that one of Sedgwick's greatest difficulties in this dispute was that "he had the misfortune . . . not to command so numerous and influential a following as his more socially fortunate opponent."[18] Murchison lived in Belgravia, had a brilliant and charming wife, and his home was famed for scientific soirees.[19] To this, however, it can be added that, though Sedgwick was well known for his eloquent public displays and his lively conversation at the club, he was also a bad writer.[20] Sedgwick was one of the only leading members of the Geological Society never to publish a geological book, though he repeatedly promised to write a history of the Palaeozoic strata. He wrote only one review in his life, despite repeated requests to do so. Part of the reason was snobbery. He refused to write even half a column reviewing Murchison's *Silurian System* in the *Times* in 1839, despite the fact that terms were still amicable between the two men, because he claimed he had "not time to write to newspaper editors any letters fit to be read," nor to read the book before the deadline, and would "not condescend to write in a paper which I think calumnious and dishonest, though very clever." Furthermore, he added, Murchison already had "the assistance" of the Sussex geologist and popular writer Gideon "Mantell, who having a little followed the trade of puffing can blow a better blast than myself."[21] Never mind the *Edinburgh Review*, Sedgwick here sounds more like Swift or Pope satirizing the puffs and hacks of Grub Street. Indeed, when he came to review "the foul book" *Vestiges of the Natural History of Creation* in 1844, "it brought Swift's satire" to his mind, filling him with "inexpressible disgust."[22] Bad science and cheap print constantly left a foul taste in Sedgwick's mouth. "Reasons!" Sedgwick exclaimed in his letter to Murchison, "Why, I have sent you reasons enough to sink a three-decker!"[23]

But it wasn't only snobbery about the ever-expanding range of print outlets that hindered Sedgwick's literary career. The majority of his publications are to be found in the *Transactions of the Geological Society*, whose circulation was limited to the organization's members, and even then he often relied on co-contributors to finish the articles for him, or delivered his contributions late and unpolished. Considering himself "constitutionally incapable of sedentary exertion," he saw his aversion to pen and ink as consequent upon his love of the hammer.[24] He even counseled his fellow geologists not to bother with good prose. Not only did he refuse to review Murchison's *Silurian System* for the *Times*, but he had already written to Murchison to tell him not to worry about the style of that sort, for "after all your book is only for geologists. Natives of the country will read and pick out parts of it, as Jack Horner ate Christmas pies; but, as a whole, it is far too good and deep for any but a true geologist. And what geologist will care, one grain of trap, how a sentence is written provided he understands it."[25] The masses, and the fashionable world, might well read the *Silurian System*. But they were nothing but plum-pluckers. Their reading tastes were trifling concerns to the business of elite geological conversation.

Stylish writing might be extraneous to good geology in Sedgwick's view, but when it came to geological works that fell outside the gentlemanly remit of the Society, writing still mattered, as it had to John Playfair a generation before. In his presidential address to the Geological Society of 1830, Sedgwick rounded on Andrew Ure, a chemist who had shocked the nation in 1818 by galvanizing the corpse of an executed murderer so that he seemed to come back to life, and then shocked the nation again by conducting a very public divorce from his adulterous wife. Ure was a member of the Geological Society, but his book, *A New System of Geology* (1829), an attempt to reconcile geology with a *literal* interpretation of scripture, was anathema to Sedgwick. This was not the kind of man, nor the kind of science, the Geological Society wanted to promote, and Sedgwick condemned the work as the latest of a long line of "monuments of folly" and "dreams of cosmogony" to have recently "issued from the English press."[26] In the published version of the address, Sedgwick wrote:

> The best of [Ure's] narrative is made up of successive extracts, often taken word for word, yet without the marks of quotation, from various well-known works on geology. Many of these extracts, although in themselves admirable, appear in the book before us but as disjointed fragments, in the arrangement of which the author has but ill performed the humble duties of a compiler. For in the secondary formations, we find enormous faults and dislocations of which there is neither any written record, nor any archetype in the book of Nature. Thus we find the lias sometimes below the oolites, sometimes between the oolites and the greensand.[27]

Thus Sedgwick demotes Ure from the status of author to the "humble" position of "compiler," and a bad compiler at that, one who fails to cite his sources and, unable to make a logical "arrangement," turns even "admirable" extracts into "disjointed fragments" of prose. This literary fragmentation in turn gives form to the natural world Ure describes, which, uncannily, is also characterized by disjunction, by "faults and dislocations" and the incorrect arrangement of its materials, the strata. Mangling his text, Ure has mangled his geology. Indeed, he has damaged the repute of geological science itself, having "pulled down and reconstructed" the "goodly pile" that the "Gentlemen" of the Geological Society "have helped to rear, after years of labour."²⁸ Thirty years before Sedgwick's address, John Playfair had claimed that elegant literary construction could be proof of the philosophical completeness of a scientific theory. Now such completeness was rejected as a fictionalizing arrangement of parts into suspiciously plausible wholes. But bad writing, clearly, could still signify "bad" science.

As it had for Playfair, this sense of the scientific value of well-arranged literary form had its roots in mathematics. Sedgwick's Cambridge background was fundamental to his particular brand of geology.²⁹ When he was awarded the Copley Medal by the Royal Society in 1863, Sedgwick was praised for his unique powers of conducting "mountain geometry," "that geometry by which we unite in imagination lines and surfaces observed in one part of a complicated mountain or district with those of another, so as to form a distinct geometrical conception of the arrangement of the intervening masses."³⁰ Sedgwick's field notebooks were filled with the workings-out of the dip, strike, and structure of the strata to "an unusual degree," a technique particularly useful to him as he spent his career examining the barely fossiliferous masses of slate in North Wales.³¹ It was this geometric ability that Sedgwick had been taught at Cambridge in the earliest years of the nineteenth century, a university in which "the focus and ultimate end . . . of mathematics" was the understanding of traditional Newtonian mechanics, and in which geometry was privileged (much to Playfair's chagrin) over the algebra perfected by such men as Lagrange and Laplace on the continent.³²

But in the years immediately following his graduation, Sedgwick also witnessed the analytic revolution at Cambridge. He was close friends with the group of young men who had formed the Cambridge Analytical Society in 1812 in the hope of addressing the British "decline of science" Playfair had lamented in the *Edinburgh Review*, and of introducing to Cambridge the analytical mathematics he had championed.³³ Algebra, these men argued, was the purest language in existence, for its signs and symbols did not refer to material objects, and, unlike descriptive prose, it did not need to communicate concepts to its readers. Algebraic symbols made pure, unadulterated sense in relation to one another, not in some messy, half-baked fit with the physical world. Nonetheless, though by 1810 calculus had made it into the Cambridge textbooks, and in the 1830s the Tripos came to include advanced algebra, integrals, and coefficients, worries about the spectral nature of this kind of mathematics remained. Largely under the influence of William Whewell, philosopher of science at Trinity College, Sedgwick's close friend, and a president of the Geological Society, analysis was accepted only when it could be applied to the study of the natural world and never studied as a pure science as it was in France.[34]

Sedgwick was more comfortable than Whewell with analytic methods pursued "without any reference to . . . immediate use or application." "Its reward," he said, was to be found in "the delights of discovery" and "the enjoyment flowing from the pure unmixed love of truth."[35] But, typically from within the Cambridge system, he also saw "analysis" in Newtonian terms. In his *Discourse on the Studies of the University*, Sedgwick quoted Newton at length: "As in mathematics, so in natural philosophy," Newton had written, "the investigation of difficult things by the method of analysis ought ever to precede the method of composition. This analysis consists in making experiments and observations, and in drawing general conclusions from them by induction." Though such a method would not demonstrate "general conclusions," it could slowly, inch by inch, reconstruct enough evidence to finally reach systematic laws and general principles. "This is the method of *analysis*," Newton had continued. "And the *synthesis* consists in assuming the causes discovered, and established as principles, and by them explaining the phenomena proceeding from them and proving these explanations."[36] Induction, of course, was the catchword of geological science.

Sedgwick glossed, "The former of these methods, when applied to the investigation, has long been known by the term *induction*: the latter . . . is called the method of *deduction*."[37] And while deduction was the "crowning fruit" and "consummation of exact science," geology, he believed, was still in its infancy. It had not yet conducted enough experiments, or gathered enough observations, to make deductions from general laws and systems.[38] "It is a humiliating fact," Sedgwick reminded his colleagues in 1830, "that the material combinations we investigate and attempt to classify are too rude and ill defined to be regarded as the appreciable results of any simple law of nature."[39] In this commitment to the observations and experiments of analytical methods, Sedgwick's notebooks were "crammed with discrete observations" to an even greater extent than those of his colleagues, and he was the most reluctant of the leading geologists to move from individual observations to more general classifications, sections, and correlations of the strata.[40] The study of geology was perfect for a Cambridge mathematician, in fact, for not only was it particularly amenable to analytic experiments and calculations, but it was also a physical, rather than a pure, science. "As the historians of the natural world," Sedgwick decreed in 1831, geologists must accept that "the earth has been brought into its present form by countless causes of which we know nothing," "conditions... infinitely too complex and ill defined to come within the grasp of any exact *analysis*" (my italics). "Our subject," he reminded the gentlemen geologists, "will never be so far abstracted from the materials which weigh it down, as to rise to the rank of an exact science."[41] Using the techniques of analysis associated with a young generation of reforming scholars at Cambridge, geology offered the double certainty that new and speculative methods would never succumb to the fictions and pure symbols of continental algebra.

Sedgwick's lack of ability as a writer, and his firmly held vision of geology as an inductive science, nonetheless, went hand in hand. Even at the end of his career, Sedgwick believed that geology was not ready to proceed to deductive methods, and he criticized Lyell for converting to a belief in evolutionary theory in the 1860s. Though Lyell was "an excellent and thoughtful writer," Sedgwick considered, he was "not, I think, a great field-observer." Writing and observing, making plausible prose and seeing the world accurately, might be mutually exclusive. Lyell's "mind is essentially deductive, and not inductive," Sedgwick continued, and "geologists have not yet sufficiently unrolled the records of the earth to reach a starting point of knowledge from which to reason deductively with perfect safety. They may varnish it as they will; but the transmutation theory ends (with nine out of ten) in rank materialism."[42] Obviously Sedgwick's concern about "the transmutation theory" is partly motivated by its theological implication. But his distaste for it also reminds us that the problem with geology as an inductive science was, as William Whewell wrote in 1831, that it did not produce the kind of "synthetic sweep" that would capture the public imagination and appeal to the ever-expanding reading public.[43] The question remained: how to rewrite the story of the earth, without ending up with nothing but a clutch of speculative fictions.

The Threat from Vestiges

In one of the most notorious episodes in the history of the science—the anonymous and "sensational" publication of *Vestiges of the Natural History of Creation*, in October 1844 (and subsequently, in ever-cheaper formats)—cosmology seemed to threaten to break down the doors of the elite geological community, and to throw all it had achieved out onto the streets. Though the identity of its author was not disclosed until the 1880s, it was immediately apparent that *Vestiges* could not have been written by any leading member of the Geological Society. For a start, it was riddled with what the gentlemen geologists saw as damning errors. Even more shockingly, it argued along Lamarckian lines that species were not eternal and fixed by God but developed along evolutionary lines, and that humans were the product of this evolutionary scale. The ensuing controversy, to which Sedgwick's truculent review of *Vestiges* in the *Edinburgh Review* belongs, sharply reveals just what was at stake in the writing of geology as a gentlemanly science.

Secord has suggested that the book employed the methods of Scott "on a cosmic scale," utilizing "the traces of the generic conventions of historical fiction" in order to reinscribe "Enlightenment cosmology as evolutionary narrative" "in a form appropriate to a Victorian readership."[44] Certainly, intoxicated by its cosmological sweep, with all its threatening evolutionary implications, readers widely acknowledged that the book "read like a novel."[45] But in the context of the 1840s it did not necessarily, or not only, read like a Scott novel. In the decades between the publication of *Waverley* in 1814 and the publication of *Vestiges* thirty years later, both the novel and the periodical culture that had fostered Scott's literary achievements had changed almost beyond recognition. Though you couldn't guess it from the records of the gentlemen geologists, the *Edinburgh Review* and the *Quarterly Review* had had their crowns toppled by a host of rival quarterlies, including the *Westminster Review* (founded in 1823), *Fraser's Magazine* (1830), and an avalanche of monthly and weekly periodicals with cheaper price tags and more readily accessible content.[46] While Scott had commanded record prices for fiction, reaching 31s6d with *Kenilworth* (1821)—a rate that would become standard for Victorian triple-decker novels—the groundbreaking Magnum Opus edition of his entire works reprinted each volume in a five-shilling, small octavo format. More worrying trends were entering the middle-class novel from the other direction: chapbooks, ballads, broadsides, and salacious accounts of murders were sold in their millions to the "semiliterate masses" in the 1830s, and the unstamped penny periodicals, with their shocking criminal content and such titles as *Calendar of Horrors! A Weekly Register of the Terrific, Wonderful, Instructive, Legendary, Extraordinary, and Fictitious* (1835–1836), were widely perceived to have wound their way into the mainstream in the form of the so-called Newgate novels of Edward Bulwer-Lytton, Charles Dickens, and Harrison Ainsworth.[47] In the new quarterly *Fraser's Magazine*, Thackeray was outraged.[48] This was not the literary culture of which Scott had been king, or of which a man like Thackeray might hope to take the crown. Indeed, as

Paul Thomas Murphy has demonstrated, the newly literate, working-class readers of the unstamped papers hated Scott for his elitism and lack of social realism, and hated fiction by association with Scott's novelistic output. Even when attitudes toward fiction changed in the working-class press during the 1830s, Scott remained a target of contempt.[49] At the same time, many illegal unstamped papers were condemned for circulating radical politics, corrosive fiction, and materialist science, including evolutionary theory, to vast readerships. When *Vestiges* delivered this fare in a seemingly respectable five-shilling format, the intrusion of popular radicalism into the literary sphere seemed complete. No matter what literary techniques it may have borrowed from Scott, this was not the gentlemanly world of clubs and learned societies to which Scott had belonged, and neither was it the gentlemanly world of the Geological Society, whose members imitated drinking games from Scott's fiction, knew his friends, and continued to read the publications with which he had been associated.

In this context, of course, it is unsurprising that those geologists who continued to self-identify with an older, literary culture, were nonetheless unable to share Playfair's breathtaking wonder at the possibilities of evolutionary theory. "As old species perish," Playfair had speculated in the *Edinburgh Review* in 1812, "do new species rise up? Is there some secret law of animal reproduction by which there is a succession of species in the course of ages, as there is of individuals in the course of years!"[50] By the turbulent 1820s, there were signs of unease among the gentlemanly community, so that even the open-minded and youthful Charles Lyell, who had "devoured Lamarck" with "pleasure," was forced to add several caveats:

> His theories delighted me more than any novel I ever read, and much in the same way, for they address themselves to the imagination, at least of geologists who know the mighty inferences which would be deducible were they established by observations. But though I admire even his flights, and feel none of the *odium theologicum* which some modern writers in this country have visited him with, I confess I read him rather as I hear an advocate on the wrong side, to know what can be made of the case in good hands.[51]

Now the theory was not a beautiful challenge for the excited imagination of a philosopher. It was a faulty fiction, a well-argued case on the wrong side of the law, and its address to the imagination had transformed its pleasures into those of a novel—presumably one of those novels with which it jostled for space on the pages of the unstamped press. By the 1840s "the uncertain lights and gigantic images" that the theory had represented to Playfair seemed, to some, nothing short of sinister.[52]

For Sedgwick, sinister wasn't the half of it. Within a few months of the publication of *Vestiges,* he had published a scathing eighty-five-page review in the *Edinburgh Review*. Sedgwick collected and bound every issue of the *Edinburgh Review* from its inception until his death in 1873, along with every edition of the *Quarterly Review*. When he wrote to the editor of the *Edinburgh Review* to accept the job of reviewing *Vestiges*, he said that "the principles of the Edinburgh Review are the principles on which I have acted and thought, since I was capable of thinking for myself."[53] These values are clear to see on every page of his review. Why had the "false assumptions" and "philosophic jargon" of *Vestiges* found so "much favour in London"? "Because of the shallowness of the fashionable reading world," Sedgwick wrote, "and because of the intense dogmatic form of the work itself. He who asserts boldly and without doubt, will be sure of a school of followers."[54] For the Sedgwick who had found writing difficult all his life, the bold assertion and dogmatic form he attributed to "shallow" readers would not come easily. "I could write a volume," he said and, punning on the book's use of the nebular hypothesis, acknowledged that he "must condense the nebulosity, and try to make it give out its latent heat. But it is hard for me to tell, in three or four pages, a story which has cost me a quarter of a century and more in reading."[55] "Oh that I could bring back again to my elbow, and obtain the help of our friend Sydney Smith," he lamented. "With all his faults, he was a very kind man, and a better companion and a better exposer of folly and nonsense never lived. I shall long remember his broad, good-humoured face, as he showed himself in our London club-houses, where I often met him."[56]

The culture of the "London club-houses" and the quarterly reviews seemed under threat from this novelistic text. And Sedgwick reviewed it as a novel masquerading as a work of science. Drawing on the familiar metaphor of the earth as a "book of nature," and its strata as "pages" or "chapters," Sedgwick accused the author of *Vestiges* of having "work[ed] up an historical tale from his imagination—using the old documents now and then to eke out an hypothesis, or to give a savour of reality to a fictitious narrative."[57] Forgetting the scientist's humble role as a reader of the earth's "documents" he had instead assumed the creative role of author. Moreover, the "fictitious narrative" and the cosmological vision of the earth were directly related through their formal features. The cosmological form of *Vestiges,* as it deformed the book of nature, and the novelistic form of the text curiously mirrored one another. "It is no doubt true," Sedgwick wrote in the *Discourse*,

> that a book of natural science (like certain chapters of *The Vestiges*) might be drawn up in the language of an hypothesis; and that the animal kingdom, including both the extinct and living forms of life, might be arranged in some preconceived order of natural development. And if this arrangement were accompanied by a new dogmatic and technical language—implying and taking for granted that, on the whole ascending scale . . . each part of the ascending series was derived, in the way of common generation, from that which went before it—our narrative would have at least the symmetry and external form of a true history. But, after all, it would be but a physical romance, and a work of imagination; and its language would be no better than a downright cheat.[58]

Sedgwick claims that "true history" takes the form of an "ascending scale" with "common generation, from that which went before it," that a "preconceived order of natural development" is an appropriate arrangement for historical writing. Clearly Sedgwick has in mind the Enlightenment vision of history associated with the Edinburgh philosophers. But this historical plot is not an appropriate lens through which to consider the natural world. The author of *Vestiges* has shaped the material of the natural world according to the forms of history, so that both the narrative line of the text and the narrative line it ascribes to nature collapse into one another. Cosmological explanations of the universe turn the natural world into "physical romance, and a work of imagination." "How widely different is the history given in the unerring book of nature from the fabulous narrative in some written records of creation," Sedgwick wrote. "Men have ever delighted to repose in the luxury of speculation who could ill broke the repulsive toil of analyzing phenomena, and cutting their way, step by step, through the solid rock on which stands the temple of material truth." This all might be "meet matter for the imagination of a poet," "grateful to the neophytes of a bad philosophy," but is more conducive to faulty "schemes of Cosmogony" than to geology itself.[59] No wonder, then, that Darwin, who read Sedgwick's review of *Vestiges* in broad agreement, but "with fear and trembling!," structured his evolutionary arguments in *Origin of Species* in resolutely nonnarrative form, its chapters arranged under abstract headings.[60] His evolutionary theory might imply that life on earth had a *story*, but his readers, unlike the readers of *Vestiges*, would not experience it that way. They could not read *Origin* as if it were a novel.

Sedgwick's voice was the voice of the *Edinburgh Review* author who claimed to safeguard, as Ina Ferris has put it, "the generic young female reader of novels who must be protected by the disciplinary authority of the reviewer" in the earlier part of the century.[61] Sedgwick worried over "susceptible young women readers" in the metropolis, particularly in Whig literary circles. "You have no conception what mischief the book has done & is doing among our London blues, & God willing I will strive to abate the evil," he wrote.[62] Again, Sedgwick asked, "How are we to account for the popularity of the work and the sudden sale of edition after edition? Men who are fed on nothing better than the trash of literature, and who have never waded beyond the surface of the things they pretend to know, must needs delight in the trashy skimmings of philosophy."[63] In short, novel readers—the young, the idle, the weak, and the otherwise susceptible—were responsible. Indeed, so stupid was the book that it might even have been written by a woman. After all, "the ascent up the hill of science is rugged and thorny, and ill-fitted for the drapery of a petticoat."[64] Women did not possess the intellectual or physical stamina required to pursue geological fieldwork. Thus unable read the earth's textual archive, women were relegated to the writing of fictions requiring no specialist knowledge. Sedgwick never married, as stipulated by the terms of his professorship, and considered himself "wedded to the rocks."[65] Though he enjoyed the company of women, not for Sedgwick was the passionate belief in female intellectual equality espoused by Playfair, that older geological bachelor often to be found "philandering at the Needles," as Francis Jeffrey had put it.[66] Sedgwick, more earnest, more impassioned, and more evangelical than Playfair, nonetheless attempted to appropriate the voice of the *Edinburgh Review* in its glory days. He constructed his defense of geology from the world of the clubs, learned societies, and periodicals of the elite, the literary culture he, and the geological community, still associated with Scott, Playfair, and the Enlightenment in which they felt their science had been born.

Scott on the Rocks

As might by now be expected, Sedgwick was ambivalent about the novel. Though he read the poetry of John Milton, Abraham Cowley, and John Dryden enthusiastically, Secord notes that Sedgwick handled books roughly, viewing them as utilitarian objects rather than sources of pleasure.[67] And when it came to the novel, he rarely gave it much of his time. He owned a few volumes of Dickens, even fewer of Thackeray, whom he admired, and

Harriet Beecher Stowe's *Uncle Tom's Cabin*.[68] His bookshelves contained no fiction-carrying periodicals of any sort. The only exception to the rule, of course, was Scott—Sedgwick owned the forty-eight-volume Magnum Opus edition of the Waverley novels, the twenty-eight-volume *Miscellaneous Prose Writings*, a twelve-volume edition of *Poetical Works*, *Letters on Demonology and Witchcraft*, and John Gibson Lockhart's seven-volume *Memoirs of the Life of Sir Walter Scott*. His editions of Swift, Defoe, and Dryden were edited and annotated by Scott.[69] His biographers tell us that "he was never tired of reading and re-reading, till he must have known them almost by heart, the novels of Walter Scott, with which, as he often said, 'he had been driven half-mad as a young man.'"[70]

But in later life, Sedgwick considered even Scott a guilty distraction from his geological work. In a typical letter of 1837, Sedgwick complained of the flu: "Even old Shakespeare and Walter Scott acted only as sounding-lines for the depth of my intellectual vacuity," he complained. "But now I can read them for an hour together, and fancy, at least, that I am refreshed and built up by them."[71] Over and again, Sedgwick claimed to read fiction only when he was so stupid with sickness that he could not manage anything more. In the mid-1850s, under doctors' orders, Sedgwick "hardly looked at a Geological book" for two years, banishing all "books (except 'Siluria')." "Of course," he added, "I mean geological books."[72] No geological books, but, a month later, again "oppressed by a horrid cold—living on cathartics; sudorifics, & slops," he wrote, "I have been incapable of reading anything harder than Walter Scott, & one or two essays of Dean Swift. In short, I have been, and am still, a sleepy, slimy, snivelling, slattern snob."[73] In 1847, again, while apparently stuck in a deserted out-of-term Cambridge with a heavy cold, Sedgwick reported that he had been confined to his rooms for several days. "Several times I went out and tried to resume my studies," he claimed, but "each time I was seised by the throat, and driven back to my solitary fireside. So at length I struck work absolutely, took to slops and thin potations, and treated my brain with things as light as I put in my stomach." "But for old Daniel De Foe," he wrote, "I should have died." "Only think!" he wrote to another friend during the same illness, "to fill up the time I have been reading *Robinson Crusoe*! For in the stupid soddened state of my brain I can think of nothing worth thinking about."[74] Novels were the embodiment of the "synthetic sweep" lacking in inductive geology, the very opposite of scientific practice. Geologists needed to spend hard days tramping through the wilderness, hammering at the rocks in all weathers and conditions, bringing the vast and chaotic earth into a semblance of conceptual order. Geology required physical and intellectual strength in all of its forms. Novels, on the other hand, were little more than seductive digressions from the hard labor of life and work, and though Sedgwick could not refrain from repeated bouts of indulgence, for him they could only be understood as the reading matter of the shallow, the sick, and the idle.

Sedgwick's rigid separation of stories and science, albeit hyperbolic, is an expression of his interpretation of the ways in which the Geological Society increasingly divorced narrative or causal patterns from the structural analysis of the earth's rocks and fossils, until at least the publication of Darwin's *Origin*. But Sedgwick's case also testifies to the ongoing power of literature to shape the culture and practices of the geologists themselves. For Sedgwick was also famous for regaling his dinner companions and his lecture audiences with waggish anecdotes on his journeys into the earth's oldest rocks. On one occasion, "geologising on the cliff-side" of the Cornish coast, Sedgwick encountered "two or three smugglers." The smugglers apparently took Sedgwick for "too sharp a looking gentleman to be looking after them stones," and offered to show him "the nicest little cove where you can land a keg o' brandy." Sedgwick declined the offer, to which the smugglers replied, "with a disappointed air," "Us sees you've no confidence in we."[75] One day in Wales, Sedgwick often told his audiences, he "had had a long tramp on the hills, and was walking home, dog-tired, and very dirty, when I caught sight of fossils in some stones laid by the wayside." Hammering at the rocks, he was approached by a lady who asked him for directions and information about the local area. "She thanked me for my information, and added: 'Poor man, you must find this very hard work.' 'Yes, indeed I do,' I replied; whereupon she took out her purse, and gave me a shilling." Sedgwick continued:

> Next evening, to my great amusement, she came to dine at the house where I was staying. I recognised her at once, but she did not know me, in my altered dress. She was visiting Wales for the first time, and was full of enthusiasm for the scenery and the people. "They are so obliging, and so communicative," she said; "only yesterday I had a long conversation with an old man who was breaking stones on the road. He told me all I wanted to know, and was so civil that I gave him a shilling." I could not resist the pleasure of saying, "Yes, Ma'am, you did, and here it is!"[76]

Greasy landladies served Sedgwick leftover potatoes and bacon but refused to allow him to tip them for their pains because, on account of his appearance, they thought he'd "seen better days"; dirty Scottish landlords whose whiskey-addled "noses look like the ends of hot pokers" were irritated with the early hours kept by Sedgwick and his inexplicable hammer.[77] Sedgwick's anecdotes very often repeat this central joke—a joke his colleagues also loved to tell and retell—that fieldwork took all good geologists into rough terrain and rougher company, temporarily obscuring the gentlemanliness that lay beneath their dirty clothes and knapsacks.[78] Of course, the gentleman-in-disguise is a standard romance motif, running right through Middle English romance and Shakespeare, not to mention the historical novels of Scott, from *Waverley* and *The Antiquary* through the medieval romances of *Kenilworth* and *Ivanhoe*. In the stories he told about his fieldwork, then, Sedgwick used the tropes of disguise and mistaken identity to make the preservation of the gentlemanliness of the geologist a source of narrative delight, and to situate the activities of the geologist within a long history of romantic quests.[79] The image of the knightly geologist was a popular one, assimilating spiritual and scientific values under the aegis of a knightly endeavor to enlighten beliefs, rescue treasures from the depths of the soil, and carry the burden of scientific discovery on their strong, masculine shoulders. Not only did Sedgwick prove his simultaneously inductive and gentlemanly mettle in these stories, then, detailing the hardships of his labor in the field, but he also shifted the desire for narrative and "synthetic sweep" in science *from* the rocks and fossils and onto the man of science himself.

As the dominance of the quarterly reviews, and of the gentlemanly brand of sociability valorized by Scott and Playfair, came under increasing threat from a proliferating print culture in the decades surrounding the coronation of Queen Victoria, the story of the earth and the story of the geologists' reconstruction of the earth thus became ever more rigidly separated in public discourse. We have to look at private letters and anecdotes to reconstruct the ways in which such stories may have helped produce "geology" as a modern science in this period. In 1875 Archibald Geikie recalled a paper read by Sedgwick at a meeting of the Geological Society in 1831, which he described as "one of those luminous efforts which by a few broad lines served to convey, even to nonscientific hearers, a vivid notion of the geology of a wide region, or of a great geological formation." The trouble was,

> embalmed in the society's printed publications, the paper, as we read it now, bears about as much resemblance to what it must have been to those who heard it, as the dried leaves in a herbarium do to the plant which tossed its blossoms in the mountain wind. The words are there, but the fire and hu-

mour with which they rang through that dingy room in Somerset House have passed away.[80]

Sedgwick's biographers claimed that it was in his charming conversation and in his thunderous lectures and public addresses that much of his force was felt by his colleagues.[81] As such, we find the man who could not write a geological book being described as a master storyteller again and again in the records of the Geological Society. "Few have ever told a story so vividly as he did," we are told, "or with such a marvellous combination of the dramatic, the humorous, and the pathetic."[82] It is, then, as much in the "fire and humour" of the geologists' conversation in meetings and at clubs, in the stories they told around the dining table about their exploits in the field, in the ways in which they imagined themselves, and were imagined by the public, as in the scientific tomes they produced, that we can recapture much of the tenor of their geological work.

For Murchison, Sedgwick's anxieties about the novel were an irrelevance. As Sedgwick's frequent counsel that he ignore the novel-reading public as he wrote *Silurian System* suggests, Murchison's social station meant that he had little to fear, and more to gain, from courting the fashionable reading public than did the Cambridge professor. His geological "style," as it has been described, was also very different.[83] For Murchison, moreover, Scott was a favorite novelist, and more. Scott's Jacobite tales formed a textual tapestry so imaginatively potent, indeed, that Murchison would spend much of his life claiming a Highland ancestry he barely possessed. Having spent the first three years of his life on the family estate in Ross-shire, he "was proud of being a Highlander, and seldom lost a chance of proclaiming his nationality."[84] One can only pity his wife, Charlotte—on their honeymoon, and on numerous subsequent geological excursions, he took her to the estate and visited the scenes of famous battles, his face "kindle[d] with the old martial fire as he went over again the events of 'the '15.'" He commissioned Edwin Landseer, the most prominent artist of the day, to paint him dressed up as his "relation" Donald Murchison, a Highlander who, after the battle of 1715, had collected rents on behalf of his exiled chief in defiance of the English seizure of such estates. Murchison even raised a monument to Donald. Landseer was both an inspired and an obvious choice for such a commission. The painter had spent ten days with Scott at Abbotsford in 1824 incessantly sketching Scott and his dogs, and the next eleven years of his life were dominated by travels in the Highlands, staying with the Duke of Atholl, the Marquess of Breadalbane, and the Duke of Bedford, hunting, and painting sumptuous oil scenes of sporting and Highland subjects. He had painted Queen Victoria on horseback, Prince Albert as a huntsman, and the royal couple as Edward III and Queen Philippa, embodiments of stock-in-trade chivalric values. Landseer's mature work, including the painting Murchison commissioned, *Rent Day in the Wilderness*, now hanging in the Scottish National Gallery in Edinburgh, drew extensively on Scott's romantic atmospherics, antiquarian detail, and vision of the Highlands as a product of history and landscape.[85] The two men, who knew one another socially, shared this love of the Highlands, of foxhunting, of chivalry, and of Scott.[86] Landseer, who had provided illustrations for the Magnum Opus edition of Scott's novels in 1829–1833, was thus the perfect choice for Murchison's rewriting of his own family history.

And a rewriting it certainly was. As Murchison's protégé the geologist and historian of geology Archibald Geikie was forced to admit, the connection between Roderick and his forebear was not direct. In a rather bathetic footnote Geikie admits that Roderick's "grandfather was a third cousin of the Colonel."[87] This barely constitutes Highland ancestry, and it hardly warrants the commissioning of a sumptuous oil painting by the leading painter of the day linking Donald Murchison with a famous historical event. What does justify it is the ideological work the painting performs. The painting bears a detailed, realistically described landscape, both of which work to efface the genealogical distance between Murchison and the Jacobite past. Densely realizing a key moment in Scottish history, drawing on the historic romances of Scott both for atmosphere and in order to indicate the scene's historical authenticity, and using Murchison as a model, the painting immerses the geologist in an ancestral history he barely possesses and in a historical world drawn straight from the novels of Scott. For Geikie, indeed, there is aesthetic gain to be made from the gap between Murchison and his sort-of ancestor. Their relation, he writes, "was of that shadowy kind in which Highland genealogists delight."[88]

Geikie is now best remembered as the author of *Founders of Geology*, an early history of the science that created and perpetuated many myths about the "heroic age" of geology, not the least of which being a hagiographic view of Murchison.[89] He also played an instrumental role in the ongoing classification of the Palaeozoic rocks according to Murchison's nomenclature until the turn of the twentieth century, defending the expanded "Silurian System" against all comers for the length of his career.[90] Like his mentor and intellectual hero, Murchison, Geikie had a penchant for Scott. "There is something thrilling to even the calmest imagination," he wrote in first geological book, *The Story of a Boulder* (1858),

> in contemplating the results of vast and sudden upheavals, in picturing the solid crust of the earth heaving like a ground-swell upon the ocean, in tracing amid "Crags, knolls, and mounds confusedly hurl'd, the fragments of an earlier world," and in conjuring up visions of earthquakes, and frightful abysses from which there ever rose a lurid glare as hill after hill of molten rock came belching up from the fires below.[91]

The quotation is from Scott's poem *Lady of the Lake* (1810).[92] There it describes the "ruined sides" and "summit hoar" of Ben Venue, a mountain in the Trossachs, which "Down to the lake in masses threw / Crags, knolls, and mounds, confusedly hurl'd, / The fragments of an earlier world." Scott's sensitivity to the geological processes of erosion and weathering, and to the potential violence of the geological world, serves Geikie's purpose perfectly, not least because Scott is describing the scene as it is encountered by a huntsman. Advancing from the pack alone, in a chase that is itself described in martial terms—as the stag springs "from his heathery couch in haste" "As Chief, who hears his warder call, / 'To arms! The foemen storm the wall'"—the huntsman has lost his horse and wanders to find his friends, enraptured by the mountain scenery.[93] The passage works through a series of echoes: the mountains echo a distant thunder to the huntsmen, whose shouts and calls echo the cries of battle that have once resounded in the same caverns and glens, so that "To many a mingled sound at once / The awakened mountain gave response."[94] The scenery has been created through violent geological action, creating chasms and caverns that capture and return the calls and shouts of the huntsmen; their own violent actions simply echo and return the violence that has taken place geologically and historically in the same space. Geology, hunting, and war create and spur the other on—as they did for Murchison, as soldier, foxhunter, and geologist. Moreover, in these opening stanzas of *Lord of the Isles*, when the huntsman finds himself alone and without horse, his echoes are not returned. Only then does he stop and admire the breathtaking beauty of the surrounding scenery, and only then do we as readers step back from the thrill of the chase and contemplate the effects of that scenery on the actions that take place within it. Like the geologists, he is inspired by war, romance, and hunting. Like them, however, he must pause from those pursuits to linger longer and alone in the wild and tempestuous world into which they have led him. This is the perfect image of the "style" of Murchisonian geological practice: the headlong, heroic dash and sweep through remote and isolated landscapes—on horseback, or as a warrior—the conquest and capture of geological territories for one's own geological "system," and then the slower, more detailed contemplation of the rocks and fossils that comprised the "system" he had conceived.

The geologists' indebtedness to Scott is unmistakable in their private correspondence. From the 1830s on, Sedgwick and Murchison undertook extensive stratigraphic work on the so-called Primary or primitive rocks, which in Britain tend to be located at what were then the remotest and least populated reaches of the landscape, in Cornwall, Wales, and Scotland. Taking post chaises and steamships from their comfortable English homes and venturing into these untamed landscapes, often staying at miserable alehouses in uncomfortable circumstances, it was not a stretch to imagine themselves as so many Waverley heroes embarking on their very own "romance." As Katherine Haldane Grenier has shown, nineteenth-century tourists to Scotland—even those who proclaimed their distance from the images Scott's fiction had originated or perpetuated—found it almost impossible to see the landscape or its people unencumbered by the particular vision he had produced of it.[95] The Highland Tour, the standard narrative device of the Waverley novels, became increasingly accessible to middle-class holidaymakers, whose vision of Scotland as a "romance" destination was inseparable from Scott's writing. Scott became "Scotland's foremost tourist guide."[96] Even when he was as far afield as Bavaria, for example, Sedgwick was reminded of Scott by the close of a grand royal hunting party. "The sport was nearly over before we arrived," he explained to his father, "but we saw the company, and the manner of the chase, which was all we wanted. The scene altogether reminded me of some of Sir Walter Scott's finest descriptions, but was upon a scale more grand than Scotland could ever boast of."[97] Again, recounting a particularly happy field trip to St. Abb's Head in 1834, Sedgwick remembered that he and his party had been "all excited to the highest pitch; sometimes speculating on the strange frolics dame Nature had loved to play thousands of years before strathspeys were thought of; then talking of Walter Scott, the Master of Ravenswood, and the Middle Ages."[98] Romance, history, "strathspey" dances, the frolics of nature, all blend here into a single, evocative vision. In 1841 Sedgwick visited and "was greatly charmed with" Wigton Bay on the Solway Firth, a "noble coast" presenting "a succession of beautifully snug bays and creeks," and noted that they have "another source of interest, being the scene of one or two of Scott's novels."[99] The geologists clearly found in the landscapes Scott had described for them the kind of holiday release into a temporary, exotic location with which the most "romantic" aspects of his fiction are identified.[100]

During this period Scott quickly became a tourist destination in his own right, too. On a geological excursion to Scotland in 1825, Sedgwick, like a latter-day Mannering, immersed himself in Edinburgh's intellectual scene, meeting savants, craniologists, and Scott: "What we saw of him," he wrote, "made us long for his better acquaintance. He talks exactly as he writes, and before you have been two minutes in his company he begins to tell good stories."[101] In 1855, while on a Highland tour with his nieces, Sedgwick again dined at Scott's home with the novelist's granddaughter Hope, though his "joy was now and then clouded by sorrowful memorials of the effects of time." "Just a quarter of a century before I dined in the same room with Sir Walter Scott and a most happy family of eight, including his sons and daughters. Of that party I am now the only remnant."[102] In 1822 Murchison visited Edinburgh at the same time as a grand procession of George IV intended to demonstrate English forgiveness for the Jacobite rebellions, but orchestrated by Scott so that it also represented Scotland in tartans and kilts, as a space culturally (if not politically) distinctive from the British Union in which it was merged.[103] Murchison watched the procession from Calton Hill, being especially pleased to spot "the beaming face and white hair of Walter Scott as he marched jauntily along in the front of the carriage."[104] In this procession Scott simultaneously cultivated a fictional image of Scotland, an image of heroic and chivalric masculinity based on the image of the martial Highlander, and symbolized Scottish national identity and its vibrant intellectual traditions.

A Highland Tale

In 1827, long before their friendship was ruined by the "Cambrian-Silurian controversy," Sedgwick and Murchison conducted their first collaborative geological tour together, visiting the geologically well-trammeled Isle of Arran off the west coast of Scotland, one of the principal settings of *Lord of the Isles* and the source of much of James Hutton's evidence for the intrusion of granite veins into preexisting rock formations. Opposed to the Huttonian hypothesis, Robert Jameson had visited Arran and reported in his *Mineralogy of the Scottish Isles* (1800) that, far from being intruded in veins from below the existing rocks, the granite was stratified and had therefore been deposited from an aqueous solution like all other rocks.[105] In the ensuing years Playfair and Jameson offered further contrary proofs about the existence of stratified granite, often utilizing evidence from Arran.[106]

Sedgwick and Murchison set about reworking the island's stratigraphic classifications, and they took with them the leading text on the stratigraphic formation of the island, MacCulloch's *Description of the Western Islands of Scotland* (1819). MacCulloch, as chemist to the Board of Ordnance, had spent many years on geological fieldwork attempting to find limestone for use in gun manufacture; he was also a geologist for the trigonometric survey of Scotland and president of the Geological Society of London from 1816 to 1818. Lecturing the military cadets of the East India Company at Addiscombe in Surrey, MacCulloch also offered one of the relatively few lecture courses on geology outside Oxford and Cambridge in the 1810s and 1820s.[107] *Description of the Western Islands*, the first systematic account of the geology of the Scottish islands, was an extended version of a series of papers MacCulloch had given before the Geological Society of London. At the time it was published, its methods were typical of those promoted by the Society. Describing the formation and deposition of rock masses largely on the basis of mineralogical classifications of the rocks, MacCulloch's work shared much with the Wernerian descriptions of the strata, but it rejected the theory of a universal primordial ocean on which those descriptions had rested. MacCulloch has been characterized as a broadly Wernerian geologist, despite the fact that he controversially rejected Werner's "transition" class of rocks.[108] But in *Description*, MacCulloch refused to take theoretical sides. The trap veins near Swishnish point on Skye, for instance, were probably "in most instances connected with a principal mass," and the erosion of those masses over time would leave those veins independent, "and thus perhaps induce future geologists to attribute a protrusion from below that which may equally have entered from above."[109] For Hutton, veins were thrust up in molten form from the earth's core; from Werner, they had trickled down from an aqueous solution above. On the one hand, MacCulloch argued against Jameson that the serrated granite mountains on the north of the island were not stratified. On the other, he claimed that they should still be classed as "primary" rocks, a classification that could not exist in Hutton's cyclical scheme.[110] And talking of Hutton's unconformity, a junction of schist and granite at Loch Ranza, MacCulloch tentatively agreed that the cracked and tilted secondary strata had been "displaced" by the granite mountains, elevated from below, so "that the granite is posterior to them," as Hutton suggested.[111] Nonetheless, "the solution of these difficulties must be referred to future times," for the movements of the granite could still not be adequately known.[112] "The question is not visionary," he wrote, "but it must not be pursued with visionary arguments, and I shall therefore leave it for future and increased experience."[113]

MacCulloch's writing, nonetheless, was imbued with the visionary and the romantic at every turn. Of the northern coast of Arran, MacCulloch wrote:

> The sun seldom penetrates these deep recesses, which exhibit to the painter all the sober and harmonious tints of reflected light as it is reverberated from rock to rock and from the clouds that occasionally rest on their lofty boundaries. It is in Glen Sannox, above all, that the effects arising from magnitude of dimension, combined with breadth of forms, and with simplicity of composition and colouring, are most strongly felt.[114]

"Brodick bay is no less beautiful," offering "a picture approaching to perfect composition in a degree rarely seen in Nature." Arran's antiquarian riches, its "rude sepulchral pillars, urns, stone chests, cairns, dunes, circles and cromlechs," excite his imagination and invigorate his prose, along with meteorological description and accounts of its castles and fishing traditions.[115] The writing exhibits all of the atmospheric effect and antiquarian detail of Scott's fiction.

This vision of the landscape and its history imbues MacCulloch's geological method too. The third volume of *Description of the Western Islands* consists entirely of illustrations supporting his textual descriptions of the islands. Opening the book, the reader is presented with a dramatic vision of "curved gneiss in Lewis," a ninety-degree tidal wave of rock cresting across a cluster of cliffs and crags (fig. 2.1). Turning from the title page we are given a "View of Brochel Castle in Rasay" (fig. 2.2). The ruined castle is framed by a broken swirl of light emanating from the dark and thickening clouds behind. In the foreground, three figures stand before the staggering sweep of rock on which the turret rests, significant only for the sense of scale they impart to the scene. The stunted ruins of Brochel castle, and the short, mossy cliffs on which they really stand, are sacrificed to the romantic distortions of the Gothic. No fewer than ten plates, each mingling detailed representations of actual landscapes with the conventions of the sublime and the picturesque, amplifying and contracting the scale at will, precede the text, maps, and sections that follow. Even then, MacCulloch blends mimetic representation with romantic visions. "It must not be imagined," he advises, "that geological maps, even though accompanied by every necessary section, can, in the present state of the science, supersede the use of geological descriptions ... at present they can

Fig. 2.1. "View of a rock of curved gneiss on the north eastern shore of Lewis." In John MacCulloch, *Description of the Western Islands of Scotland* (1819), vol. 3, plate I. Reproduced by permission of the syndics of Cambridge University Library.

only be considered as topographic records of unconnected facts, or as aids to the imagination."[116] It is impossible to transcribe the complexity of the rocks without resorting to "a scale of inadmissible and extravagant dimensions" for a map. Just as the painterly visions of the landscape that preceded them were disproportionately large, these sparsely drawn sections are disproportionately small. In this sense, the scenes' romantic grandeur offers a mode of orientation for the reader of geological maps and sections: they reinstate the panoramic breadth inevitably lost in cartographic convention.

For MacCulloch, all geological knowledge relies upon the imaginative dimension embodied by the maps, for "innumerable circumstances render it impossible to examine the rocks which are near the surface, and, still less, those beneath it," so that only "a previous knowledge of the laws by which rocks are related to each other," of "observations much more widely generalised," can make sense of those sections that are observable.[117] Here, however, the imaginative is tethered securely to the rational. In his plates on Arran (fig. 2.3), MacCulloch gives imaginary and ideal sections of the strata, on their vertical

Fig. 2.2. "View of Brochel Castle in Rasay." In John MacCulloch, *Description of the Western Islands of Scotland* (1819), vol. 3, plate II. Reproduced by permission of the syndics of Cambridge University Library.

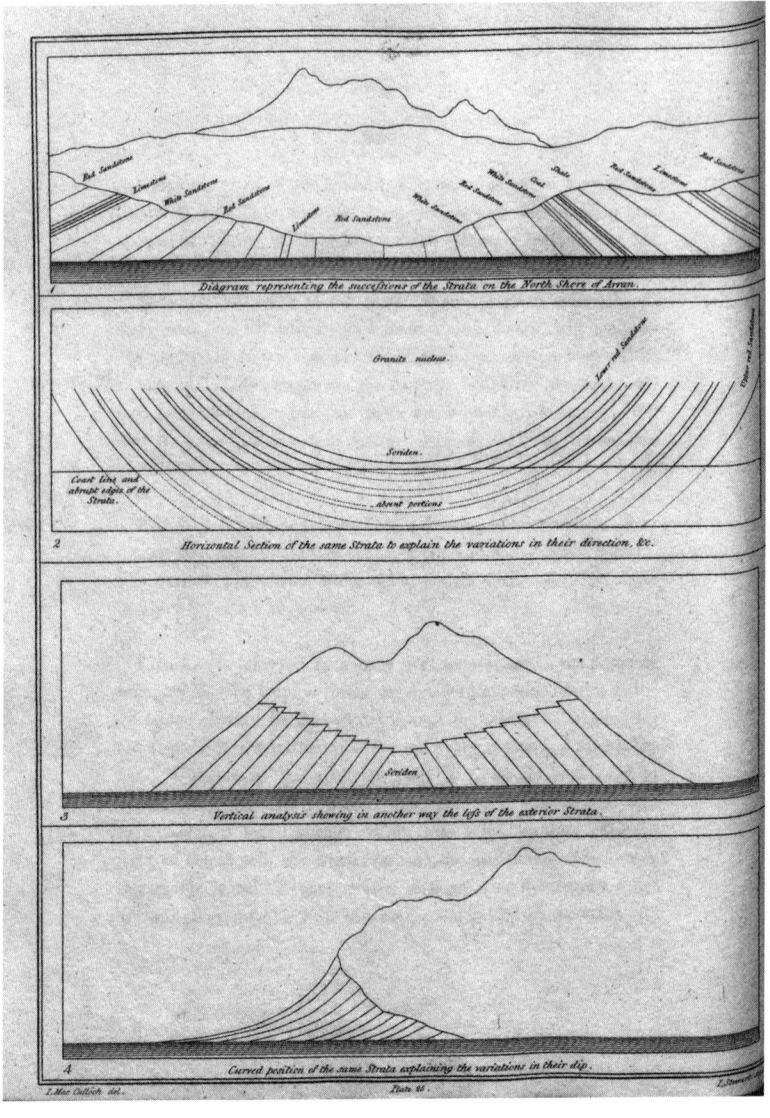

Fig. 4.9 "Various sections illustrating the sandstone on the north-east shore of Arran." In John MacCulloch, *Description of the Western Islands of Scotland* (1819), vol. 3, plate XXV. Reproduced by permission of the syndics of Cambridge University Library.

and horizontal planes, all of which must be combined to create a three-dimensional model of the strata as they dip away from the granite mountains. Having already given us several romantic views of that same land, replete with recognizable literary and artistic convention, MacCulloch has ensured that we see the entire view by which these partial representations may be given meaning. Immersing us into a romance vision of the landscape before pulling us back into rationality, MacCulloch evokes the dramatic Highlands of fiction and canvas so that he can train his reader to understand the geological map correctly. In this sense his maps are not tools of orientation but examples of disorientation, requiring orientation by the paintings that precede them. The geological vision is inserted into, and made sense of by reference to, the Highland mists and castles with which they share the landscape.

MacCulloch's next book, *Highlands and Western Islands of Scotland* (1824), was addressed in four letters to Walter Scott. The political import of the geological vision is clear here. It was widely criticized for its support for the Highland Clearances, in which the Highlanders were driven from their land in the name of political and economic progress. In *Description of the Western Islands*, MacCulloch had described "the agricultural system of Arran" as having "hitherto been of the worst and most antiquated description," its inhabitants having lost "some of the characteristic virtues of the Highlanders" and "acquired in return, only the vices of that civilization" in nearby regions "to which they have so free an access."[118] But by the time he wrote *Highlands and Western Islands* in 1824, MacCulloch claimed that Arran and its people had been transformed. Arran, he wrote, had "changed under my very pen; thanks to the excise, the steam boats, and the Duke of Hamilton," and, with all the speed of a scene change in a diorama, had seen its filthy hovels and uneducated populace replaced "with sheep, and ribbands, and excisemen, and shoes, and muslin, and rents, and taxes, and absentees, and kelp, and English, and cod, and herring, and lobster smacks, and justices of peace, and breeches, and shops of all wares, and schools, and roads, and cockneys travelling in gigs, and innkeepers who have learnt little from their instructors but the art of making them pay for what they do not get."[119] Though MacCulloch's unusual parataxis appears to lend reproach to this description, mingling the English and cod, or roads with the cockneys who travel on them, in fact this is a bustling celebration of "the picturesque beauty and the variety" of the island's topography and its newly modern, newly vibrant population.[120] The rapid transformation of Arran under the influence of taxes and steamships made it an elegant model for the modernizing program MacCulloch was calling for. Geology, moreover, clears Fingal and the fairy tales (like the Highlanders) from the surface of the landscape, orienting the reader to look deeper, at the processes of history rather than the individual lives and communities that are sacrificed to it.

As Peter Garside has written, MacCulloch's text operates on a "deterministic" view of history, shared with Scott, in which "it is impossible to retain virtues peculiar to primitive society and to enjoy at the same time the fruits of fully civilised society." As Scott himself wrote in a letter to Maria Edgeworth in 1830, the clearances were "the inevitable consequence of a great change of things." But both writers were aware of the dissonances in this "inevitable" march from primitivity to civilization: "Like Scott, MacCulloch believes there is a natural tendency to sentimentalise the past," and he also indulges in Scott's nostalgia for the losses incurred in the march of civilization.[121] Writing of the region of Aberfoyle, one of "the retreats of the noted Rob Roy," MacCulloch describes a "narrow pass" that might be "intended in the skirmish described in" that novel.[122] He continues:

> If Ithaca, Segovia, Bagdad, the Sierro Morena, and Datchet Mead, are classic ground, and if we cannot visit without interest the scenes of those fictions which have, for centuries, formed the delight of youth and age, neither can we now easily contemplate the pass of Aberfoyle or the wilds of Loch Cateran without viewing them through that atmosphere of a new enchantment which has been lately spread over them.[123]

Nonetheless, for Macculloch Scott's novels did not only describe and sentimentalize a lost past. They also sped up the processes by which that past was consigned to history. "Whatever enchantment that pen of yours," MacCulloch tells Scott, "whether wielded by yourself or your shadow"—the still-unidentified author of the Waverley novels—"there is a compensation of evil in it."[124] "Time was," MacCulloch continues, "when I strayed about these wild scenes, and, as I listened to the endless tales of Rob Roy and his Mac Gregors, could imagine myself glorying in past times, as if I also had been sprung from the children of the Mist." But Scott's writing has changed all that. Now those tales

> have found their way to every circulating library, Brighton and Margate flaunt in tartan, the citizen from Pudding Lane talks of Loch Hard; and recollections of Miss Stephens, Diana Vernon, and Liston, with the smell and smoke of gas lights, and cries of "Music, Off, Off," confound the other senses, and recall base realities where there was once a delicious vision.[125]

"Who can even hope to tell a Highland tale, in the teeth of such company as this?" MacCulloch asks.[126] Scott's stories, popular as they have been, have spread romance beyond the preserve of the gentlemanly collectors of folk tales and ballads. No longer are they the aesthetic property of elite, intellectual culture. No longer, in that sense, are they romantic at all, cheapened by the circulating libraries, the gaslights of the London stage, the unavoidable whiff of the fashionable seaside town. This popularizing process has had wonderful effects on Arran, conjuring prosperity with the flick of a wand, but still, in MacCulloch's book, Scott's fictions have destroyed the romance that is their very subject by co-opting it into the forces of modernity.

This dissonant tension between romance and modernity reenacts itself in MacCulloch's shifts between literary genres. MacCulloch's nostalgic digressions into his own romantic boyhood and into "the new enchantment" that Scott has spread over Scotland, pave the way for a lengthy disquisition on the military and political history of Arran. Because Scott has removed the Highland "mist" from the stories of *Rob Roy* by making them so vulgarly accessible, "I must even strip poor Rob and his oppressed clan as naked as ever the law did, since I dare not pass over such important personages, and exhibit them to you in the style of the Newgate calendar," he says.[127] Detailing the "antiquity of" Rob's clan through "the days of James III, and some time before that period," MacCulloch discusses with Scott the difficulty of inconsistent historical sources on this topic, draws on statues and writs, and charts the history of the clans alongside official royal documents.[128] Just as MacCulloch plunged his readers into romantic and artistic representations of the landscape, only to make it possible to pull them back into the rational task of reading stratigraphic maps and sections, on the level of his prose he plunges us into Scott's Highland mists as a counterweight to the authoritative legal and historical disquisition that also forms part of this landscape. As he has proven, this is a technique borrowed directly from Scott himself, whose literary narratives plunge their readers into an archaic past but are eminently part of the modernizing forces of Enlightenment culture.

Three years after MacCulloch's *Highlands and Western Islands*, Sedgwick and Murchison visited Arran. In their written texts there would be no trace of the tension between the plunge into a romance world and pull back into rationality that MacCulloch had borrowed from Scott. Their papers on Arran, delivered before the Geological Society, complete with stratigraphic sections, are models of plot-free description.[129] But their letters from this period reveal a different story. It was not the cockney-ridden, lobster-smacked, shop-filled Arran of MacCulloch's 1824 text that they saw. MacCulloch had reminded his readers to "take care to wipe the Highland mist well off from their eyes" when they visited the island, and to learn instead to see the island's people in all their "variety."[130] Sedgwick and Murchison ignored his advice. To be sure, they did not believe that the geology of Arran had been shaped by the work of the mythical giant Fingal, as MacCulloch urged his readers to understand, but their correspondence reveals that they did, nonetheless, "expect to meet a Helen Mac Gregor, a Dugald, or a Captain Knockdunder, at every corner."[131] Without any sense of embarrassment Geikie depicts the two men in Arran drinking whiskey, dancing jigs, and eating haggises. Sedgwick takes positive joy in the traces of the older, squalid Arran when he comes across them, reveling in the sight of the "good-humoured, high-minded, well-informed, racy, and dirty" Highlanders.[132] This was "the first place in which I saw the Highlanders in their native habitations," Sedgwick later wrote, and, luckily for him it seems, they "indeed are none of the best. Those of the lower orders have often neither chimney nor window, and from the distance of two hundred yards might be mistaken for peat-stacks."[133] While MacCulloch had claimed that Arran had *once* been the scene of "ruinous huts where it ought to have had houses, and wretched villages where it might have maintained towns," Sedgwick continued to see the ruinous huts.[134] Whatever the truth was, his description far more closely mirrors that of "the miserable little *bourocks*" in which the Highlanders live in *Rob Roy*, "composed of loose stone, cemented by clay instead of mortar, and thatched by turfs, laid rudely upon rafters formed of native and unhewn birches and oaks from the woods around," and whose "roofs approached the ground so nearly, that Andrew Fairservice observed we might have ridden over the village the night before, and never found out we were near it, unless our horses' feet had 'gane through the riggin.'"[135] Sedgwick and Murchison's response to such landscapes, structured by their pleasurable experiences of Scott's novels, is one of unmitigated pleasure.

For MacCulloch, this kind of pleasure was "the indulgence of the fantastic wishes" bought at the expense of the happiness of the Highlanders themselves. As he put it, "What the poor people themselves have to gain by remaining what they were two centuries past, it would be hard to say, and it really is far beyond the bounds of fair indulgence to the romance of their betters, to consent to see the Highlanders at large, suffering any inconveniences from which they might be relieved."[136] This kind of "romance" emerges as a construction shared by men of science, men of letters, men whose culture is defined by their shared participation in a class, and Scott, Sedgwick, and Murchison are all alike sharers in its aesthetic pleasures—even if Sedgwick, more publicly, was ambivalent about it.[137]

Furthermore, as MacCulloch had said, the island offered "to the geologist an epitome of the structure of the globe; forming a model of practical geology for the instruction of the beginner and for the study all."[138] For him, the nature of its "epitomising" powers was broadly topographical, for the island possessed "rocky and rugged mountain," "swelling hill," "open valley," and "green retired glen," "all that diversity of surface which is rarely found condensed into so small a compass, and, more rarely still, combined with an insular situation." His map of Arran, published in the third volume of *Description of the Western Islands,* built upon a scant few existing maps and gave the most detailed shape to the island yet in terms of scale, hachuring, and use of color to delineate the strata (plate 2).[139] For Sedgwick, paraphrasing MacCulloch, Arran was indeed "a geological epitome of the whole world, and is, moreover, eminently picturesque."[140] But for him this phrase carried much more theoretical import. MacCulloch had presaged against the method of correlating or comparing rocks from diverse geographical locations in order to form an ideal sequence of the strata, for, he argued, "every deposit of secondary strata must, to a greater or lesser extent, be considered partial; there being no proof that they have ever possessed universal continuity." For Sedgwick and Murchison it was precisely this correlative potential embodied in the rocks of Arran that motivated their research.[141] Sedgwick wrote that the work would "not only assist in completing the natural history of Arran" but would help "to fix the true epoch of all those interrupted fragments of secondary formations" found along the west and north of Scotland. It might even help "fix" the secondary formations of England itself.[142] Such an action was possible because these new geologists used fossils as well as lithology to determine the structure of the rocks—a practice deeply opposed by MacCulloch, who would bitterly contest Sedgwick and Murchison's work on the island. As Lyell put it, MacCulloch would "treat in a tone bordering on ridicule some entire departments of science connected with geology, such as the study of fossil conchology."[143] But it was also the new means by which those "sober geologist[s]" who did not now dare "to give an ideal history of the revolutions of the earth" could hope to regain synthetic sweep for their science, conjuring up imaginary romance worlds from a single fossil or object, suggesting the evocative vision that MacCulloch had blended into the heady mix of his prose.

The geologist, Sedgwick was later to write in his appendices to William

Wordsworth's *Guide to the Lakes* (1842), "studies phenomena, groups them together, contemplates them in all their bearings, and so attempts to rise from phenomena to laws."[144] In doing so, however, he would have to fully immerse himself in the chaos and drama of former worlds, accepting that no story could yet be constructed from their ruins. Only in the process of that immersion, whose details we find in the literary and sociable records of the geologists rather than in their *Transactions*, could the gentleman geologist hope to reinvest the world with some kind of order. This the geologists had learned partly from Scott. Nostalgic for an "essentially eighteenth-century world of order," for a "heroic past," geologists like Sedgwick and Murchison also felt themselves to be participating in the program of modernization and enlightenment that MacCulloch, however controversially, had called for. As in Scott's novels, where romance is often viewed ironically, in the *Transactions* and *Proceedings* of the Geological Society, this romance is held at arm's length. But it operated there nonetheless, in conversations, in the club room, in speeches, and in the mind's eye of the geologist as he quested into rocky terrain.

The Roaring Linn

The significance of Scott to men like Sedgwick and Murchison is brought into relief when we look at the geologist who effectively solved the "Cambrian-Silurian dispute": Charles Lapworth. Lapworth began his career not as a geologist but as an English teacher. He cared about his prose. "Time and again," his biographers note, "he would tear up and burn manuscripts until by much re-writing they met the high standard he imposed upon himself. 'Either the readers or the writer of a paper have got to have a bad time;' Lapworth once remarked, 'and it is better it should be the writer because he is in the minority.'"[145] Certainly he was not averse to the epic sweep: in a presidential address to the British Association for the Advancement of Science in 1892 he careered "from the microcosm of earth to the macrocosm of creation-at-large," moving "*in imagination*, to the fiery eddies of the sun, and thence to the glowing swirls of the nebulae; and so outwards and upwards to that most glorious septum of all the visible creation, the radiant ring of the Milky way."[146] By the 1890s cosmology was back in vogue and could be employed with much more abandon than it could in the 1830s. But Lapworth's language could be tactful too. The term Ordivician referred to "the Ordovices, people who inhabited the part of Wales in which some of his work on the Ordovician rocks were carried out."[147] In choosing this name, he gave paid respectful homage to Sedgwick and

Murchison, whose Cambrian and Silurian classifications had also referred to the Welsh lands and peoples who had once lived on the rocks they studied.

Moreover, even once that cosmological sweep had found its way back into elite geological discourse, the practical skills of the geologist and his patient, geometrical approach to complex stratigraphic questions remained firmly embedded in a more popular image of the geologist. Charles Kingsley described Lapworth with "wire-rimmed spectacles on his nose, geological hammer in hand and hung all over with collecting boxes, telescopes, barometers, ordnance maps, photographic apparatus for finding out everything about everything and a little more, too."[148] In doing so, he picked up on Lapworth's self-fashioning as a geologist of the absentminded, antidomestic type, a type enshrined in literary texts such as *Bleak House*, with its monomaniacal geologist Professor Dingo, or in the 1880 novel *Wooers and Winners* and its negligent geologist-hero Archibald Thorpe.[149] As late as 1961 his daughter-in-law remembered him "as an ideal father-in-law, though very absent-minded," for "during pauses in mealtime conversation at home, he would draw geological sections on the white linen tablecloth." "Years after his death," his students recorded, "while sorting through boxes of geological specimens, Birmingham geologists discovered an unopened packet of Lapworth's sandwiches perfectly petrified."[150] The image gave concrete form to popular views about the *practice* of geology, moreover, and it helped sustain Lapworth's identity as a stratigraphic geologist bent on fieldwork. His biographers praise him primarily as a stratigrapher and a mapmaker—subordinating palaeontology to the role of "a handmaid, never a mistress," to the study of the structure of the rocks—and for the increasingly large scale of his maps, as he resisted the abstractions required to render natural immensities in pocket size. "In addition to a keen eye for country" his biographers noted, his natural geological faculties

> included a wonderful perception of minute changes in both horizontal and vertical divisions of the rocks, and the special significance of these in his stratigraphical work. He also had an exact and retentive memory for special localities and critical places, which rarely played him false. . . . He was also a good geometrician and could "see solid" (i.e. in three dimensions) into a map or the ground it portrayed. When he "saw solid" he seemed to see through the surface of the ground by means of that surface and to visualise the conditions that might have controlled the observed phenomena.[151]

But "above all, he possessed the power to realise vividly the conditions that might have given rise to the observed phenomena, so that in imagination he saw them all at work and studied their results. This characteristic was responsible for the saying that when once Lapworth had made up his mind on a subject, it was no use to contradict him, for *he was there when the rocks were made.*"[152] Here again was that process of imaginative immersion that nineteenth-century geologists mythologized as a necessary preliminary to the more sober task of inductive observation.

Lapworth moved to Galashiels on the Scottish Borders in 1869 to take up a teaching post near the home of his favorite author, Scott. The manuscripts of two of his lectures given at Edinburgh in 1879 reveal that this imaginative identification with the "romance" of landscape, and the pull back to appropriate a slightly fusty antiquarian persona as a sober inductivist, had its roots in his reading of the famous poet and novelist. In those lectures he describes Edinburgh from "the summit of Arthur's Seat" as "a view of which every inhabitant of this beautiful city is proud," "Sea and shore, lowland and upland, sunken plain and lofty peak," "looming" and "glittering" and "rolling" over every horizon while, quoting the Scottish poet and physician David Macbeth Moir, "To the west, the floor of the country rises stage by stage into 'Pentland's gloomy ridge / Belting the pale blue sky.'"[153] "The landscape," he continues, "is corroded with spots of old renown. The very plain is dotted over with the sites of historic battle fields. Locality after locality crowds upon the sight and fills the mind with memories of victory and death, of glory and pity, of poetry and song."[154]

As Lapworth takes his students through a variety of Scottish landscapes for the purposes of geological explication, he is haunted by these "thousand associations with the past. Every locality has its memories of old." In each spot, "Artist and poet, practical man and scientist, natives and strangers are all fascinated in time by familiarity with" the "old world country" of the Scottish borderlands. Geological exploration simply adds to the "historical and legendary appreciation" of such sights.[155] It is unsurprising, then, that the second of Lapworth's 1879 lectures takes his students directly into the heart of Scott country. "There are many charming routes through the heart of the Southern Uplands or Borderlands of Scotland," he writes:

> But there is one, however, which has long been acknowledged to stand preeminent. It is that which runs through the ancient forest of Ettrick and is followed by the country coach, that plies two or three times a week, in early spring and summer between the towns of Selkirk and Moffat. There is no other upland route that can be compared with this. It bears the palm

in variety of interest, not only to the lover of nature but to the artist, to the poet and to the geologist.[156]

The path between Selkirk (of which Scott was sheriff deputy from 1799) and Moffat now forms part of the Walter Scott Way. Scott is far from being the only poet associated with this landscape of course—the poet James Hogg, for instance, known during his lifetime as the Ettrick shepherd, is mentioned by name in Lapworth's lecture, which also includes a quotation from the Scots poet David Macbeth Moir. This is country steeped, in Lapworth's account of it, in literary and legendary lore. But Scott does infiltrate Lapworth's prose as he winds his students "for mile after mile, through the deeply wooded glen of the lower Yarrow," noting "the parks of sweet Bowhill and fatal Philiphaugh," past Newark Castle, described in *Lord of the Isles*, to the "breezy uplands, near the Dowie Holmes of Yarrow," and is beset by "sad and tender thoughts" of "these old familiar names," each filled with "memories of poetry and song," "legends of love and sadness—'all wet with tears and pale with wae.'"[157] The source of the quotation is obscure, but these are scenes recounted in "The sang of the outlaw Murray" in Scott's *Minstrelsy of the Scottish Border*. Scott explained, "It is certain that, during the civil wars betwixt Bruce and Baliol, the family of Philiphaugh existed, and was powerful" and may have "refused allegiance to the feeble monarch of the day, and thus extorted from him some grant of territory or jurisdiction."[158] Nonetheless, Lapworth says, "we have no time to dwell on" this now, but must instead climb "onward and upward" until "we stand upon the highest point of the roadway, upon the watershed of the region, upon the boundary line between the basins of the Tweed and the Annan," where the waterfall known as Dobb's Linn used to be a "favourite hiding place of the hunted covenanters."[159] As Scott described the waterfall in *Marmion* (1808), it was the place

> Where deep deep down, and far within
> Toils with the rocks the roaring linn;
> Then, issuing forth one foamy wave,
> And wheeling round the Giant's Grave,
> White as the snowy charger's tail,
> Drives down the pass of Moffatdale.[160]

It is in "this strange spot," "its waters ... more than once stained of a deeper tint with the blood of martyrs," that Lapworth commences his study of the rocks. For "the key that will unlock all the mysteries of the geology of the

Uplands lies in this mysterious glen, and we must not leave the place until we have discovered it."[161] In this "mysterious glen" Lapworth reveals his pioneering work on the structure of the southern uplands of Scotland. For it was here that Lapworth found an abundance of "graptolites" in the stratified black shales of which it is comprised. Establishing that the graptolites, fern-shaped imprints on the rock, were the fossilized remnants of organic, possibly crustacean, creatures, Lapworth had been able to establish three distinct "zones" of these fossils within the shale. Here, form was important in two senses. First, graptolites were evidently a rapidly evolving species, so that within closely related strata they exhibited *different* formal characteristics. Second, those different organic forms could be best seen and understood by the geologist if he had a cabinet with a drawer for each stratum being investigated:

> In our spare time on an evening we name and identify these Graptolites but we are always careful to replace them in the proper drawer. Long before we have collected our hundred species, we shall have made another and more startling discovery. We find that our drawers contain 3 distinct sets of Graptolites. The 2 upper sets of rocks contain one set of graptolites. Those answering to the 2 middle sets of shales contain a second set and those answering to the flinty beds at the bottom of the section contain a third set. None of these species of Graptolites found in one set passes into the other.[162]

So, "Instead of 5 divisions of the dark shales we make three" now: the Upper Shale, the Middle Shales, and the Lower Shales.[163] Here, the evolution of the graptolite enables the geologist to divide the rocks into three stratigraphic bands, analogous to the drawers of a cabinet. Correlating the cabinets he has made at Dobbs Linn with the thousand-foot-high sections of shale exposed at the Craigmurchan Scaurs, which "lies among the heights above the source of the Ettrick," Lapworth reveals that, in fact, rocks that had been deposited in a single group of layers had subsequently been smashed and folded and faulted against one another in every possible way, until they had been broken up and thoroughly reshuffled, like a deck of cards.[164] The graptolite faunas proved this because the three distinct groups of fossils repeated themselves over the sequence. The rocks had been shuffled one over the other and compressed into a smaller area, so that "only *hundreds* of feet of strata are involved, not *thou sands*, as had previously been believed to be the case."[165] Most importantly, this enabled Lapworth to resolve the Cambrian-Silurian controversy. The rocks under dispute could be divided, through the use of graptolite zones, into not *two* systems—the Cambrian and the Silurian, whose boundaries were at issue—but three, with the "Ordovician" dividing the other two.[166]

Of course Lapworth's Ordovician system took form without Scott's intervention. But the literary setting of his lectures, describing the geologist's movement through the landscape to Dobb's Linn, steeping him in poetic and historical legend, does more than simply add atmospherics to his scientific description. It claims for Lapworth's nomenclature a place *within* that Scottish history and myth, drawing from it evocative and explanatory power. And it situates Lapworth within an intellectual tradition stemming back through Sedgwick and Murchison, through the geological writer Hugh Miller, whose writings had reached a broad public and had been filled too with quotations of Moir's poetry, to MacCulloch, a tradition indebted to the "romances" of his literary hero, Walter Scott.[167] Finally, it invests the graptolites and the strata they explicate with the mythic and historic status accorded to the Covenanters or to the heroes of Scott's historic narratives. They become part of the world that Scott described.

Conclusion

In MacCulloch's *Description of the Western Islands*, the generic tensions that had given Scott's novels their vitality and cultural authority were plain to see on every page. Murchison and Geikie, seeking to gain for science the same level of popularity and prestige as had been accorded to the novel by Scott's success, appropriated the novelist's "romance" style, dashing at the strata, capturing the rocks as terrain. Sedgwick, much more ambivalent about novelistic encroachments upon geological form, nonetheless admired the writer and shared with him his antiquarian digressions and discursions, his almost comically inductive approach to knowledge. Lapworth's scientific conciliation also took form from Scott, as he presented his geological world in the historic and mythic terms of Scott's poetry and prose. What Scott gave each of them was a language for imagining his own geological practice. These men created their self-images, and their definitions of the practices of a good geologist, from Scott's vision of the past. Like Scott, who pulled back from romance into plotless antiquarian digression and lengthy descriptive prose as a means of giving novel a gentlemanly and respectable flavor, the gentlemen geologists would move into these landscapes like so many Waverley heroes, before pulling back to the hard intellectual tasks of contemplation, organizing their collections on the page or in the cabinet. Hammering at the uncouth strata in the remotest regions of Britain, venturing into "romance" landscapes like Scott's Waverley heroes, they produced a new set of descriptions of the world which, via publication in the gentlemanly forums of the quarterly reviews and the scientific learned societies, might form the basis of national pride and self-definition. Scott, in part, gave shape to the world that they described.

Lyell's Mock Epic { **CHAPTER THREE**

On November 11, 1817, three days before his twentieth birthday, an aspiring poet sent his father his latest literary effort from his undergraduate lodgings at Exeter College, Oxford. The poem, written in modish Spenserian stanza, a form that had recently been revived by Lord Byron in *Childe Harold's Pilgrimage* and with which John Keats and Percy Bysshe Shelley were also experimenting, romantically recalled his trip to the Hebridean island of Staffa during the university vacation two months earlier. But it was also written in the wake of a series of disappointments. In both 1816 and 1817 the young man had entered verses for Oxford's prestigious Newdigate Prize, won later in the century by John Ruskin, Matthew Arnold, and Oscar Wilde.[1] But his poems on the annually set topics—the horses of Lysippus in 1816, the Farnese Hercules in 1817—had not come close to clinching the honors, and even his father had pronounced that his son's work was "not above mediocrity and certainly ought not to gain a prize," though he added in paternal tone, that it "is not destitute of *estro poetico*."[2] Checked but undaunted, the undergraduate would submit his verses for the prize once again, and again unsuccessfully, in 1818, this time some Latin verses and some lines on the Coliseum.[3] In November 1817, then, he was already beginning to doubt that he would ever make it as a poet, and it was not without some self-consciousness that he sent his father his newest piece of verse. The poet, of course, was the geologist Charles Lyell.

The narrative of this poem (see appendix) sees "deluded Fancy's child" spellbound by the "wizard" magic of an "Eastern tale," related to us in the first stanza in sensuous terms. These "fairy worlds" ring with magical effects, from the intoxicating alliteration of rhyming phrases—"beings bright," "realms rich with unborrowed light," "noon ... which never knew the night"—to the repetition of negatives, four in four successive lines, to evoke the ethereality of that paranormal light emanating from sunbeams that "never fell," "unborrowed," "unquenchable," in an unknowing noon. Turning from the visionary dreams of Arabian fiction to the real world, the protagonist feels himself, like many a Romantic figure before him, a Dantean hero, exiled in "a desert bleak and wild." But the speaker of the poem quickly tells us a different story. Nature, summoning the world into being through her voice, has a secret that the child of Fancy might hear if only he were aesthetically competent to detect her "strain," a secret the poet knows and shares with the reader in conspiratorial fashion. The secret is Staffa, a "massy pile" of pillars raised from the ocean by an omnipotent but artistic goddess. The goddess's world operates in a different, more picturesque aesthetic register, but its hidden charms, slowly perceived by the speaker of the poem, softly echo the aural effects of the first stanza with the alliteration of the fricative consonants /s/ and /f/, the cavern's "sculptured side," the "favoured foot of Mortal yet to come," and another, more muted pair of negations, "no shapes of Terror," "no guardian dragon"—reflections of those more spectacular sounds and more uncanny negations in the eastern tale. Staffa might not have the glamor of Arabia, but, repeating a word from the first stanza in the last, it is no less "enchanted." There is the hint of a familiar moral dimension to this conclusion: the speaker of the poem and his ideal reader overcome the childish delusions of fiction to developa more intellectual sensitivityto the north, superior to the merely sensory pleasures of the east. And with moral achievement comes power: the mighty voice of the goddess Nature may be read as an allegory for the poet's force as he too summons up the natural world with his powerful voice.

With its narrative of maturation, it is tempting to read "Lines on Staffa" as a somewhat earnest performance of Lyell's approach to adulthood. Sending the poem to his father, he attempts to show that at Oxford he has learned a gentlemanly sensitivity to the natural world and attained the poetic accomplishment to compose stanzas with the most fashionable of contemporary poets. It was not the first poem he had sent home seeking approval from a father who had taken great care over his son's education, nor would it be the last reference to poetic matters in that correspondence: while visiting Venice the following summer, for instance, he wrote of seeing Lord Byron himself rowing in his gondola every morning after breakfast.[4] But if the poem partly expresses the undergraduate's gentlemanly panache, it also tentatively expresses his growing interest in a new pursuit, geology. Between 1830 and 1833 Lyell would publish arguably the most important geological work of the nineteenth century, *Principles of Geology*. *Principles* sold some fifteen thousand copies in eleven extensively revised editions before Lyell's death in 1875, and it was read by many more.[5] In "Lines on Staffa" those budding geological interests found their first, cautious expression: "Whatever you may think of the poetry," Lyell added in a note to his father, "you will agree with me in regretting that [the geologist Abraham Gottlob] Werner should have died without the knowledge of this geological discovery concerning the origin and formation of basalt."[6] Werner, who had only recently passed away, certainly would have disagreed that Nature had "bade uprising lift the massy pile" from beneath the ocean. For him, basalt had slowly settled from the thick aqueous solution that had constituted the primeval globe, as had all of the rocks. An "uprising" motion suggested the contrary view, that it had been thrust upward by the power of the earth's internal heat, not by water but by fire. More to the point of Lyell's self-conscious joke unveiling his poem as a geological discovery, Werner certainly would have been surprised to learn that those controversial basalts at Fingal's Cave in Staffa were not necessarily the products of either fire or water but rather an artistic creation of the fairy goddess Nature.

It is hardly surprising that Lyell, as a young man at Oxford, should have been considering these fashionable geological questions. The geological lectures of William Buckland, a leading member of the Geological Society of London and the university's Reader in Mineralogy, were then attracting hordes of undergraduates and dons and considerable attention further afield for their powerful and often witty evocation of the aeons of geological time that had preceded the advent of man.[7] Lyell was in the crowd. Excited by Buckland's showmanship and by the worlds he unfolded in the lecture hall, Lyell borrowed Robert Bakewell's *Introduction to Geology* (1813) from his father, a prominent botanist, who had found it "an interesting and popular volume, notwithstanding a peevish preface," and in the summer of 1817 they together visited Robert Jameson, founder of the Wernerian Natural History Society, in the hope of viewing his renowned collection of specimens at the museum of the University of Edinburgh.[8] Disappointed in that hope, they had nonetheless dined with Jameson and professed to have enjoyed his company.[9] And it was partly on a geological errand for Buckland that Lyell had visited Staffa later in the same vacation, verifying for him that Fingal's Cave had "broken tops" rather than smooth ones, disconfirming the theory of a former student of Werner, Leopold von Buch, that the cave had been formed by deposition along a dyke of softer lava that had since been eroded.[10] The serrated tops of

Fig. 3.1. "View of the entrance of Fingal's Cave in Staffa." In John MacCulloch, *Description of the Western Islands of Scotland* (1819), vol. 3, plate VII. Reproduced by permission of the syndics of Cambridge University Library.

the columns (see fig. 3.1), suggested instead that they had, as "Lines on Staffa" agrees, been formed by rapid intrusion from below rather than gentle deposition from above.

But if Lyell's father was encouraging of his son's new scientific interests, he would be less than pleased when, ten years later, Lyell abandoned a career in the law in order to dedicate himself entirely to geology and to attempt to make a living at it.[11] For Charles Lyell senior, science was a disinterested gentlemanly pursuit but not an appropriate source of income or a paying career. Tentatively framing a scientific speculation in fashionable Spenserian stanza and within a Romantic "fairy scene," then, may have allowed Lyell to hint at his new geological interests without explicitly vocalizing whatever deeper ambitions the science may have been awakening in him.

Whether or not Lyell already knew that he wanted to devote himself to geology, at Oxford high aspirations for the science were quickly developing, and poetry was a key medium by which those aspirations could be voiced within a potentially hostile environment.[12] As part of a wider program designed to give the physical sciences a place in the curriculum, in 1817 Buckland was busy persuading the Prince Regent to establish a second scientific readership at the university, this time in geology. His bid would be successful, and he himself would hold the post from its inauguration in 1819.[13] Part of his strategy for recruiting undergraduates and dons to an interest in the science, however, was poetic in a somewhat less ideal sense than Lyell's father may have approved of. A well-known lithograph, entitled "A Coprolitic Vision," produced by Buckland's Oxford colleague W. D. Conybeare in 1830, captures something of the notoriety Buckland gained for his study of what he called "coprolites," or fossils containing petrified excrement, and also, in its oxymoronic title, captures the parody of poetic "vision" by which Buckland presented his findings.[14] The image shows the professor, precariously balancing on a rock and dressed in subfusc, entering a filthy cave, geological hammer in hand. The extinct hyenas whose habits and habitats Buckland had famously described fight in the foreground, and pterodactyls, a species "discovered" by Mary Anning in 1829 and presented by Buckland in a paper at the Geological Society that same year, swarm overhead.[15] If anything, the image claims intellectual ownership over those flying monsters for the gentleman professor in his academic dress, etching their female, working-class discoverer out of the story. But it also demonstrates the extent to which that academic and gentlemanly "vision" of the primeval past went hand in hand, at least in Buckland's formulation, with conspicuous impropriety: flying pterodactyls drop dung all over the cave in midflight, and fat, spiraling columns of bezoar-stone, a concretion of fecal matter found in the stomach and intestines of ruminant animals, support the cave.[16] Buckland even had a mahogany table inlaid with coprolitic specimens that, much to his own amusement, he frequently encouraged his unsuspecting guests to admire. The effect could be disorienting. Lyell, for instance, described Buckland's "usual style" as so "strange" a "mixture of the humorous and the serious that we could none of us discern how far he believed himself what he said."[17]

As ungentlemanly as Buckland's coprolitic vision of geology might seem, it may have worked to guard scientific territory for a male elite, keeping out those too delicate to hear its dirty secrets. It was the combination of robust physical humor and a sense of entitlement that proved Buckland's most effective weapon in recruiting new students and in claiming the gentlemanly authority to describe a geological past that could still be seen, in some quarters, as heretical. It was this too that prompted the strongest antipathy to geology as an object of university study within Oxford. In the 1880s the prominent Oxford clergyman Thomas Mozley recalled with contempt that Buckland was "always coarse" and occasionally "emboldened to unwonted profaneness," but it was not only this that irked the leading members of the Oxford Movement.[18] According to Mozley, though John Henry Newman, John Keble, and R. H. Froude responded differently to the discoveries of the geologists, they were united in their resentment of "the moral tone ostentatiously displayed by many men of science, flattering themselves that they had beaten 'Revelation' out of the field."[19] Mozley was prone to exaggeration, and it is not difficult to detect some bitterness here, but that moral tone was certainly an intrinsic feature of Buckland's geological rhetoric. His caves were not only profane. They were also sublime.[20] His prehistoric caves and underground caverns were presented as mysterious realms inhabited by latter-day dragons and beasts, spaces that could only be safely interpreted by gentlemen geologists conceived of as epic guides and heroes on dangerous quests into their subterranean depths. Buckland's writings and performances, and those of many other geologists in his wake, were filled with allusion to the epic poets Homer, Virgil, Dante, and Milton.[21] Associating geology with classical and Christian epic gave it perceived compatibility with the textual scholarship of classics and the Bible on which Oxford's role as a teacher of clerics was established.[22] If Milton had rewritten Genesis, and if Dante had provide Virgil as a guide to the esoteric underworld, geologists educated in those traditions, Buckland intimated, were now ready and waiting to take up both those quests for the nineteenth century.

Most importantly, historians are beginning to argue that epic performed real scientific work for geological writers like Buckland in the early nineteenth century, as it did for subsequent popularizers of evolutionary theory.[23] The teleological narrative line of epic form, and particularly of Miltonian epic, with its claim to describe the total history of a nation or a people from beginning to end, could give dramatic shape to the history of the earth. The geologist-guide too unveiled "a cosmic narrative tracing a progression from brute matter to humanity over millions of years," a world whose past was awe-inspiring enough to require epic treatment and that derived classical and biblical authority from that most prestigious of poetic forms.[24] In his 1836 Bridgewater Treatise, Buckland made the terms of this titanic history explicit:

> The deeper we descend into the strata of the Earth, the higher do we ascend into the archaeological history of past ages of creation. We find successive stages marked by varying forms of animal and vegetable life, and these generally differ more and more widely from existing species, as we go further downwards into the receptacles of the wreck of more ancient creations.[25]

Based largely on the work of Jean Baptiste Fourier, geologists widely, though tentatively, considered that the earth had once been a molten, fluid mass, which had gradually cooled and contracted until it had reached its current equilibrium.[26] This meant that the earliest periods of earth history were likely to have been much more volatile than later periods, and that geological processes in the present were of relatively minor interest, and it dovetailed neatly with the evidence of the fossil record, which seemed to reveal "successive stages" of ever more complex life forms. In its deepest depths, and therefore in its oldest rocks, was a greater abundance of large, simply structured species analogous with those now found in tropical climates. As the geologist "ascend[ed]" to dizzying philosophical heights as they *descended* through the layers of the earth, in earthly imitation of Dante "ascending" through Hell and Purgatory toward Heaven, those life forms became increasingly more complex and increasingly resembled those now living.

Of course, there was still much work to be done in fleshing out this hypothesis, and Buckland did not claim that the epic of earth history had yet been written. To do so, indeed, would have been to claim the kind of totalizing, all-encompassing vision of the earth that the Geological Society explicitly rejected. Buckland's work was not a presentation of an epic history of the earth but was largely based on the reconstruction of the habits of extinct animal populations and of the specific habitats in which they lived. As such, it took a synchronic more than a diachronic approach to earth history, describing in detail geological worlds at particular points in time rather than speculating on the development of those worlds *through* time.[27] Even once those geological worlds were assembled along a chronological line of development they could not easily be read as plotted stories of earth history, at least without a very active and engaged reader to interpret them as such.[28] In print at least, even Buckland's most cosmological speculations were confined to evocative concluding paragraphs in suggestive, rather than explanatory, fashion.[29]

Still, he was keen to present geology as a science that would, however long it took, finally reach the all-encompassing view. "The season has not yet arrived," he "candidly" admitted, "when a perfect theory of the whole earth can be fixedly and finally established, since we have not yet before us all the facts on which such a theory may eventually be founded." But "the first, and second, and third story of our edifice" of geological knowledge could be "soundly and solidly constructed" independently of one another, and it was only a matter of time before "the roof and pinnacles of the perfect building would be completed."[30] Buckland's architectural metaphor worked to suggest both that geology was to reveal a fixed body of knowledge that could be produced as if by architectural plan and completed within a fixed time frame, and that such knowledge was close to completion. As such, while "during the infancy of Geology" it had been "prudent ... not to enter upon any comparison of the Mosaic account of creation with the structure of the earth, then almost totally unknown," the question had now become paramount. Once geologists completed the edifice of their science, cosmogony and geology might meet again. It was in this treatise that Buckland recanted his earlier belief that there was *geological* evidence for a global flood coinciding with the flood described in Genesis, but he also described a world in which every natural phenomenon, from the symmetry of crystals to the laying up of coal in the depths of the earth's strata to the structure of the eye of a trilobite, could be evinced as "an integral part of one grand original design."[31] For Buckland, seeking to establish geology as "a legitimate branch of inductive science" from within conservative Oxford, the relationship between geology and the Mosaic cosmogony had been put on hold, but it was with confidence that he asserted that the two might one day be reunited.[32] "Although time must elapse before the roof and pinnacles of the perfect building can be completed," he wrote, "we protest against the rejection of established parts, because the whole is not yet made perfect."[33] And it was with this conviction in the underlying "whole" that Buckland ended his work: "The Earth," he rhapsodized in his conclusion to the Bridgewater Treatise, "from her deep foundations unites with the celestial orbs that roll through boundless space, to declare the glory and shew forth the praise of their common Author and Preserver."[34] Buckland's "celestial orbs" are stars, but the phrase also recalls the classical, medieval, and early Renaissance notion of celestial orbs as physical spheres encircling the earth or the sun, within which the stars sat like jewels. More importantly, the phrase echoes Milton's *Paradise Lost,* itself littered with celestial orbs and spheres, and situates geology within the grandeur of Milton's cosmological epic.[35] If geology had gained authority through caution, if it could not dare boast a new cosmology of epic scale, proportion, and totality, it nonetheless carried a powerful epic promise.

Lyell's "Lines on Staffa" indicates at least that he was sensitive to the potential poetry of the science as it was revealed to him in the lecture halls at Oxford, and reveals too that if he was beginning to think of dedicating more than his spare time to geology, then poetry was a useful medium in which to present it, at least to his skeptical father. And it was this sublime, epical language that appealed to him rather than Buckland's burlesques. By 1827 he had adopted its terms in an article in the *Quarterly Review* designed to convert educated readers to geology: though geologists had not gathered enough data to create an epic story of earth history just yet, once all the data had been gathered and compiled, the geologist's students would be able, he wrote, to "confide themselves to his guidance, as Dante in his sublime vision followed the footsteps of his master, and beheld, with mingled admiration and fear, in the subterranean circles environing the deep abyss, the shades of beings" who could "recal from oblivion the secrets of the past."[36] The key difference, however, and one that would come to dominate the differences that would soon emerge between Lyell and his former mentor, was that for Lyell, geology's epic promise was much further from realization. By the time he wrote *Principles of Geology*, indeed, Lyell would declare that it was a promise that geology could never hope to fulfil.

Poetry and Science

Lyell the undergraduate, puzzled but inspired by the aeons of geological time unveiled to him in the Oxford lecture halls and enchanted by the western islands of Scotland, grew up, of course, to become one of the most famous geologists of his age. His biographer implies that as he did so the childish delusions of Fancy described in "Lines on Staffa" and replaced by the adolescent fantasy of a poetic vocation finally gave way to the mature, scientific apprehension of the natural world with which he would stake out his place in the world.[37] Such a narrative equates Lyell's maturity with a rejection of poetry for science, retelling in more forceful terms the same conventionalized story Lyell edges toward in "Lines on Staffa" and in his letter to his father. It also recapitulates in Lyell's life story the familiar Enlightenment narrative of progression from a primitive age of poetry to an enlightened age of reason, in which poetry is superseded as an instrument of knowledge by science and philosophy as civilization advances or, in this case, as the hero of the tale reaches maturity. And this was precisely how Lyell presented his "principles" in *Principles of Geology*. Famously beginning with five chapters outlining a "historical sketch of the progress of geology" Lyell claimed to debunk the mythic and fallacious content of the classical, oriental, and recent cosmologies that had hitherto "retarded" the science.[38] As he put it:

> In an early stage of advancement, when a great number of natural appearances are unintelligible, an eclipse, an earthquake, a flood, or the approach of a comet, with many other occurrences afterwards found to belong to the regular course of events, are regarded as prodigies. The same delusion

prevails as to moral phenomena, and many of these are ascribed to the intervention of demons, ghosts, witches, and other immaterial and supernatural agents. By degrees, many of the enigmas of the moral and physical world are explained, and, instead of being due to extrinsic and irregular causes, they are found to depend on fixed and invariable laws. The philosopher at last ... rejects the fabulous tales of former times, on the ground of their being irreconcilable with the experience of more enlightened ages.[39]

The philosopher at last, in fact, "becomes convinced of the undeviating uniformity of secondary causes," the very principle that Lyell's volumes are designed to expound. Seeking to put geology on a firm methodological footing, to enable it to "rise to the rank of an exact science,"[40] Lyell's *Principles* was a sustained argument for the study of *vera causae*, causes that could be directly observed in nature, as the only means by which the millions of years of geological history that had never been observed by man could be "found to depend on" those "fixed and invariable laws."[41] The unobservable world was to be understood by analogy with the observable world. This meant, in the first instance, that only *kinds* of causes that had been witnessed within the course of human history were adequate explanations for the geological past.[42] "Demons, ghosts, witches, and other immaterial or supernatural agents" were out of bounds, along with natural phenomena like comets, whose potential to affect the course of earth history was merely hypothetical. And with this all leading geologists concurred. In his well-known critique of Lyell's uniformitarian geology, William Whewell, for instance, described the history of geology in terms that also relegated supernatural and extraordinary explanations for geological events to the science's mythic or poetic prehistory. "A few appearances hastily seen, and arbitrarily interpreted," he wrote,

> are enough to give rise to a wondrous tale of the past, full of strange events and supernatural agencies. The mythology and early poetry of nations afford sufficient evidence of man's love of the wonderful, and of his inventive powers, in early stages of intellectual development. The scientific faculty, on the other hand, and especially that part of it which is requisite of the induction of laws from facts, emerges slowly and with difficulty from the crowd of adverse influences, even under the most favourable circumstances.[43]

Whewell's retelling of the growing-up of geological science as a sloughing off its poetic prehistory, again cast in the language of "stages" of human intellectual advancement, is marshaled not in the service of Lyell's *vera causa* methodology, but of enumerative induction, the method around which the Geological Society had been built, the accumulation of masses of data from which legitimate theories would emerge, almost of their own accord. Geologists would gather information about the geological past not by analogy with geological processes observable in the present, but from the archive of evidence that had been left behind, from the fossils and strata buried deep beneath the earth's surface. By the 1820s, and with the work of a generation of pioneering geologists including Buckland, Conybeare, and Sedgwick, there had been a relaxation of such stringent methods: the "strata" and their "formations" were necessarily theoretical categories, extrapolated from the relatively few rocks that were exposed at the surface, and fossils deemed characteristic "types" of each strata were being used for dating.[44] But the overarching paradigm was still overwhelmingly inductive. Whewell's story of the history of science performs the same trick performed by Lyell's biographers, of recapitulating the move from poetry to science in the body of the geologist himself. In Whewell's account, the pattern of discovery of this method, in which "the induction of laws from facts" is a slow and painful process, is also the pattern of its development in the history of human thought. Slowly, painfully, the method emerged only with much time and labor "from the crowd of adverse influences" that sought to retard geological science. The methods, and the enlightened men of an enlightened age who could carry them out, symbolized one another's achievements. In a neat twist, for Whewell, it was not the story of earth history but the sheer labor of geological research and the ability of its practitioners to resist their "inherent love of the wonderful" in the pursuit of the "scientific faculty" that possessed "the charm of an oriental fiction."[45] Read in this light, the protagonist of "Lines on Staffa" need not have cast off his Arabian fancies: his intellectual journey, his learning to see the world in chastened scientific terms, was a tale all of enchantment all of its own.

In arguing for *vera causa* methods, however, Lyell attempted to redraw the conceptual boundary between the "poetic" and "scientific" ages of the science. Lyell argued not only that events in the earth's past could only be understood if they were assumed to have been of the same *kind* as those currently operating, but also that they should be considered to have operated at the same *intensity* as they now did.[46] If those causes that could now be observed by geologists operating on the earth's surface were to form an adequate analogy for the deep past, this supposition was essential. The method of enumerative induction had not required that the present be a template for the past in this way; it simply involved studying the evidence that the past had left behind, and those evidences had seemed to reveal a greater intensity of geological processes in the deep geological past. For Lyell, while it was not impossible that floods or earthquakes or volcanoes had ever been violent enough to devastate the entire globe in a single episode, the fact that such episodes had not occurred within the span of recorded human history made this a merely metaphysical speculation, as was the notion that the earliest preserved rocks marked the beginning of earth history.[47] The oldest rocks were merely the oldest surviving rocks, beyond which the history of the earth may have extended for an unimaginable length of time. Their apparent lack of fossils did not mean that the earth at that period had been devoid of life, which would, again, have marked a beginning, but rather that, through the millions of years that had passed since their deposition, those rocks had been crushed and melted in a process Lyell named metamorphosis until they bore no relation to their original states and all their fossils had been destroyed. Those geologists who believed that there may have been global floods (regardless of whether those floods had scriptural counterparts), or that the stratigraphic column was a complete representation of earth history, "that planets, like certain projectiles, are always red hot when they are first cast" and gradually lose heat, enabling the progressive introduction of more complex species, were now branded mere "cosmogonists" purveying "an arbitrary hypothesis," indulging their "fancies in framing imaginary systems for the government of infant worlds."[48] And enumerative induction, relying on the notion that the record of the earth's past stored in its strata was relatively complete, was an utterly inadequate model once the geologist realized that those strata represented only a few pages of a much vaster volume whose many chapters had largely been torn out and discarded. Though he did not name them directly, by this formulation many of Lyell's colleagues now appeared not as rigorously inductive men of science but as old-style cosmogonists in a poor disguise. Though the gentlemen of the Geological Society considered themselves to have ushered in a newly scientific dawn for the study of the earth, reading between the lines of Lyell's history of geology, they were now, against their will, protagonists of its poetic prehistory. The scientific age of geology, Lyell implied, had truly begun with the publication of his *Principles*.

The Wolf and the Lamb

Perhaps the commonest assumption about Lyell's "uniformitarian" geology is that it was an argument about the shape and pattern of earth history directed against existing patterns offered by his colleagues. Lyell has been imagined as "the historian of time's cycle," writing an antiprogressive history of the earth in which species recurred in a perpetual series of cycles of the same or different kinds, or of a single cycle or "great year" of earth history that shaped all of its events, or of a not quite cyclical, "steady-state" system of checks and balances in which all the earth's processes exhibited "relatively minor fluctuations around a constant mean."[49] Others have Lyell describing a "system of indifference," comparable to the shock absorbers on a car, having "no goal-seeking or goal-directed mechanism," a proposition that might end up with the result that there is no progression in the history of the earth but that is not, of itself, a statement about historical direction at all.[50] In literary studies, the influential work of critics including Gillian Beer and George Levine has so powerfully demonstrated the ways in which literary plots reflected the shapes and structures of the evolutionary world, with its lack of origins and its lateral and entangled modes of connection, that by a backward projection geology has often been read as providing Victorian culture with a set of distinctive plots.[51] As outlined in the introduction, in this literature "catastrophism," represented by Lyell's adversaries, is commonly held to be a defense of the reality of the catastrophic events narrated in the Bible and of the young earth as dated by biblical chronologists, a natural-theological attempt to retain faith in the fixity of species, with new species introduced after each mass extinction, and a belief that geological change *always* operated via catastrophic means as a plot on which to hang the events of earth history.

In this literature Lyell's "uniformitarianism" is still often considered a secular view that geological change was always gradual and incremental, taking place over an immense span of time, and not admitting of any unpredictable, large-scale, or sporadic disasters. This has been brought into significant question by historians in the last thirty years.[52] The "central business" of geology was not the formulation of laws and patterns of change and causation, but stratigraphy, the working out of the order and structure of the earth's rocks, and nobody ever speculated that the earth operated via only catastrophic means.[53] Even Buckland, the most powerful geological storyteller of his generation and the geologist who held longest to the notion that the Mosaic flood was a genuine geological event, believed that the earth was "millions of millions of years" old, as he put it in his Bridgewater Treatise, and utterly rejected the attempts of "scriptural literalists" to square geology with Genesis.[54] More to the point, there was good empirical evidence for at least one global catastrophe accompanied by mass extinction in the earth's past, the great break in the secondary strata now seen as accompanying the extinction of the dinosaurs, and the fossil record did suggest that species had become more complex through time. Far from believing that geological change had always been catastrophic (a view that would, paradoxically, have been "uniformitarian" in suggesting that the rate and intensity of change had held constant across earth history), many of Lyell's critics simply held that the visible evidence afforded by the strata suggested that the deep past had witnessed at least one global geological event. There was no *a priori* notion that all earth processes had been catastrophic, or that gradualistic explanations might threaten such a world picture.

Indeed, Adam Sedgwick wrote to Lyell in 1835 in some annoyance at a paper given by the Swiss geologist Louis Agassiz at the Dublin meeting of the British Association for the Advancement of Science. "Among other marvels he told us that each formation (e.g., the lias and the chalk) was formed at one moment by a catastrophe," Sedgwick wrote incredulously, "and that the fossils were by such catastrophes brought from some unknown region, and deposited where we find them." Agassiz's ichthyological research was the important thing, not his wildly theoretical catastrophic speculations. "I hope we shall before long be able to get this moonshine out of his head," Sedgwick concluded, "or at least prevent him from publishing it."[55] There was no "catastrophism" and no school of catastrophists antagonistic to Lyell's work. While Sedgwick claimed, in his anniversary address to the Geological Society in 1831, to have enjoyed "nineteen-twentieths" of the first volume of *Principles*, the question he raised was over the adequacy of Lyell's method as a means of understanding the past. While he and most other geologists were happy to accept that the method had real value, still it seemed astonishingly anthropocentric to assume that millions of years of earth history had only *ever* seen effects that man, whose duration on earth was so comparatively short, had witnessed and recorded. The difference between Lyell and his colleagues was a point of method, and they never separated into two distinct "schools," each with its own plot of earth history.

Equally, Lyell's uniformitarian geology did not suggest that all geological change took place by the accumulation of imperceptibly gradual processes all operating at the same rate. The first volume of *Principles*, following the first five historical chapters and four chapters outlining his theory of an oscillating rather than a cooling climate, comprised a catalog of inorganic causes of geological change divided into aqueous causes (erosion by streams, rivers, and tides; the formation of deltas, hot springs, and coral reefs) and igneous causes including earthquakes, landslides, and volcanoes. From Pompeii to Lisbon to Calabria to Chile, thousands of people are shown to have died horrible deaths in a never-ending series of "terrific catastrophes" that Lyell parades before us.[56] In just a small sample of the conflagrations and disasters that had taken place on earth in the last thirty years, Lyell wrote, "New rocks have risen from the waters; the coast of Chili for one hundred miles has been permanently elevated; part of the delta of the Indus has sunk down ... the town of Tomboro has been submerged, and twelve thousand of the inhabitants of Sumbawa have been destroyed."[57] And yet, he wrote, geologists continue to "declare with perfect composure that the earth has ... settled into a state of repose."[58] The world in its present state was infinitely more catastrophic than those we have been accustomed to describing as the visions of "catastrophists." And its effects could be sporadic, variable in their intensity, and irregular. Speaking of a flood at Tivoli in 1826, Lyell reminds his readers, for instance, that "for four or five centuries consecutively this headlong stream, as Horace truly called it, has often remained within its bounds, and then, after such long intervals of rest, at different periods inundated its banks again, and widened its channel":[59]

> The destroying agency of the flood came within two hundred yards of the precipice on which the beautiful temple of Vesta stands; but fortunately this precious relic of antiquity was spared, while the wreck of modern structures was hurled down the abyss. Vesta, it will be remembered, in the heathen mythology, personified the stability of the earth; and when the Samian astronomer, Aristarchus, first taught that the earth revolved on its axis, and round the sun, he was publicly accused of impiety, "for moving the everlasting Vesta from her place." ... If the days of omens had not gone by, the geologists who now worship Vesta might regard the late catastrophe as portentous. We may, at least, recommend the modern votaries of the goddess to lose no time in making a pilgrimage to her shrine, for the next flood may not respect the temple.[60]

Here is Lyell provocatively transforming the so-called temple of Vesta, spared by the last of a convulsive and unpredictable series of floods, into a symbol of what does not exist, "the stability of the earth," and of the simultaneously heathen and unscientific worship of that stability by her "modern votaries," our so-called catastrophists. These men cling to ancient beliefs, to the worship of a Roman goddess, and to a belief in portents, a view almost as outmoded as the belief that the sun revolves around the earth, and to a faith in the stability of the earth rather than in its catastrophic potential. Lyell's tone, addressing a depersonalized reader who is assumed already to share Lyell's ironic distance from those votaries of Vesta, enables his final flourish, the warning that the temple of the earth's stability may be washed away in the next breaking of the river's banks. Unraveling the complex history of another Roman temple, the Temple of Jupiter Serapis, whose image formed the frontispiece to the first volume of *Principles*, Lyell claimed that fossil shells found in different layers of the temple's architecture proved it to have been submerged beneath the ocean and upraised again by violent means three times in the course of fifteen hundred years.[61] Much more widely accepted by his peers was Charles Babbage's explanation that those submergences and reemergences had been far more gradual.[62] The real debate between Lyell and his colleagues, then, was not about the overall nature of geological change, but about the method by which geology could come to be considered an "exact science" fit for modern purposes. Lyell's *Principles*, as this chapter is arguing, and as James A. Secord has put it, was not a cycle or a series of cycles or an incremental plot of earth history but an "anti-narrative" of the highest order.[63]

Lyell's volatile earth meant that the records searched for by geologists wedded to the method of enumerative induction probably did not even exist. In his second volume, dedicated to describing the processes of organic change, Lyell argued that the chances that any species would be fossilized were so low that geologists were lucky whenever they found any evidence for past life on earth. Such a lack of evidence posed serious challenges for the notion that life on earth had followed a progressive plan. It was certainly not that Lyell believed in the gradual transformation of species over time, what would come to be called "evolution." He agreed with all his British contemporaries that species had a real and fixed existence in nature, arguing that the mummies of animals collected from Egypt during the Napoleonic Wars revealed no perceptible species changes in the four thousand years that had elapsed since their mummification.[64] He exempted humans from his argument that all classes of species could have existed throughout all periods of earth history, at least partly motivated by a concern to retain human dignity in the face of transmutationist theories.[65] Hybrids and monsters, he said, were often infertile and were always degenerate forms of their parent species rather than new species of higher organization.[66] And observed variations in species caused by domestication or by a sudden change in environmental circumstances invariably took place not developmentally over a course of ages, but quickly, even catastrophically.[67] In the case of even tiny environmental changes, such as a slight increase in temperature in a particular region, the effect was often the rapid extinction of species not favored by those changes at the hands of those that were, so that there was no time for gradual variation to occur.[68] But this very volatility meant that the fossil record was radically incomplete and its overall pattern and direction was indecipherable. Not only was it possible that millions of years had preceded the formation of the earliest surviving strata, or that all their fossils had been destroyed, but it was no surprise that the stratified secondary strata contained fish, mollusks, and amphibious reptiles but very few land mammals. The secondary strata were marine and freshwater deposits, formed on the beds of seas and lakes, and were therefore much more likely to contain aquatic rather than land species. Research was revealing, too, a "very few exceptions" to the posited line of progress. One or two fossils of complex plants had been found in strata of the Carboniferous Era, whose two or three hundred known plant species were otherwise of simple organization, and two species of "warm-blooded quadrupeds" had been discovered in the fish- and reptile-abundant oolitic series. These few species, Lyell argued, were "as fatal to the doctrine of successive development as if they were a thousand."[69] One species out of progressive order pulled the whole chronological edifice to the ground. It was at least possible that, with the single exception of human beings, all classes of species, mammals, reptiles, amphibians, fish, birds, insects, and plants could have existed at any period of earth history. The implication was clear: if the geologic record was so mutilated and fragmentary, geologists could never hope to know the pattern or shape of earth history, epic, progressive, or otherwise. Essentially, in Lyell's account, while the earth certainly had had a beginning and likely would have an end, the geologic record itself offered no evidence of that beginning, a middle that was mutilated beyond all comprehension, and no discernible clues as to its ending.

This meant that all those geologists who were engaged in the construction of what they called a geological "alphabet," a full exposition of the order and structure of the strata, were engaged in a project that was doomed to failure.[70] Again borrowing directly from the language of his colleagues and redefining it to his own ends, in his third volume Lyell claimed that, in his first two volumes, he had written a new "alphabet and grammar" of geology, and moved into a syntactic exposition of the exact science that geology had, under his tutelage, now become.[71] Using what he called the earth's "tertiary" strata, its most geologically recent deposits, as a test case for the methodology he was proposing, Lyell identified four distinct periods. Crucially, he could not do so by the common method of identifying the most characteristic fossils of each group of strata. The most characteristic species of any epoch, he pointed out, may not have been fossilized at all. Marine fossils had the highest chance of preservation in any period, and the relative *percentages* of extant and extinct species, at least within the recent, tertiary strata, would offer the most reliable evidence of the age of the rocks in which they were embedded. The greater the percentage of extant species in a given sequence of tertiary strata the more recent its date. In the "Newer Pliocene" strata, the loose sands and gravels on the earth's surface, 90 to 95 percent of the fossilized species Lyell identified remained extant. Moving backward through the strata, from the observable world to the unobservable, he calculated that the "Older Pliocene" contained 30 to 35 percent extant species, the "Miocene" 18 percent, and the "Eocene" only around 3 percent.[72]

Such a method presumed that there was an overall uniformity to the rate and intensity of change of earth history and to the distribution of species through time, but it also allowed for large fluctuations and catastrophic change within that generally uniform pattern. Likening the geologist to a census collector, Lyell imagined the kinds of discrepancies that might produce irregular results: "at some periods a pestilential disease may lessen the average duration of human life," he wrote, "or a variety of circumstances may cause the births to be unusually numerous, and the population to multiply, or, a province may be suddenly colonised by persons migrating from surrounding districts."[73] If the census taker returned to the same spot only every ten years, it might look to him as if such changes had been catastrophic. It was only on average that the earth, like the rate of population change, could be considered uniform.

Similarly, an antiquary would be fatally misguided were he to assume, on finding "two buried cities at the foot of Vesuvius, immediately superimposed upon each other, with a great mass of tuff and lava intervening," and on discovering that the older town was Greek and the newer Italian, that there had been a sudden change from Greek to Italian.

Suppose he afterward found *three* buried cities, one above the other, the intermediate one being Roman, while, as in the former example, the lowest was Greek, and the uppermost Italian, he would then perceive the fallacy of his former opinion, and would begin to suspect that the catastrophes, whereby the cities were inhumed, might have no relation whatever to the fluctuations in the language of the inhabitants, and that, as the Roman tongue had evidently intervened between the Greek and Italian, so many other dialects may have been spoken in succession, and the passage from the Greek to the Italian may have been very gradual, some terms growing obsolete, while others were introduced from time to time.[74]

It was the same, Lyell argued, in geology, a science that had long borrowed from the methods of the antiquaries in order to explain the deep past. Catastrophic events could happen, rates of change could accelerate and slow down, and violence was a fact of life, but all this could be overcome and understood if an analogy between known and unknown worlds was taken for granted.

As Lyell's image of the census suggests with its administrative overtones, this was not a geology that would well fit with the language of epic. Its patterns, if patterns they could be called, were statistical rather than teleological, offering not a totalizing historical explanation of the earth and its processes but a set of possible averages gathered from a limited and provisional data set. Indeed, both his census collector and his puzzled antiquarian stand not for a particular model of population or historical change but for a specifically modern epistemological problem. The census collector and the antiquarian operate as familiar, readily recognizable figures whose knowledge relied upon the soundness of their methods for filling in the inevitable gaps in their evidence. Lyell's project was not to plot the history of the earth, but to find a reliable method for speculating on the unknown worlds that must have existed in an unobservable past.

Heroes and Villains

For the last four decades historians have argued that the "historical sketch" with which *Principles* begins is responsible for giving the impression that Lyell really did create geology anew, that uniformitarianism marked the inauguration of a science opposed to the half-baked catastrophism of his predecessors. Acknowledging the achievements of Lyell's colleagues and the importance of their work in the history of geology, historians have described those chapters as an act of rhetorical "distortion," a "Whiggish heroes-and-villains history" "at its most opportunistic," "propaganda; not history, but a 'historical romance.'"[75] In some sense, of course, they did offer a distorting history of geology. But this revisionist approach has overstated the case, for it continues to emphasize the notion that there was a fundamental and deep-seated opposition between Lyell and the existing geological community, though it puts the blame for that opposition on Lyell's rhetoric rather than on any lack of expertise among his contemporaries. But we should not forget, as Lyell certainly did not, that it was with Buckland's support that Lyell was appointed secretary to the Geological Society in 1819, or that it was with Whewell's help that he named the Pliocene, Miocene, and Eocene strata of his tertiary epoch, or that during his career he gave forty-five papers to the Geological Society, stretching long after the publication of *Principles,* of exactly the descriptive and enumerative kind that the Society was renowned for.[76] Indeed, in his only direct mention of the Geological Society in *Principles,* Lyell claimed that his work was the direct outgrowth of his involvement in it, a point that, however tactical, was reiterated at length in many of its early reviews.[77] Whatever their qualifications about Lyell's theory or its implications, the reaction from his peers was generally positive. Lyell was pleased to note, on bumping into a friend at the Athenaeum, that people were "full of astonishment and admiration" at his "having, among other things, been the first to take the bull by the horns, about the antiquity of the earth, and contrived to do it in so inoffensive a way."[78] It wasn't so much that Lyell was perceived to have made a startlingly original contribution to the science, as that he had found a language by which to safely and effectively communicate the consensus of his colleagues to a broader public. Lyell's desire to present himself as bringing geology out of its poetic epoch and into its scientific age tells us more about his indebtedness to the values of the geological community, which had formulated that story on different terms, than his real difference from it. What is more, for Lyell as much as Buckland, poetry was not something to be grown out of, but was a key medium through which to think through problems of geological method, and to signify and strengthen one's inclusion in the community of geologists who had taken custodianship of the science.[79]

This desire to placate his readers and colleagues even as he criticized them, to remain a part of their community, was central to Lyell's geological vision in *Principles.* On leaving Oxford and moving to London to pursue a career in law, Lyell quickly became an acquaintance of John Gibson Lockhart, son-in-law of Walter Scott and editor of the conservative *Quarterly Review,* and went on to write several essays for that publication, on such topics as scientific institutions and the state of the universities.[80] He saw the *Quarterly* as the most powerful platform through which he could reach fashionable but conservative readers and convert them to a belief in the immensity of the geological timescale. The *Quarterly Review* was *the* journal that could "free the science from Moses," he wrote in 1830.[81] But it had to be handled cleverly. Lyell counseled his friends not to "irritate" anybody with irate prose. The battle for hearts and minds could be won not by triumphing over "the bishops and enlightened saints" in pompous fashion, he wrote, but by complimenting "the liberality and candour of the present age," flattering *Quarterly Review* readers into submission. "The decision of the Q.R. would settle the doubts of many of those good people, who only want to be told from authority what they *may* think," he surmised.[82] And it was always better to gently cajole one's opponents than to rant and rave and risk alienation.

Despite his punchy and provocative prose in the opening chapters of *Principles*, this conciliatory approach was deeply important to Lyell as he engaged in controversy while writing the first volume of *Principles*. Quite independently, the Scottish naturalist John Fleming had come to conclude, like Lyell, that Noah's flood had no basis in geological evidence, and that the only causes that had ever acted on the earth's surface were those currently in operation. From the mid-1820s he set about making as many scientific enemies as possible on these grounds.[83] First he deliberately offended Buckland, still a leading figure in the science. Fleming said he should be "consigned to occupy a page in the History of Prejudice, along with the persecutors of Galileo," and dubbed his book an "idola specus" or "idol of the cave," punning on Buckland's cave research to suggest that he was guilty of indulging in one of Francis Bacon's idols of the mind, modes of thought that inhibited true scientific reasoning.[84] A year later Fleming attacked another Oxford don, Buckland's friend William Daniel Conybeare, on the ancient climate. Though Lyell disagreed with Fleming's views on climate. he found himself, at first enjoyably, drawn into the fray. In a series of debates at the Geological Society, Buckland and Conybeare styled themselves "Diluvialists," believing in powerful floods as important agents of geological change, and termed Lyell, Fleming, and Roderick Murchison "Fluvialists" for their belief that the action of rivers, rainfall, tides, and currents was enough, given a long enough span of time, to carve out new valleys and demolish continents.[85] These attributions were clearly playful, echoing both the "Neptunist" and "Vulcanist" tags and their claims to all-encompassing explanatory power, against which the Geological Society was supposed to be opposed. But unlike these earlier appellations, they are at once more narrow and less heroic, stripped of classical content and more obviously self-ironizing. The aesthetic terrain on which scientific argument was fought had, to some degree, already shifted.

Lyell wrote of this argument in a letter of May 1829 that tells us much about the place of irony at the Society. "A splendid meeting last night," he wrote, with "Sedgwick in the chair":

> Greenough assisted us by making an ultra speech on the importance of modern causes. No river, he said, within times of history, has deepened

its channel one foot! It was great fun, for he said, "Our opponents say—'Give us time, and we will work wonders.' So said the wolf in the fable to the lamb: 'Why do you disturb the water?' 'I do not; you are further up the stream than I.' 'But your father did.' 'He never was here.' 'Then your grandfather did, so I will murder you. Give me *time*, and I will murder you.' So say the Fluvialists!" Roars of laughter, in which Greenough joined against himself. What a choice simile! Murchison and I fought stoutly, and Buckland was very piano. Conybeare's memoir is not strong by any means. He admits three deluges before the Noachian! and Buckland adds God knows how many *catastrophes* besides, so we have driven them out of the Mosaic record fairly.[86]

Describing George Bellas Greenough as an "ultra," Lyell makes a topical reference to the faction of Tories who, less than two months before, had broken away from their party in opposition to the passing of the Catholic Emancipation Act. Greenough, Buckland, and Conybeare, all founding members of the Geological Society, appear here clinging to old beliefs and traditions in the face of a new reality, represented by the new generation of younger, more liberal-minded geologists, Adam Sedgwick, Roderick Murchison, and Lyell. Two things are important to note here. First, it is clear that other geologists within the Society agreed with Lyell that the gradual operation of water across a long enough span of time could enact vast geological changes. Lyell was far from alone on this matter. Second, though Greenough and Buckland did not agree on this issue, this did not make them biblical literalists or apologists for a young earth. It was taken for granted among all members of this debate that the earth was millions of years old. But for the Diluvialists, *no* amount of time could ever be enough for a stream to demolish a continent. As Henry De la Beche put it in his *Geological Manual*, "If a mouse be harnessed to a large piece of ordnance, it will never move it, even if centuries on centuries could be allowed; but attach the necessary force, and the resistance is overcome in a minute."[87] Greenough presumably "joined against himself" in the "roars of laughter" that ensued from his use of Aesop's fable of the wolf and the lamb because, in casting his opponents in the role of wolflike tyrants, determined to win no matter how contrary were the facts to their arguments, he nonetheless acknowledged his own hyperbole. The pugnacious spirit of the Geological Society was not so violent that it could not support a laugh.

Greenough's ability to laugh at his own error and his mode of indirect debate, couched in puns and fable, was central to the gentlemanly code of the

Geological Society and to its safeguarding of the respectable name of the science. Such a style was important to Lyell, who began to worry that Fleming's incendiary style would prove counterproductive. "I sincerely hope you will keep in well with some party," Lyell wrote to him, "but if you go on much longer, with an anti-Bucklandite, anti-Mackayite, anti-Lamarckian warfare, I begin to fear the odds ... your indiscriminate attacks are ... injurious. To show that you can strike home may save you from many an adversary but I am sure many new ones will be called forth."[88] As a direct consequence of this dispute Lyell resolved that, no matter how the first volume of his book was received, he would "not begin pamphleteering," or give "time to combat in controversy."[89] Instead, he would argue in precisely the poetic and classical terms that had been so successful at the Geological Society.

In the *Principles*, in fact, Lyell sought to place the research program of the Geological Society on the firmest public footing possible. But while O'Connor argues that "by the 1830s geologists had assembled a large enough fossil repertoire to begin" to write the epic of earth history, here I want to argue that that futuristic projection of a science that could one day gather enough evidence to furnish materials for an epic was seriously undermined by Lyell's work even as the decade dawned.[90] In defending his science and in creating principles that would bring the project of the Geological Society to fruition, as well as make his own career, Lyell would attempt to expose the hidden narrative aesthetic he saw operating in its midst, and to create a new literary mode in which to bring forth its most systematic vision of the earth and its past. In this, he would not abandon poetry, as the standard story goes, and neither would geology become an unpoetic science. Forging a language by which to reform his science from within, poetry, as it had been for Buckland, would be among his most powerful weapons.

Charles Lyell, Antihero

As he penned *Principles*, Lyell continued to take a lively interest in poetry, both in print and in private. In 1829 this interest hinged directly on a modern rereading of epic, when his father asked him to review Gabriele Rossetti's controversial *Comento analitico* on Dante's *Divine Comedy*, partly in order to defend it against the vitriolic reviews it had already received in the major quarterlies.[91] Rossetti had argued that Dante's poem was not an epic in the true sense but a coded allegory designed to communicate secret antipapal messages between members of an ancient heretical sect preserved in the rites of the Freemasons. The sect communicated, often in poetry, via "a secret conventional language," its "gergo," of which the *Divine Comedy* was a master-text and to which it could therefore function as a key.[92] The epic content of Dante's poetry was an irrelevance, and epic itself a fallible historic document, not the perfect and all-encompassing expression of a completed historical epoch.[93] Worst of all, the implication of Rossetti's argument was, as he later made explicit, that Dante was not only opposed to the temporal powers of the Holy See but also to its spiritual authority. Dante, the greatest Catholic poet in Western history, was at best a heretic, at worst a faithless fraud.[94]

Worse still, if Dante could be proved a member of a secret heretical sect on the evidence of an allegorical reading alone, ungrounded in historical proofs, where might such readings end? Why not conclude "that theology [itself] has been always a Masonic trick," as one reviewer put it, and that the sacred texts were nothing more than coded documents by which heretics had communicated?[95] Despite acknowledgments that he had thrown some light on many inexplicable passages of Dante's work, Rossetti stood accused of what, at first glance, appear to be mutually exclusive offences. The first was of an excess of unlicensed imagination, a wild temptation to indulge in fantastical theories ungrounded in "inductive" proofs; the second a sneering "skepticism" that threatened to destroy not only the affective value of Dante's poetry but the very fabric of the faith on which that poetry rested, reducing Dante's vision of "the things of eternity" to a "spiteful satire under a fictitious covering."[96] These were criticisms that rang loud and clear with Lyell, for it was the same accusation that he and his colleagues had leveled at "theories of the earth," for instance, and that they had had leveled at them from equally hostile quarters. In his review, Lyell accepted "the strong propensity, prevailing in Dante's age, to indulge in allegory, and his own express avowal of the use he has made of it in the poem in question," as well as the fact that Dante "openly advocated, in the 'Commedia,' the reform of many political and ecclesiastical abuses."[97] But he would not go so far as to abandon inductive proof. There was not a single reference to the "gergo" in any historical record; its very existence was at best unproven, Lyell claimed, and he found ridiculous Rossetti's speculation that extant dictionaries of the gergo existed but were undiscoverable because they had been written in the same secret language they were designed to decipher.[98]

On the question of "skepticism," Lyell argued that Rossetti had displayed a distinct *lack* of the kind of synchronic historical imagination that had made Buckland's evocations of the geological past so stunning and so convincing.

Unable to immerse himself fully in the culture of the poet, Rossetti had produced an ingenious but unlikely reading of his works. For, "even in criticising the works of contemporary nations, especially those of the German school," Lyell wrote,

> we must allow largely for differences of national taste and sentiment, or we shall unjustly condemn an author as absurd for passages admired by a public whose judgment, on the whole, we cannot but respect. But when we are carried back to the age of chivalry and romance, of allegory and metaphysics, of crusaders, astrologers, and miracle-workers, we ought to prepare ourselves to encounter almost any anomaly.... Writers of that age believed the obscure to be as appropriate an element in an amatory poem, as Burke declared it to be in the sublime.[99]

The obscurity and unintelligibility of much of Dante's poetry was part of the essence of its meaning as a medieval text, a reflection of the age in which it was produced and a function of its historical value. As Jane, wife of Humphry Davy, told Lyell at a party in April 1828, "I advise all to read Dante as I have done, three times, and as I mean to do a fourth, before they read Rossetti. There is much [in the poet] that I cannot understand, it is true, but there is much that delights me."[100] Rossetti's insistent determination to read poetry by a literalizing historical light had paradoxically led him into historical error, an inability to distinguish between the figurative and the literal dimensions of the poetry, between its sense and its spirit, and in doing so it had drained the poem of the aesthetic pleasures that were so intrinsic to its meaning.

The problem of striking a balance between inductive proof and historical imagination, and the charge of skepticism that this could lead to, particularly preoccupied Lyell as he planned the *Principles of Geology*. But there was another figure who was at that time claiming, in much more overt terms, to have written a history stripped of epic and to have created a "scientific" method by which to imagine the past, whom Lyell appears to have found much more convincing. The hint comes in his plea for culturally sensitive reading not only of Dante, but "especially those of the German school." In the historical chapters of *Principles*, Lyell would cite the German historian Barthold Georg Niebuhr, whose *Römische geschichte* had recently been translated into English by Connop Thirlwall and Julius Charles Hare, friends of Adam Sedgwick and Whewell at Trinity, and which was also causing a furore in 1829 on the pages of the *Quarterly Review*.[101] Deeply critical of classical sources, and especially of Livy and Dionysus, Niebuhr had interpreted their histories not as factual records of the Roman people but as myths and unreliable fragments. Niebuhr's treatment of those classical sources engendered the same criticisms provoked by Rossetti's reading of Dante: that he was overtly imaginative, reconstructing the Roman past from shards of evidence that he deemed reliable but many disagreed with, but also that he was deeply skeptical, raising the possibility of similarly critical treatments of the biblical texts.[102] In the light of these too-imaginative, too-skeptical new readings of revered texts by Rossetti and Niebuhr, readers of the *Quarterly Review* might have been wondering in 1829 just how they were to understand and imagine the past in a newly critical and scientific light, without destroying the beliefs that upheld them in the present.

Livy, Niebuhr said, had been "moved ... to write" not as a historian but as an epic poet. He had brought

> down the marvels of the heroic ages into the sphere of history; as was commonly done even by those who in what belonged to their own times and experience were far from credulous, at a period when the thoughtless belief of childhood continued undisturbed throughout life. Even those primitive ages when the gods walkt [*sic*] about among mankind, he would not absolutely reject.[103]

Similarly, when Lyell quoted from the *History of Rome* at the end of the fifth and final historical chapter of *Principles*, he implied that his own project had been to uncover that credulous faith in heroic myths and legends that had continued to dominate thinking in geology, itself in a hitherto primitive stage. The historical writing the Romans had bequeathed the present, Niebuhr said, constituted "an epical narrative of actions and events" and not a historical one. Niebuhr would move the history of Rome into a newly and self-consciously "scientific" age, as Lyell was proclaiming to do for geology, by ridding it of this epical dimension. "In order that the heroes and patriots of Rome may rise up before our view, not like Milton's angels, but as beings of our own flesh and blood,—we require more and something else, over and above what we find in his [Livy's] inimitable narrative," a "scientific" method that was critical of sources, stripped back their epic values and forms, and reconstructed both a richer and a more probabilistic account of the past from the fragments of evidence that survived this scrutiny.[104] It was this reconstructive aspect of Niebuhr's work that had opened him also to charges of an excessively speculative imagination, but that Lyell quoted in the *Principles*, leaving his readers with a parting shot at the end of a dramatic chapter: "He who calls departed ages back again into being, enjoys a bliss like that of creating."[105] Lyell's view that the archive of earth history represented by the strata was mutilated and lost beyond comprehension meant that his geology would need to be as imaginative as Niebuhr's reconstruction of the history of the Roman people from a radically deconstructed record. In citing Niebuhr, Lyell tethers this reconstructive imagination to the self-consciously "scientific" criticism that had been lacking from Rossetti's *Comento Analitico*, grounding imagination in a method that rang true with the values of scientific induction.

In this light it is not surprising to discover that there was in fact another literary genre with which Lyell was toying as he wrote the first edition of *Principles*. Shortly after the first volume was published he was attempting to win the hand of his future wife Mary Horner. In a somewhat formal moment of their courtship in 1831, Mary asked him to write her an account of his life, by which she might more fully judge his character. Lyell duly provided it. In this autobiographical text his poetic exploits as a young man loomed large. He told stories about the precocious young boy-poet he had been before his disappointment with the Oxford prizes. On one occasion he had submitted a versification of Scott's *Lady of the Lake* for a school prize on the theme of "Local Attachment," and had won despite having flouted an unwritten rule that all such poems would be composed in rhymes of ten syllables. Bearing in mind that Lyell was now a thirty-four-year-old man, I think we can forgive ourselves a smile at his continued pride that the ten-syllable rule had had to be formally instituted after his transgression of it.[106] With even greater *amour propre* Lyell recalled a Latin copy of verses he had written just before he went up to Oxford. The verses were "on the fight between the land-rats and the water-rats, suggested by reading Homer's battle of the frogs and mice— a mock-heroic." One of Lyell's teachers "had just drained a pond much infested by water-rats," Lyell recalled, and they had been stealing food and running amok all over the school ever since. The incident, Lyell wrote, trying very hard to impress Mary now, "convinces me that I must very early have felt a pleasure not usual among boys of about sixteen in exerting my inventive powers voluntarily." He continued:

> The plot was begun with a consultation of water-rats, to each of whom altisonant Greek names were given, after the plan of Homer—cake-stealer, gin-dreader, book-eater, ditch-lover, &c. The king began by describing a dream in which the water-prophet covered with slimy reeds appeared to him, foretelling that the delicious expanse of sweet-scented mud would

soon dry up, and foreboding woes. Part of the warning was copied or paraphrased from the Sybil's song to the Trojans in the "Aeneid" of what should happen when they reached Italy. The dream and warning ... being communicated, the others entered into the debate what they should do, and it was agreed that, as the fates had decreed the drying up of the waters, they should migrate to a neighbouring sewer, and should destroy the house-rats, who consumed so much provender in the school room, and who had usurped their rights. One passage, in which a chief was described as a great map-eater, and having at one meal consumed Africa, Europe, Asia, America, and the Ocean, was admired as a good specimen of pompous description of mighty deeds, on the first entrance of a hero in an epic poem. The verses ran on to thirty-eight, and when done, there was great discussion whether I should dare show up such a thing.[107]

In the middle of writing the mammoth three-volume, career-making book in which he hoped to "create a science," as he put it, Lyell was in private parading his abilities as an epoist, capable of constructing the most ambitious poetic form, the epic, while at the same time mocking it. And it is clear that Lyell still enjoys his own joke. He is especially keen to paint the central discrepancy of mock-epic, the incongruence of lofty style and trivial subject, in the strongest terms possible. He presents himself as a man with the panache to transgress literary boundaries in a manner that, far from alienating his peers or his teachers, might catch them out in laughter. He did "show up" his mock-epic, in fact, and remembered with self-congratulation that his teacher had been "struck, more than he chose to admit to us, with the invention displayed in the whole thing. He told the class that it was such good Latin that I deserved great credit, but he did not wish them or me to send him up more mock-heroics."[108]

Lyell may not have bestowed any more mock-heroics on his unsuspecting schoolmaster, but he didn't stop writing them as he penned *Principles*. They were, in fact, a part of the fabric by which he presented himself as an urbane reader of the geological archive and, in the light of his immersion in the culture of the *Quarterly Review* with its contemporary debates on Rossetti and Niebuhr, created an aesthetic value for geology that may have read as modern, enlightened, and erudite to a broader public. In the third chapter of volume 1, for instance, Lyell stated that he did not believe there to be an earthly location for the Garden of Eden, and feigned arch surprise that the seventeenth-century cosmologist Thomas Burnet had said in his *Sacred Theory of the Earth* that it was located in the southern hemisphere, somewhere near the equinoctial line. "[Samuel] Butler selected this conceit as a fair mark for his satire," Lyell reported, "when, amongst the numerous accomplishments of Hudibras, he says—

> He knew the seat of Paradise,
> Could tell in what degree it lies;
> And as he was disposed, could prove it
> Below the moon, or else above it."[109]

Lyell took pleasure in citing Butler's cumbersome octosyllabic couplets, with their broken scansion and imperfect rhymes, as they aurally recreate the effect of faulty reasoning in Hudibras's proof that Paradise is "Below the moon, or else above it." But he was also quick to condemn the gullibility of Charles II, "who is said never to have slept without Butler's poem under his pillow" and yet also ordered a translation of Burnet's satirized cosmology from Latin into English. Charles had evidently been seduced by Burnet's "eloquent" prose, his "power of invention of no ordinary stamp," Lyell wrote. The *Sacred Theory* "was, in fact, a fine historical romance."[110] From Butler's Hudibrastic satire Lyell begins to construct an authorial persona that is not damning or vindictive, but that can claim ironic distance from cosmology and turn its conventions on its head.

A few pages later, Lyell turned this strategy to epic account, invoking Dante to poke fun at the followers of Werner and their belief in granite as "the rock of ages," as he punningly put it, as the beginning of an ancient earth history that was nonetheless bounded by a fixed temporal scheme. Lyell's description of granite as a rock of ages, the rock "we turn to in our time of need" in Isaiah 26:4, subversively hints that there is a Christian temporality lurking beneath what is supposed to be a merely descriptive Wernerian geology. When Werner is in scientific need, Lyell argues, he hides behind the notion of a beginning. What is more, he and his followers saw written on granite "in legible characters, the memorable inscription

> Dinanzi a me non fur cose create
> Se non eterne."[111]

The translation is "Before me things create were none, save things Eternal," and these characters they "regarded as sacred."[112] Read straight, this quotation from Dante's *Divine Comedy* appears as a literal description of granite considered as the first rock, before which nothing existed and which has endured through all the ages of the earth. But as Lyell, in provocative mood, evidently expects his readers to know, this "memorable inscription" is from the *Inferno*, and is written on the gates of hell. An extended version of the passage reads,

> Through me you pass into the city of woe:
> Through me you pass into eternal pain:
> Through me among the people lost for aye . . .
> Before me things create were none, save things
> Eternal, and eternal I endure.
> All hope abandon ye who enter here.[113]

For the reader of Lyell who also knows his Dante, then, and this is perhaps the most famous of all quotations from Dante's work, there is another reading to be had. Uncovering the hidden cosmology underpinning Werner's seemingly secular geology, in which time is "eternal" rather than genuinely temporal, Lyell turns epic eschatology against this geological view: think like Werner, he suggests, and you are entering dangerous philosophical territory. Here is Werner's mode of reasoning, "All hope abandon ye who enter here."

Presenting Werner as a would-be epoist, then, Lyell revisited the geologist who had just died as he composed "Lines on Staffa" in a different genre. Once he had penned Spenserian stanzas of an inclusive Romantic strain that presumed a reverence for nature shared with Werner, even as he offered an alternative explanation of the rocks. Now, in equally gentlemanly fashion, he wrote the mock-epic that would expose him and all those geologists who believed the stratigraphic column to offer the promise of a complete story of the earth. Just twenty pages later the satiric persona Lyell develops in these early chapters of the *Principles* finds its fullest expression in a thought experiment by which he demonstrates that human beings, inhabiting only a quarter of the earth's surface, are at a disadvantage when it comes to analyzing geological processes. "There can be little doubt," he writes, "although the reader may, perhaps, smile at the bare suggestion of such an idea, that an amphibious being, who should possess our faculties, would still more easily arrive at sound theoretical opinions in geology" than we can, for that creature could see to the depths of the sea, as humans, limited to the regions of the air, cannot. But even the amphibious creature could not "reason on rocks of subterranean origin" from direct observation:

> But if we may be allowed so far to indulge the imagination, as to suppose a being, entirely confined to the nether world—some "dusky melancholy sprite," like Umbriel, who could "flit on sooty pinions to the central earth,"

but who was never permitted to "sully the fair face of light," and emerge into the regions of water and of air; and if this being should busy himself in investigating the structure of the globe, he might frame theories the exact converse of those usually adopted by human philosophers.[114]

This is of course another mock-epic, for Lyell's sprite is Umbriel, the gnome of the earth from Pope's *Rape of the Lock*, a poem that had applied epic machinery to compare the public lopping of a lock of a young woman's hair to the rape of Helen of Troy. Umbriel does not pass through the gates of hell on his epic journey to the underworld, for his underworld is not hell, but the heroine's spleen. Lyell plays here explicitly on the mixture of the sublime and the ridiculous that had long been a part of geological discourse, but he does it in order to turn against all those whose geology he suspected of being underpinned with mythological frameworks. While Buckland had claimed for geologists the status of expert Virgilian guides to the underworld, for Lyell there can be no expertise on an underworld that is essentially unobservable. No matter that geologists presented themselves as empiricists avoiding story, stories from the earliest periods of human history, Lyell suggests, had found their way into geology nonetheless. Lyell used mock-epic to imply that the geological writing with which he disagreed was "epic," to suggest that it was too heavily plotted. Mock-epic offered him a parodic language in which to distance his own project from *plot*.

Mock-epic was not simply a means of laughing at his predecessors, moreover. It was a method by which Lyell could present his geology as consonant with the very traditions he was criticizing. Mock-epic enabled Lyell to continue to associate geology with the classical and literary traditions with which Buckland had begun to forge an authoritative cultural ground for the science, to continue to claim for geology the grandeur with which it had become associated without becoming locked in the narrative machinery of epic that was so at odds with the Society's rejection of cosmology. Lyell conceived of himself not as attacking geology but as bringing it to the full realization of its aesthetic and scientific conclusions. So Milton and Dante figure prominently in the *Principles*, as we would expect from a former student of Buckland and an Oxford-educated liberal attempting to appeal to readers of the *Quarterly Review*. But they often appear in what might be considered Niebuhrian fashion. Writing on the destructive geological power of the floods of the river Po, Lyell notes, "The practice of embankment was adopted on some of the Italian rivers as early as the thirteenth century; and Dante, writing in the beginning of the fourteenth, describes, in the seventh circle of hell, a rivulet of tears separated from a burning sandy desert by embankments 'like those which, between Ghent and Bruges, were raised against the ocean, or those which the Paduans had erected along the Brenta to defend their villas on the melting of the Alpine snows.'" That last is Lyell's translation for the following passage, which he gives in the Italian:

> Quale i Fiamminghi tra Guzzante e Bruggia
> Temendo il flotto che in ver lor s'avventa,
> Farno lo schermo, perchè il mar si fuggia,
> E quale i Padovan lungo la Brenta,
> Per difender lor ville e lor castelli,
> Anzi che Chiarentana il caldo senta—
> *Inferno*, Canto XV.[115]

Dante's epic poem is quoted here not for the mythic journey of its protagonist but for its ability to yield important historical information that is significant for the geologist. "Some speculators . . . who disregard the analogy of existing Nature," Lyell later writes on the earthquake at Calabria in 1783, "and who are as prodigal of violence as they are thrifty of time, may suppose that Calabria 'rose like an exhalation' from the deep, after the manner of Milton's Pandaemonium"—thus briefly invoking the poetry of hell yet again. "But such an hypothesis will deprive them of that peculiar removing force required to form a regular system of deep and wide valleys, for *time* is essential to the operation."[116] While Milton's epic machinery is an effective tool for the geologist who needs to describe the unobservable underworld, it is also a mythic machinery whose power of evocation is much greater than its power to frame a complex understanding of the natural world and its processes.

Again and again Lyell, like Niebuhr, bemoans the inadequacy of his sources: "That Egypt was the gift of the Nile, was the opinion of her priests before the time of Herodotus; but we have no authentic materials for determining, with accuracy, the additions made to the habitable surface of that country since the earliest historical period," he writes, complaining of the lack of geological information in ancient texts, though "we know that the base of the delta has been considerably modified since the days of Homer," because Homer's text, like many epic texts, can be read for useful historical detail that is often lacking from historical accounts.[117] Or again, and most strikingly, "It is most singular that Pliny, although giving a circumstantial detail of so many physical facts, and enlarging upon the manner of his uncle's death, and the ashes which fell when he was at Stabiae, makes no allusion whatever to the sudden overwhelming of two large and populous cities, Herculaneum and Pompeii," in AD 79. He continues:

> All naturalists who have searched into the memorials of the past, for records of physical events, must have been surprised at the indifference with which the most memorable occurrences are often passed by, in the works of writers of enlightened periods; as also of the extraordinary exaggeration which usually displays itself in the traditions of similar events, in ignorant and superstitious ages. But, of all omissions, the most inexplicable, perhaps, is that now under consideration; and we have no hesitation in saying, that had the buried cities never been discovered, the accounts transmitted to us of their tragical end would have been discredited by the majority, so vague and general are the other narratives, or so long subsequent to the event.[118]

Tacitus, Suetonius, Martial, Dion Cassius—all ignore events or record "without discrimination, . . . facts and fables" as if each were as reliable as the other. But out of these mythic histories, epic poems, and superstitious traditions, Lyell reconstructs truth, divining fact from fable in the textual record, in an analogy for the methodical imagination that can be relied upon to reason sensibly about evidence that has been lost in the earth's archive as well as in the historic one. Lyell, like Niebuhr, rids a discipline of its faith in heroes, but retains imagination as its principle epistemological mode. In doing so, he brings into alignment both his colleagues' rejection of grand narrative and their impulse to claim epic grandeur for the science.

Lyell's choice of poet for the task of reenvisioning the world in this way is Byron, whose experiments in Spenserian stanza Lyell had imitated in his undergraduate days at Oxford. Geologists, Lyell tells us, had formerly believed that it was the sea that altered its level over time, while the land remained stagnant. "But it is time that the geologist should in some degree overcome those first and natural impressions which induced the poets of old to select the rock as the emblem of firmness—the sea as the image of inconstancy. Our modern poet, in a more philosophical spirit, saw in the latter 'The image of Eternity,' and has finely contrasted the fleeting existence of the successive empires which have flourished and fallen on the borders of the ocean, with its own unchanged stability.

—— Their decay
Has dried up realms to deserts:—not so thou,

> Unchangeable, save to thy wild waves' play:
> Time writes no wrinkle on thine azure brow;
> Such as creation's dawn beheld, thou rollest now.
> CHILDE HAROLD, Canto iv."[119]

Byron, the "philosophical" poet who can tell that the sea, perhaps paradoxically, is the image of constancy, while the land is utterly fickle, later helps Lyell to describe the manner in which seeds can be flung by winds prevailing in one direction, or by hurricanes, almost without limit, thus diffusing themselves over the earth so that their points of origin are indecipherable: "All are familiar with the sight of the floating sea-weed 'flung from the rock on ocean's foam to sail, / Where'er the surge may sweep, the tempest's breath prevail,'" he writes.[120] Byron breathes of submission to forces beyond one's control, self-abandonment, and self-forgetting in this passage from *Childe Harold's Pilgrimage*, a passage that is emblematic of the directionless, anti-epoizing quality that had given his poetry such potency in the years of Lyell's adulthood. It was "the present moment" that was "Byron's strong suit," as one historian of epic poetry has recently put it, not the diachronic and cosmological plotting of the past that belonged to epic.[121] Instead, the Byronic hero made "a black hole in narrative," "subordinating opinion, description, and momentum to the advancement of a central self without evident motive, bearing, or goal. Here ... was a venue where epic need not apply."[122] Writing of the past, Byron made it vivid and present. Indeed, it has been argued that, in the wake of Byron's achievement, epic poetry became a heroic impossibility for all but the apocalyptic and deluge poets whose work flooded from the presses in the 1820s and from which Lyell, for obvious reasons, would have been sensible to distance himself and his science.[123]

It is on the model of Byron's seaweed, flung about the ocean with abandon, on the model of Byron's aesthetic of inconstancy, that Lyell fashioned his own vision of an earth history that was modern, advanced, and all grown-up in the late 1820s and early 1830s. In his second volume Lyell catalogs the dispersal of species across the earth's surface in a bid to demonstrate that their points of origin are often undetectable. Describing the dispersal of seeds over the surface of the earth, Lyell suddenly breaks into the present tense:

> A deer has strayed from the herd, when browsing on some rich pasture, when he is suddenly alarmed by the approach of his foe. He instantly plunges through many a thicket, and swims through many a river and lake. The seeds of the herbs and shrubs adhere to his smoking flanks, and are

washed off again by the streams. The thorny spray is torn off and fixes itself in his hairy coat, until brushed off again in other thickets and copses. Even on the spot where the victim is devoured, many of the seeds which he had swallowed immediately before the pursuit may be left on the ground uninjured. . . . A tempestuous wind bears the seeds of a plant many miles through the air, and then delivers them to the ocean; the oceanic current drifts them to a distant continent; by the fall of the tide they become the food of numerous birds, and one of these is seized by a hawk or eagle, which, soaring across hill and dale to a place of retreat, leaves, after devouring its prey, the unpalatable seeds to spring up and flourish in a new soil.[124]

From the drama of the deer pursued by his foe the seed emerges "uninjured," the "fall of the tide" is counterbalanced by the "soaring" of the eagle, and the little heroic seed emerges to "spring up and flourish" from the trail of death and destruction that precedes it. A perpetual check and balance manifests itself in Lyell's diction here, but it is a directionless check and balance: continuous present tense and a lack of geographical specificity locate the scene in any time and any place. Lyell's world is thronging and alive with change, change in all directions, at all times, by all possible means. Throughout *Principles*, Lyell asks us to imagine worlds we have never seen. While the New Zealander cannot imagine the proportions of land and sea to be different in the northern hemisphere than they are in the southern, the modern European reader has no trouble speculating "on the state of America in the interval that elapsed between the period of the introduction of man into Asia, the cradle of our race, and that of the arrival of the first adventurers on the shores of the New World. In that interval, we imagine the state of things to have gone on according to the order now observed in regions unoccupied by man." We move backward through the geological record, from the worlds we know to the worlds we don't, or outward from the geology of Britain, or the French and Italian landscapes familiar to so many through the Grand Tour, to the Africas or Ganges or the New World. But these are the patterns by which we can know the world, and not the pattern of the world itself. Lyell's *Principles*, like the dispersal of seeds or the Byronic hero, offer us a world whose patterns can never be known.

Lyell's aerial viewpoints do not, then, as they might in an epic poem or a pageant of earth history, offer scenes from the geological past moving past us in narrative procession.[125] Instead, we constantly enter and reenter a dazzling array of perspectives in *Principles*, each of limited range: an alien being, a satiric gnome, a seed passing through the body of a deer, or a piece of driftwood turning into an island as it moves toward the land.[126] The progress implied by the fossil record is a sham, Lyell insists, for only a very limited number of species are preserved, and the epic potential of the rock record, its promise of a total vision of earth history, is a myth. The record is simply too fragmentary for its full story ever to be reconstructed. Alternatively, Lyell sometimes suggests, progress in the fossil record is a sham not because of the record's incompleteness, but because the slab of time it represents is only a tiny speck in a much larger history whose overall pattern is unknowable. In this view, epic vision is parochial, limited by the vagaries of human perspective. There is always a larger view that could be taken, and a smaller one, a view in short that would imply a different story altogether. Lyell's world was a Byronic "anti-narrative," absorbing the grandeur of Buckland's geological vision even as it exposed the narrative logic that underpinned it, a mock-heroic fit for a cosmopolitan geologist in the early 1830s seeking to make his mark on the world, win a wife, and bring his science to feel the force of its own conclusions. Lyell did not offer a rival plot of earth history to his geological opponents, and he did not push poetry from his science. Instead, in *Principles,* Lyell finished the job, already started by his colleagues, of unmaking the story of the world in prose.

Maps and Legends **CHAPTER FOUR**

As large-scale, totalizing plots were pushed into the background by the gentlemen of the Geological Society, a new form began to dominate the science.[1] This form offered a concrete and compelling means of arranging the rocks and fossils in comprehensive order, and it did not require speculation on the causal laws that drove historic change, or on the beginnings or endings of the universe. The form appeared to be safely empirical, at least by comparison with geological plots of progress or Creation, and it had the added advantage of being preeminently useful.[2] It could release coal, minerals, and building materials from deep in the earth's crust, and could offer critical information on the structure of the land to mine owners, engineers, and architects.[3] This form, of course, was the geological map.

Geological maps were a distinctively nineteenth-century form. In the late eighteenth century, mineralogical maps had used spot symbols to denote sites of geological significance, such as mines or quarries, or color washes to indicate rock exposures and sites of mineral resources.[4] But though these maps could be accompanied by vertical "sections" showing slices of the landscape as it stretched down into the earth at the particular locations denoted by spot symbols, they did not correlate the information yielded by the various exposures and excavations dotted across a given region, and so did not offer a general picture of the strata beneath the surface of the land. It was not until the nineteenth century that European geologists would evolve a powerful visual vocabulary with which to render an entire region in three-dimensional terms, from "transverse sections," side-on views of the rocks displayed like the side of a slice of layer cake (plate 3), to the "traverse section" (fig. 4.1), to columnar

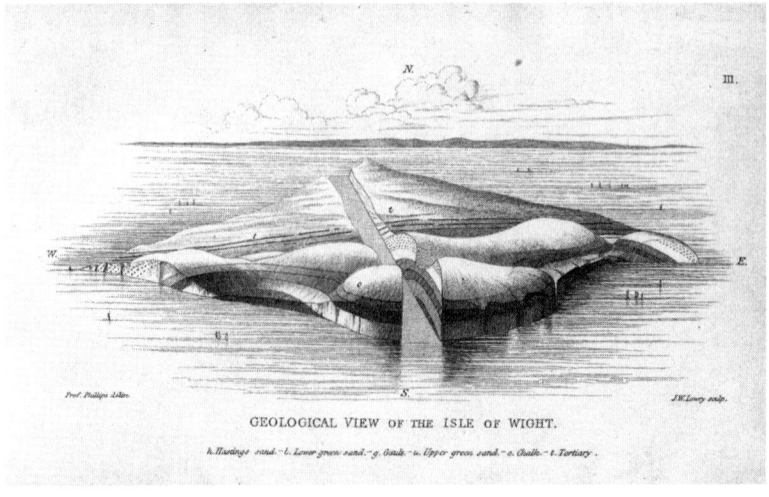

Fig. 4.1. Thomas Webster, traverse section of the Isle of Wight. Frontispiece to John Phillips, *Guide to Geology* (1834).

sections presenting the complex and overlapping layers of rock in the map as a chronological column showing the order in which they had been deposited (plate 4).[5] These three-dimensional images represented the structure of the land as it had been created through the immense span of geological time.

In 1815 the Yorkshire canal surveyor William Smith published his pioneering geological map of England and Wales (plate 5), the first geological map of an entire country, and he followed it with a series of more detailed county maps.[6] In 1820 his work was extensively plagiarized (and improved upon) by George Bellas Greenough for the Geological Society in a document designed to symbolize the empirical research methods of the new science and to galvanize its new research program.[7] Though historians have often emphasized the gentlemanly character of geology in the early nineteenth century, especially as it was practiced by members of the Geological Society, it is now clear that the concerns of those members significantly overlapped with the miners, engineers, and surveyors from whom they distanced themselves rhetorically, and this is nowhere clearer than in the production and use of geological maps.[8] As the eminently gentlemanly Greenough well knew, it was with maps that geologists could make their most successful lobbies for public support. The state-funded Ordnance Geological Survey was inaugurated in 1835 to produce what has been called a "census of the strata," and the Geological Survey later comprised the Mining Records Office and the Museum of Economic Geology, whose very architecture illustrated the stratigraphic arrangement of the earth's rocks and fossils and symbolized the powerful conceptual order geologists had bestowed upon the earth (see fig. I.3).[9]

By the 1840s, as printing technologies, design strategies, and new oil-based inks and watercolors improved, color-printed geological maps became available not only in the transactions and proceedings of scientific societies but commercially, too, as sheet maps and in general atlases, periodicals, and travel books.[10] Ambitious maps like Roderick Murchison and James Nicol's *Geological Map of Europe* in 1856 were reviewed in periodicals and lauded for their "high philosophical interest" and "real commercial value," and maps of British colonies garnered similar reception.[11] Cheaper maps for educational and commercial purposes reached ever-wider audiences. In 1860 J. M. W. Reynolds, for instance, published a 7.5-by-5.25-inch *Geological Atlas of Great Britain*, "just the size for a... side pocket," so that travelers and tourists could study the strata both while they were traveling by rail through the cuttings made by navvies, and once they arrived at their destinations. Many maps were published including information on railway timetables, stations, and stops. Indeed, if "the railway timetable became as vital a part of the geologist's toolkit as the hammer and the lens," it might also be said that the geological map was almost as likely to form part of the general traveler's luggage as his yellow-back novel.[12]

Nonetheless, this three-dimensional form was not without its problems. Traveling across wide topographical areas, geological mapmakers would often need to extrapolate a general three-dimensional view from only a few exposures of rock, and even in those exposures they would need to untangle the rocks from their twisted, contorted, smashed, and contradictory appearances in nature in order to ascertain the order in which they may have been originally deposed. Rocks, out of sequence and lying at all angles, even upside down, might overlap other rocks in equally complex arrangements, so that naturally demarcated groups rarely existed. In particular, this meant that the boundaries of a map, often determined by administrative county boundaries, were arbitrary. Mapping only the north of a particular British county could produce a very different column of rocks than would mapping its western regions, despite the fact that the territory represented might overlap. In the three-dimensional business of geological mapmaking, the marking of boundaries on the surface of the land could determine the significance and understanding accorded to the layers of rock that stretched beneath it. In order to overcome the idiosyncrasies of local landscapes, geologists correlated their findings from one region with those of another. But in the individual region, the drawing of conceptual or natural boundaries was inevitably a complex and tricky affair.

It is the argument of this chapter that the emergence of the "region" as a means of thinking about and categorizing place was fundamental both to the development of the novel in the nineteenth century and to geological mapping on a local level. The geologist's conceptualization of uneven natural and political boundaries was irrevocably tied to the literary, historical, and cultural associations of the land. In turn, the definition of those boundaries along cultural lines shaped the interpretation of the strata as they descended deep into the bowels of the earth. This is not to say that we can tell a neat story about the geological map shaping literary representations of place in the regional novel, for instance, or about novels transforming the geological understanding of a region. In fact, the relation between the literary and the geological conception of place was of a much less directly causal nature. Nonetheless, by the beginning of the nineteenth century there was already a long and prestigious tradition in geological literature of linking rocks and the peoples who lived upon them. Abraham Gottlob Werner's theory of *Gebirge,* or geological formations, for instance, defined the strata both by rock type and by the character of the peoples who lived upon the rocky surface. Alexander von Humboldt too had linked the characteristics of peoples with the geological structure of the land on which they lived. Geology was, at least implicitly, seen to determine the characteristics of the people living on the land. Less concretely, perhaps, peoples, cultures, and rock formations were often assumed to share characteristics, and definitions of the one could help shape understandings of the other. It is the business of this chapter to think harder about how those connections between people and place might have been felt to operate.

This chapter takes as its case study the mapping of the West Riding of Yorkshire, an area of Britain particularly rich in such associations. The West Riding sits at the heart of Britain and was renowned for its heavily faulted, uplifted, and eroded secondary rocks, which gave it a wild and picturesque demeanor. It was equally renowned for its historic significance as a key region in the English Civil War and as the site of Roman and Viking invasions. By the 1840s the West Riding of Yorkshire was also of central industrial significance, containing the cities of Leeds, Bradford, and Sheffield and famed for its wool, iron, and coal production. Studded with mines and quarries, it sat in the county that possessed the most famous of local natural history societies, the Yorkshire Geological and Polytechnic Society, and it was becoming known as the turbulent landscape in which the controversial novels of the Brontë sisters were written and set. As this chapter will explore, Emily Brontë's novel *Wuthering Heights* had a peculiar connection with the geological unraveling of the Yorkshire landscape. Here was a site in which grand-scale geological features, a complex mix of geological processes, both violent and calm, and wild scenes of beauty, literary import, and industrial development sat in complex juxtaposition. The West Riding of Yorkshire was therefore particularly rich in symbolic and scientific significance in the nineteenth century. If it is possible to argue that the strata of a region were given definition, and their boundaries mapped and understood, by drawing on the cultural and historic associations of the land, then this would certainly have been the case in Yorkshire's West Riding. Looking at the history of geological cartography in the county, alongside literary, topographical, and linguistic interpretations of it, with a special focus on *Wuthering Heights*, this chapter explores the ways in which mapmaking was partly determined by the imaginative, as well as the scientific, lay of the land.

Maps and Metonyms

That the dimensions of the geological map were at least partly literary does not mean that such maps were integrated into large-scale plots. The map, in fact, subordinated story to structure. And the literary figure by which people and rocks were most insistently linked reflects this structural, rather than plotted, mode of relation between them. Geological writers in the nineteenth century commonly linked people, cultures, and rocks not so much by telling stories about the ways in which different types of rock formation had produced different kinds of people, but rather by metonymic association. The features of one could be used to read or interpret the features of the other. In the *Edinburgh Review*, for instance, J. F. W. Johnston, cofounder of the British Association for the Advancement of Science, wrote that "A common unity" could be found "pervading all":

> The various colours represent, not only the various rocky formations, but diversified mineral productions also, and different agricultural capabilities and tendencies. They indicate where great cities establish themselves, and why; what brings masses of people together in particular localities, of what special class this population is composed, and what are likely to be its moral and social dispositions; why one manufacture takes root on this

spot, and another on that; why here corn waves, or cattle fatten, or sheep crop the springing herbage; why here the rich proprietor and the wealthy farmer live together in comfort, and encourage each other in progressive improvement—why there husbandry is backward, the proprietor in difficulties, and the cultivator wasting life and means in a heartless struggle."[13]

"The natural reason for the growth of large towns and crowded populations," Johnston continued, "is to be found in the geological structure of the rocks on which people live," with crowded spots near coal formations breeding "pestilence and immorality," while regions supported by tin, copper, or lead mines, such as those in Northumberland, Durham, the higher Yorkshire dales, and Cornwall, were ubiquitously "the home of a peculiar race of people—higher altogether in mental habits, in morals, and in enterprise, than what other and perhaps neighbouring spots are nourishing." Unlike coal mining, metal mining produced "ingenious, hardy, and persevering people."[14] Of course, Johnston's argument depended upon a series of elisions, not least of which is the fact that the existence of a coal formation or of rich metal deposits in a particular location did not inevitably mean that those resources would be actively mined by the overlying populations. Clearly the relation between overcrowding and "immorality" is more complex than Johnston allows, too. But it is the *form* of Johnston's argument that is important, for though he partly suggests a causal relation between the geological features of a given region and the characteristics of its inhabitants, he offers no detail on just how that relation might work. Instead of articulating the narrative mechanisms by which people, place, landscape, architecture, culture, and economics might be linked he relies on postulated metonymic associations among them. The argument is simple enough when it comes to explaining the establishment of a city near a river, for instance, but it is less obvious precisely how the existence of metal beneath the soil might influence the moral and social disposition of the inhabitants of that city, except insofar as the black dust of underground collieries and the debilitating working conditions imposed upon coal miners might conjure up black, hellish, and immoral connotations, while the metal mines of Cornwall, which might be equally dangerous but had been mined for centuries both on the surface and underground, accrue historic and healthful meaning from the landscape and the people who had traditionally lived upon it. Theirs is a metonymic relation, one set of associations bleeding into another, rather than a causal or narrative relation. There is a "fund of thought, hidden, so to speak, beneath the varied colours of a geological map," Johnston wrote.[15]

Most importantly, this "fund of thought" was—like the geological treasures of the underground, which were also "hidden" beneath the colors of the map—indicative of the physical and spatial structure of the world in the present rather than the purveyor of a story of its past. The maps of the physical atlases produced in 1848, a year, of course, of much political consternation, offered solace for the future, since read correctly the geological structure of the landscape could be made to reveal the political and economic destinies of the world's largest nations:

> There are some among us who of late years have delighted in holding up Russia and the United States, as objects of our political apprehension. When they learn to decipher the tints of the map of which we are speaking, they will probably think themselves entitled to draw from them still more alarming prognostications. Judging from the wealth and power which her small patch of blue has given to England, we may augur a lofty afterhistory to the empire of the Autocrat, as well as to our relatives beyond the Atlantic. But this lofty future England *hopes* to see and share; she does not fear it. Mental and moral culture are now inseparable, we think, from physical and material development; and we have the consolation of believing that the freaks of power in past ages will become impossible among our posterity.[16]

The "small patch of blue" refers to the blue that was used to color coal formations on geological maps. Climatic differences, made intelligible by two rain maps (one of the world, and one of Europe), combined with geological structure to create favorable or unfavorable conditions for the production of different crops.[17] "Two maps on the geographical distribution of plants" highlighted zones in which certain crops could not grow (at certain heights above sea level, for instance), and "two ethnographic maps" revealed "the influence of variety of races in the development of the resources of a country": the Celtic communities of Portugal, Spain, France, Ireland, Wales, and the central highlands of Scotland were deemed "all generally deficient in agricultural skill," while areas in which "Teutonic blood predominates"—"the low country of Scotland, England, Flanders, Switzerland, and Germany— ... take agricultural precedence of the Celtic countries," for instance. Such "facts" could be used to explain the success of some races on difficult soils and the lack of success of others on seemingly fertile ground. The many maps of the globe, each on identical scale with the other, invited the reader to lay a variety of physical features one over the other and to read overlapping significance into each. The geological map, Johnston argued, underpinned this layered vision of the natural world and its various histories, both because geological features underlay all the natural and historical events that took place on the earth's surface and because the very image of a layered present, which structured the *Atlas*'s global view, derived from the stratigraphic shape of geological maps and columns. We lay meteorological, botanical, ethnographical, and geological information over one another in layers that accrue meaning through their contiguity, just as the rocks are laid one atop the other in a stratigraphic sequence. Such a structure does not explicitly tell us the story of the earth through time (though a story, or stories, or part of a story, might be implied or suggested by it), but offers a different kind of total view than the geological grand narrative by uniting several branches of knowledge in a single, layered perspective, for which the appropriate literary figure was metonym. Mapping deep time, we might argue, the Victorians gave it contiguous or place-bound as well as temporal and sequential organization.

It was because the boundaries of the map as selected on the surface of the land could produce very different vertical columns of rock for the same regions, and because no region was fully "typical" or "representative" of any rock sequence, that contiguous relationships between people, places, buildings, cultures, and rocks were important to the geological mapmaker. Yorkshire, partly covered by a coal-mining district, in the geographical center of Britain, was an especially fertile source of such relationships. As it was described in 1859 in a leading article in the *Westminster Review*, which summed up a wide range of important writings on the county published between 1730 and 1858, Yorkshire was a county in which "Nature and man have worked together to give her the pre-eminence":

> Her place in our island is in the very heart of It—in the vital part of Britain. Her land, rich in every form of native beauty, in mountain and sea-coast, in valley and moor, river and rock, and wood and pasture, in all that can delight the eye and gladden the desire, is the home of all that is most precious in the national sentiment. Her men—a sturdy, shrewd, and stalwart race, hard-headed and hard-fisted—have so notably done their day's work in all time as to have left their mark upon our English history, mainly contributing to make that history what it is. For two thousand years has Yorkshire held her foremost place among the counties, and during all that time has played a chief part in our transactions. Briton, Roman, Saxon, Northman—she has been the theatre of all their most remarkable achieve-

ments—a witness to every process by which out of those jarring elements has been wrought the England as we have it.[18]

The county offered a perfect palimpsest of geology, archaeology, history, culture, and economics in precisely a metonymic rather than causal manner. The Yorkshireman here is a synecdoche for the Englishman, just as "Yorkshire is the epitome of England."[19] Those multitudinous processes that have "wrought . . . England" are both geological and historical, for "the Englishman, said De Foe in spite, is the mud of all the races," made up from Briton, Roman, Saxon, and Northman, "and from that useful conglomerate called the Yorkshireman it is as difficult to detect and draw out the component elements as to a given quantity of Humber alluvium to assign every parent rock and every creative force."[20] The geological complexity of the county, in the sense that it is "rich in every form of native beauty," not only serves as a metaphor for the complex history of the county but has also shaped its historic destiny, by providing shelter, agricultural land, and hunting ground for the waves of new peoples who had populated it. But again, that seemingly causal relation between geology and history slips quickly into metonym. In the West Riding, the place of wool and iron manufacture and industrial development, "the process which makes wealth to accumulate makes also men decay," so that they become "stunted, shambling," and "hollow-breasted." "In the extreme West" of the West Riding, "a region of wild romance, the cradle of Wharfe, Nid, Aire, and Ribble," lived "bold, active, and laborious" people with manners of "an unhewn roughness, almost picturesque in its hideous abnegation of grace or feeling," and a "rude and monstrous . . . passion for money."[21] The people are "picturesque" like the "region of wild romance" in which they live, and they are "unhewn," rough as the unpolished rocks that surround them.[22] In this context even the use of the geological term "conglomerate" to describe both the mixed pebbles of the Humber river and the jumbled-up racial and cultural identity of the Yorkshireman is metonymic, suggesting an intuitive "fit" between people and place.

The article ends with a geological nod to the Brontës, by then the region's most famous literary exports, whose writings had, at least indirectly, helped give popular shape and definition to this image of Yorkshire and its people: "The one original genius which the Riding has produced, in Charlotte Brontë, owes nothing to the local sentiment. More Irish than Yorkshire, the Brontës derived nothing but their ruggedness from their wild dwelling among the Haworth moors."[23] Nonetheless, the article continues to assert a contiguous relation between the novelists and "their wild" Haworth "dwelling." The Brontë sisters are rough, rugged, and "unhewn," like the uncultivated inhabitants of that "region of wild romance" in which they lived, and the coarse and wild characters and forms of their fictions are suggested by the rugged landscapes of Yorkshire. Indeed, as the novels of the Brontës came to shape popular perceptions of the county during the mid-nineteenth century, the controversial status of their writings drew out a range of anxious responses about the precise kinds of relations that might be said to exist between people and place. *Wuthering Heights* and its reception can help us understand more precisely the ways in which those relations were felt to operate.

In the *Edinburgh Review* the layered view of the identity of particular regions, a view whose formal characteristics resembled the layered forms of the geological map and which took geology as the underlying explanatory for all other natural and social forms, becomes a tool of literary criticism, so that geology and fiction too stand in metonymic relationship. The rugged and romantic features of the landscape, the industrial significance of Yorkshire's layers of rocks and minerals, and the size and position of the county were taken for granted as determining features of its imaginative literatures. Form, character, genre, language: the character of each could be explained by reference to the rocks and minerals of the world in which they had been created. But we should remind ourselves that that metonymic slippage between world and text could work in the opposite direction. The rugged and romantic Yorkshire literatures produced in the nineteenth century may have given visual, textual, and imaginative definition to the abstract colors and blocks of its geological maps and columns in that period, and the conceptual boundaries geologists drew for their maps were imaginative, often partly motivated by the literary, historical, or cultural associations of the land. Those relationships need not be directly causal, moreover, nor need we resort to a shadowy and insubstantial notion of "one culture" in which geological mapmakers and novelists worked in the nineteenth century. Instead, I suggest that when it came to the description and definition of a region, different kinds of knowledge about that region—the literary, the historical, the scientific—may have overlain one another, the connections between them loose and implicit but nonetheless determining. As such, this chapter will take various interconnected layers of nineteenth-century Yorkshire description in turn: the literary evocation of Yorkshire in *Wuthering Heights,* its geological and historical evocations in a variety of works by Adam Sedgwick, John Phillips's geological map of the county, and the rewriting of *Wuthering Heights* as a piece of domestic realism underscored by a geological subplot in Isabella Banks's 1880 novel *Wooers and Winners*. It charts the accumulation of knowledge and data, of affect and emotion, of associations and evocations, as they gave shape to the understanding of the land, its history, its rocks and its people, each one overlaying—though not always directly influencing—the other.

Brontë Country

An early and influential critic of the "regional novel" contended that the prolific production of regional literatures in Britain was at least partly owing to its "amazing" "geological diversity": the country could "supply almost any geological specification," and "wherever the rock changes, the soil changes, the crops and cattle change, the industry changes, and the manners and customs of the people tend to be different too."[24] On those grounds she argued that *Wuthering Heights* (1847) is not a regional novel at all, for though the setting of the novel "is superbly regional," offering "pictures of the West Riding moors in all their moods," only one character, the servant Joseph, speaks in dialect and "the plot . . . could have occurred anywhere; the theme is love turning to hate turning to revenge, about which there is nothing specifically Yorkshire."[25] Indeed, the novel trades in a Romantic ambiguity of boundaries—boundaries not only between the natural and supernatural worlds, for instance, but also between different landscapes and between Yorkshire and the rest of England. The two dominant topographical settings of the novel, the Heights, standing atop a hill in a tempestuous limestone region, and Thrushcross Grange from which it protrudes, in a lowland moorland valley, may be distinctly differentiated from one another, but they are also mired in geographical vagueness, befuddling those critics determined to map the locations of the novel against a "real" Yorkshire landscape.[26] It takes interpretative work for the reader even to comprehend the lay of the land in the novel, as shown by the many maps critics have attempted to draw of the region (see fig. 4.2 for an early example). Indeed, "nature" is very rarely described directly in the novel. The famous snowstorm of chapter 3 quickly obliterates the lime posts and quarries that are the landscape's only discernible features. We never see Catherine and Heathcliff together on the moors, the place with which they are most insistently associated, and the action of the novel takes place almost wholly indoors.[27] In addition, what is considered to be "natural" in the novel is very frequently imbued with a sense of the supernatural. As such, the potency of *Wuthering Heights* as a work of literature depends upon the vagueness of the boundaries it draws

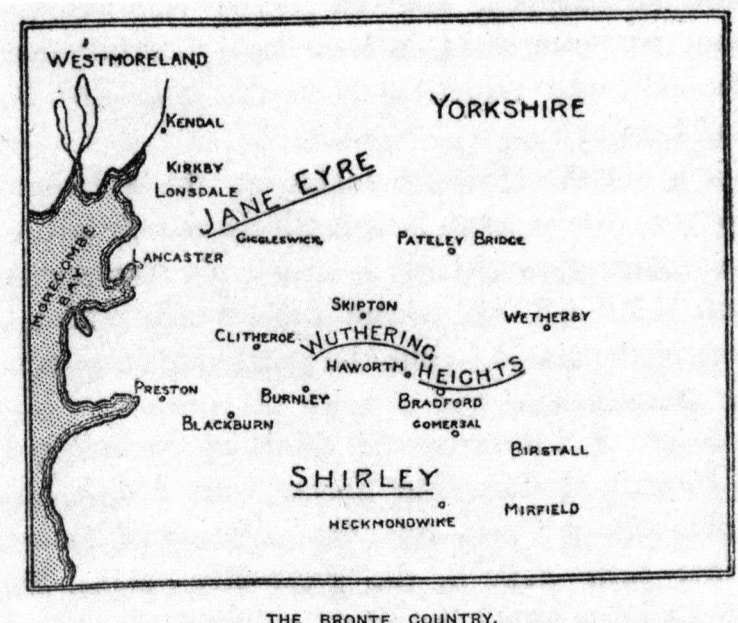

Fig. 4.2. Map of Brontë country. In William Sharp, *Literary Geography* (1904), 119. While places are vaguely mapped in relation to one another here, they are not grounded in any description of the topographical details of the county itself.

between the worlds and landscapes it describes, and is of limited use in helping the reader understand the geography of the West Riding. The novel seems, then, as if it is an antimap. And yet, its lack of topographical specificity is what helps it create an essentialist—even eternalist—vision of Yorkshire partly abstracted from history and geography, giving a kind of moral identity to the land, its rocks, and its people, which would become central to its geological mapping in the nineteenth century.

The novel was also, one critic has argued, "the first modern novel to be named for a geological process (erosion)."[28] The word "wuthering," reflecting a burgeoning interest in the scientific study of dialect in the 1840s, is glossed by the central narrator of *Wuthering Heights* the urbane visitor Lockwood, as "a significant provincial adjective, descriptive of the atmospheric tumult to which its station [Heathcliff's estate] is exposed, in stormy weather."[29] The aural similarities between "wuthering" and both "withering" and "weathering" suggest the potentially erosive effect of this "atmospheric tumult" on the buildings of the Heights, and Lockwood's antiquarian gloss hints at its possible etymological relation to these better-known terms. However, as the novel's contemporary readers understood it, the term suggested more the stormy atmosphere of the high ground in which the estate is situated than a geological process.[30] And the geological significance of the story is both more concrete and more tantalizing than a reference to a single geological process might suggest. *Wuthering Heights* was also, in part, the product of a peculiar local legend that had sprung up partly around the figure of the Cambridge geologist Adam Sedgwick, originally from Dent in the north of the West Riding. Indeed, the novel may have been culled and transformed from a local legend involving Sedgwick as a key protagonist.[31] I will recount the story here, for it begins to unravel the precise, if slippery, connections between the geological mapping of Yorkshire and the literary descriptions of the land.

Sedgwick, it appears, was executor to a complex and contested will involving two powerful families in his home village of Dent in the northeast of Yorkshire. These two families, the Sills and the Masons, both owned sugar plantations employing slaves in Jamaica in the eighteenth century, estates which went into decline at the turn of the nineteenth century, reducing some of the members of the families to poverty. When John Sill died, he bequeathed his thirteen estates in Dent and his Jamaican sugar plantation to the sons of his brother Edmund Sill, who had recently taken in a thirteen-year-old orphan, Richard Sutton, from Liverpool. The sons of Edmund Sill were cruel to Sutton, at least according to gossip in the dales, though their sister Ann was kinder. When the sons died prematurely, the property passed to Ann, who later bequeathed the family manor house and grounds to her cousin, also called Ann Sill, who lived in Rochester in Kent. It was this will of which Sedgwick was an executor. The will went into probate in 1835, eighteen years after Ann Sill from Rochester's inheritance, and in it five thousand pounds was left to Richard Sutton. With that money he bought the estate in which he had grown up, and been mistreated, for his own.[32]

Links with the slave economy were often kept under wraps in Britain, and much abolitionist and emancipationist writing was concerned with the attempt to uncover the names and localities of English plantation-owning families. Sedgwick himself was sympathetic to the emancipationist cause and supported the north in the American Civil War. A catalog of his library reveals that the only contemporary novel he owned at the time of his death was *Uncle Tom's Cabin*, Harriet Beecher Stowe's 1852 abolitionist novel, widely thought at the time to have intensified the conflict that led to the Civil War.[33] He also owned Stowe's 1853 *A Key to Uncle Tom's Cabin*, which supported with documentary evidence the novel's account of the conditions under which African American slaves lived, along with a wide variety of antislavery texts dating from the 1830s onward.[34] Sedgwick was a pronounced Whig, prominent in the call for university reform, and it is unsurprising that he was opposed in this way to slavery. But this commitment may also have been sharpened by his personal involvement with the plantation families of Dent. For, according to Christopher Heywood, the Sill estate "was shielded from attack" by emancipationists "by the long life and singular authority of its surviving Executor," none other, of course, than Sedgwick himself.[35]

The episode was first related to the public in 1838 in William Howitt's *Rural Life of England*. Sedgwick was only mentioned anonymously, but his role in the affair was called provocatively into question. There was a "most extraordinary story" "ringing through one of the dales when we were there" "from end to end," Howitt wrote.[36] Ann Sill of Rochester had been convinced by a lawyer, over the course of twenty years, that her West Indian property was worthless and that she was on the brink of poverty. Only when she was at death's door did "the hypocritical monster [the lawyer], with a refinement of cruelty perhaps never paralleled," admit to her that her funds had in fact vastly accumulated while she had been alive and that she had been rich all along.[37] The affair, in Howitt's account, was a plot by the lawyer to steal Ann's money. In addition, throughout her lifetime he ignored her repeated requests to change the will, so that when she died she bequeathed, against her own judgment, considerable funds to a young man. That man was Richard Sutton, "who first came into the lady's house as a shoe-black, or some such thing, ... and was of so base a nature that he had [willingly] chosen ... flagellation" as a punishment for his crimes.[38] "This man had now actually purchased the lady's house of the executors, and lived in it!" Howitt continued in dramatic vein. "We were told that on the will being read, the other executor," Sedgwick, was amazed at what he had unwittingly been party to:

> he seemed filled with the utmost astonishment and indignation, and abruptly said to [the lawyer]—"Why, there is nothing but damnation for you!" and with that proceeded in such piercing terms to show to the lawyer the cruelty and wickedness of his conduct, that the man trembled through every joint. It was added that the lawyer "never looked up afterwards," but was in the greatest distress of mind, and daily wasted away [until he died].[39]

The melodramatic rendering of the story by Howitt—one can almost see the lawyer, prostrating himself before the heavens, his moral condition epitomized by his wasted body—may have appealed to Sedgwick. Sedgwick appears to have enjoyed melodrama, perhaps for its didactic qualities, but mainly for the transparency of the relation it suggested between physical and moral states of being. It is not a coincidence that a culture in which geological maps could be read to determine the moral characteristics of the inhabitants of a given region, as reviewers of its physical atlases claimed, was also a culture with a penchant for the moral and physical legibility of the melodrama. Both are consistent with a worldview in which contiguous relationships between people, buildings, places, events, and landscapes were overlain, transparently, one atop the other, and in which there was frequent emphasis on the congruence between physical, moral, and material realms.

But *Wuthering Heights* is not quite the melodrama we might expect. It is a more complex story than Howitt's local legend "ringing through the dales," and though it has features in common with the melodrama it is also a Romantic novel and is deliberately unclear about what constitutes the boundary between the moral and physical worlds. It is impossible to tell whether Emily Brontë knew the detail of Sedgwick's involvement in the story. She did not know Sedgwick personally, though outside of busy towns like Keighley, the West Riding and the Dales were not densely populated. Dent was forty-seven miles from Haworth, and Sedgwick's father and then brother were successive curates of their local parish, which was only ten miles from the Clergy Daughters School attended by Charlotte, Maria, and Elizabeth Brontë at Cowan Bridge. Patrick Brontë was curate of Haworth, so that both were clergy families in relatively nearby parishes. And at Keighley Mechanics' Institute the Brontë sisters had access to the Bridgewater Treatises, and to Wordsworth's *Guide to the Lakes*, to which three letters on the geology of the region, written by Sedgwick, were appended in 1842.[40] Sedgwick and the Brontës were connected by their shared occupancy of a "place," a place in which their ideas, their texts, and their values might overlap.

Of course, Howitt's story is suspiciously similar to the plot of *Wuthering Heights*. In Brontë's novel Heathcliff is brought as an orphan, repeating "over and over again some gibberish that nobody could understand," to the Heights, where he is bullied and enslaved by his adoptive brother Hindley Earnshaw but conducts a passionate relationship with his adoptive sister Catherine.[41] The resonance with the story of the Sills of Dent is obvious: a boy of inferior social class and of supposedly wild and cruel habits is brought into a family, is bullied by all but the sister, and eventually takes advantage of the declining fortunes of those estates to rise and own them both. Heathcliff's blackness, his uncertain origins, and the enslaving acts of cruelty perpetrated against and then by him, only serve to intensify the story's connection with slaveholding plantations in Jamaica.

There is, furthermore, a second kind of economic decline behind the plot of *Wuthering Heights*, in addition to that of the slaveholding estates of Yorkshire, and with which Sedgwick might also have identified. Heathcliff has been read as a symbol of the "dispossessing bourgeosie," "a type of the historically ascendant world of capital," "a dynamic force which seeks to destroy the old yeoman settlement" represented by the Earnshaws and by Wuthering Heights.[42] The Earnshaws are of yeoman stock. Hindley drinks himself to the verge of bankruptcy and is obliged to mortgage the Heights to Heathcliff, and Brontë occasionally hints at the geological treasure trove that might lie beneath the land of the estate: the Heights is situated on a West Riding limestone dale, surrounded by quarries that are perilously hidden beneath the snow drifts as Lockwood attempts to walk between the Heights and the Grange in the snowstorm of chapter 3. From the Grange, Penistone Crags, possibly based on Ponden Kirk, an outcropping of gritstone rock on Stanbury Moor, can be seen looming behind the Heights. A mile and a half beyond them we are told that there are lime quarries visited by Joseph, a servant of the Heights, though we never directly visit them in the course of the narrative. At least potentially, the yeoman and his agricultural practices might be displaced by the rocks that underpin his land. Most importantly, the yeoman Earnshaws *are* displaced by Heathcliff and his murky wealth, as the novel progresses.

The decline of the yeoman farmer in the West Riding in the early decades of the nineteenth century was a subject of an angry pamphlet written by Sedgwick in 1868 on the renaming of Cowgill Chapel in his home town of Dent. In this pamphlet he bemoaned the decline of Dent as "a land of 'statesmen,' that is, of a rural and pastoral yeomanry," in which each small estate had a right to "large tracts of mountain pasture" and its own flock and herd (see fig. 4.3).[43] In the course of his lifetime, Sedgwick said he had seen Dent go from "a land of rural opulence and glee" to a state of penury. Many of its yeomen, "not having learnt to adapt their habits to the gradual change of times, were ruined, and sank into comparative poverty," or had emigrated. "I need not tell my countrymen," he wrote, "that there are one or two present examples of landed property in the valley which exceed any that was held by a single 'statesman'

Fig. 4.3. "Image of Dent in Pastoral Style." Frontispiece to
R. W. Thompson, *Sedbergh, Garsdale and Dent* (1910).

in the days of its greatest prosperity. But alas, these larger proprietors are no longer the resident yeomanry of the valley."[44] As the region grew increasingly isolated in the early nineteenth century, "the silken threads that had held society together ... began to fail, and lawless manners followed."[45] Sedgwick may have been speaking from experience here about the shoeblack who became the prototype for Heathcliff, a shoeblack whose usurpation of a yeoman, plantation-owning estate Sedgwick had accidentally allowed.

In her impassioned defense of the Brontës and their fictions, in her 1851 *The Life of Charlotte Brontë*, Elizabeth Gaskell also focused on the decline of the yeomen, depicting the area around the Brontës' home in Haworth as an isolated and inhospitable landscape populated by hardened and fiercely independent inhabitants. Corrupt and nonexistent magistrates allowed violent crimes to go unpunished, roads were impassable even in summer, and winter snows blocked access to villages for months at a time. In her opening chapter, Gaskell progressively intensified the stony silence and impenetrability of people, places, and buildings in this landscape, and in her second chapter, drawing explicitly on the story and setting of *Wuthering Heights*, she moved beyond the isolation of the "hill villages," which were nonetheless "in the world," to tell her reader of the even greater "loneliness of the grey ancestral houses to be seen here and there in the dense hollows of the moors":

The land has often been held by one family since the days of the Tudors; the owners are, in fact, the remains of the old yeomanry—small squires, who are rapidly becoming extinct as a class, from one or two causes. Either the possessor falls into idle, drinking habits, and so is obliged eventually to sell his property: or he finds, if more shrewd and adventurous, that the "beck" running down the mountainside, or the minerals beneath his feet, can be turned into a new source of wealth: and leaving the old plodding life of a landowner with small capital, he turns manufacturer, or digs for coal, or quarries for stone.[46]

The sequestered situation and gloomy solidity of these "grey ancestral houses" have two historical consequences. In one case, the yeoman is made obsolete and unable to function (as signified by his alcoholism, which recalls that of Hindley Earnshaw in *Wuthering Heights*) and is superseded by new owners who lack his historic relation with the estate. In the second case, he avoids obsolescence by transforming from a yeoman of Tudor stock into a member of the Victorian capitalist class that threatens to dispossess him, turning "manufacturer" or exploiting the land for coal and minerals. In either case, "the old yeomanry" is left stranded, the monument of a superseded epoch, to which the "grey ancestral houses" are monuments in stone. Gaskell leaves this much broader economic history aside, focusing on personal and internal "causes" for the decline of the yeoman ("he falls" into bad habits, or he "turns manufacturer") and not on the industrializing economy in which those choices made sense. The narrative links between people and place remain largely uninterrogated, and the effect is to intensify the remoteness of this landscape and its people even when they turn to mining or manufacture for their sustenance. Thus, they continue to appear otherworldly, out of time and place, even after the yeomen have passed.[47] The argument relies upon the drawing of isolated boundaries around the estates, in which the yeoman may live on, though it is also implied that it is, in part, the underlying geological structure of the land on which he has subsisted that has ultimately led to his demise.

Importantly, Gaskell's argument was a defense of the Brontës' novels as the outcome of their environment. In her account the Brontës are not, as their critics had it, the writers of "books, coarse even for men, coarse in language and coarse in conception, the coarseness apparently of violent and uncultivated men," but simply the accurate and passive transcribers of an inherently coarse landscape of which they were not a part.[48] This tactic had its roots in Charlotte's own defense of *Wuthering Heights* in her 1850 preface to a new edition of both that novel and *Agnes Grey*, in which she argued, as *Bentley's Miscellany* summarized it, that those who objected to her sister's novel "knew nothing of the author; were unacquainted with the locality where the scenes of the story are laid (the West-Riding of Yorkshire) and that therefore the wild moors of the North of England can have no interest for them." These readers, in their ignorance, "hardly know what to make of the rough strong utterance, the harshly manifested passions, the unbridled aversions, and headlong partialities of unlettered moorland hinds, and rugged moorland squires, who have grown up untaught and unchecked, except by Mentors as rough as themselves."[49] *Bentley's* had no time for the argument: "The truth is," it concluded, "there are no such people as Catherine Earnshaw and Heathcliff in Yorkshire, or anywhere else. It were a libel on that magnificent county to think otherwise."[50]

However unconvinced *Bentley's Miscellany* may have been, the strong metonymic association between people and place cultivated by Gaskell and Charlotte as a literary defense was fundamental to the revised reception of the Brontë novels in the 1850s. Other critics began to defend *Wuthering Heights* on the grounds of "the barrenness and solitude of its scenery, the originality of its characters, and the haggard truth of its descriptions, produc[ing] a severe unity of effect, approaching the sublime."[51] And that "unity of effect" could lead to a reemphasized association of author and place in more positive terms. The *Examiner*, for instance, extended this defense of the stories to a defense of the moral characteristics of the authors themselves. They are, the critic wrote, "of a hardy race" who "do not lounge in drawing-rooms or boudoirs," hothouses or "perfumed apartments" but breathe an air that "whistles through the rugged thorns that shoot out their prickly arms on barren moors, or . . . ruffles the moss on the mountain tops," who write of "rough characters, untamed by contact with towns or cities," speaking in "plain vigorous Saxon words, not spoiled nor weakened by bad French or school-boy Latin," and who possess "rude habits" and live in "ancient residences—with nature in her great loneliness all around."[52] In another article in the same magazine, further metonyms linked the structure of the book, the character of the author, and the manners and customs of the characters ("savages ruder than those who lived before the days of Homer") in their "considerable power," each being "wild, confused, disjointed, and improbable."[53] It was only "with difficulty that" this reviewer could "believe in the appearance of such a phenomenon" as Heathcliff, "so near our own dwellings as the summit of a Lancashire or Yorkshire moor." Nonetheless the imaginative effort to "trust ourselves with an author who goes at once fearlessly into the moors and desolate places, for his heroes" was far preferable to "the affectation and effeminate frippery" of other modern novels.⁵⁴ Gaskell and Brontë's metonymic work paid off, for it was the Yorkshire moor, "so near our own dwellings" and yet in a different sense so far, that provided justification for the violent and melancholy hero not in another continent but at the very heart of Britain, and also identified the anonymous author as the brave, masculine, and truthful survivor of the rugged world of which he wrote. The case for the literary defense rested on the abstraction of a particular Yorkshire landscape into a remote, wild, and desolate region culturally and geographically isolated from the nation in which it sits.

This defense relies on the drawing of conceptual boundaries between the world of the novel and the world of the reader. Yorkshire is another place, both part of and yet comfortably remote from the rest of England. But the drawing of such boundaries was necessary precisely because the novel itself never clearly defines them, threatening its readers into encounter with a villainous hero lurking potentially in their midst. Heathcliff, his origins uncertain and his absences from the Yorkshire landscape of the novel unchecked and unnarrated, could not easily be contained as either a "libel" on, or as a type of, a remote Yorkshire past. Where exactly did the "unhewn" world of Yorkshire end and the rest of Britain begin? Was Heathcliff the remnant of an earlier epoch now consigned to history, or a sign of the barbarism lurking within even the most seemingly "civilized" of nations and of hearts? The novel leaves these problems disconcertingly unarticulated. And the topographical vagueness of its setting only intensifies that disconcertion.⁵⁵

If, as is often contended, regional novels use place "to indicate how character grows out of certain occupied localities," suggesting a series of causal connections between places and peoples, then *Wuthering Heights* doesn't fit the mold.⁵⁶ However, as Gaskell's defense of the novel reveals, it does trade in *metonymic* connections between places and peoples, and those give the novel much of its ambiguous power. If tourists like Lockwood were, in the 1830s and 1840s, looking for "quaint customs and rugged landscapes" outside the urban center, imagining "the greater part of the British population as remnants of a primitive past that lingered on the fringes of the modern nation,"⁵⁷ then it is clear that one function of the regional novel was to assign "different temporalities to core and periphery," to make of different regions not only a distinctive place but a distinctive *time*.⁵⁸ In this context any distinction between the "place" and the "time" of the novel requires clarification: place, in a geological world, *is* time, for the rocks and land of the region are felt to belong to, or to have identity with, a specific geological period. *Wuthering Heights* measures time by place. In doing so, it is perhaps the most regional of all Victorian regional novels.

This, moreover, is explicitly linked to the structure of the land, with its lime quarries and its outcroppings of gritstone rock, and especially with its figurative association of Heathcliff with the rocks and minerals of this stony world as they appear in the novel's best-known uses of "opposable natural terms" to describe character.[59] Nelly, for instance, considers the "contrast" between Heathcliff and Linton to resemble "what you see in exchanging a bleak, hilly, coal country," like that surrounding the Heights, "for a beautiful fertile valley" like that of the Grange.[60] Heathcliff, "dark almost as if it came from the devil," responds with "flinty gratification" in his most piteous suffering and is "rough as a saw-edge," the serrated edge of a rock, "and hard as whinstone"—a dark, usually volcanic, rock that can only with difficulty be broken.[61] He is an "image of the darkness that gives off fire" that "appears in Yorkshire coal mining," and of "the subterranean dark power of which . . . Penistone Crag" is "an outcropping," "an inner and older, more primitive power than that possessed by civilized man living on the surface of his more pleasant, less demanding low lands" at the Grange.[62] As Catherine describes it in perhaps the most famous passage of the novel, her "love for Heathcliff resembles the eternal rocks beneath—hard, but necessary."[63]

If Heathcliff dispossesses the Earnshaws from their land, then, he does so not so much because he is an unnatural force, a "type" of the capitalist and industrialist economy located outside the geography of the novel, but because he is the very essence of the structure of the land on which the estates on the surface are built. Heathcliff, and the ancient, protruding coal, limestone, and whinstone with which he is associated and on which the landscapes of the novel sit, embodies *both* the natural world and the potentially industrial products embedded in its rocky depths. In geological terms, there is no distinction between them. Like the rocks and minerals, he is both nature and industry. Indeed, in his association with rock Heathcliff both precedes or even transcends the history of human habitation on the lands of Yorkshire and transcends the geological history of the land to represent something essential and eternal about the character of the "place" to which its histories and its geographic specificities are consistently returned. History is collapsed into the structures of place. Yorkshire *is* mapped in *Wuthering Heights*, but not by using geographical or historical markers. Instead, it is given definition by the rocks and the people who inhabit it, and whose overlain characteristics unite them eternally in rugged independence.

Sedgwick, we may conclude, may have had a double culpability for the decline of the yeoman, both as the negligent executor of a will and as a geological mapmaker whose efforts to unravel the structure of the land would reveal raw materials that would fundamentally alter the local economy. And yet, the layered view of the Yorkshires of Sedgwick and the Brontës, which I have begun to provide here, reveals less a narrative of geology's contribution to a shifting local economy and more a vision of Yorkshire as the permanent physical embodiment of a layer of history, of a primitive, essential, rugged sensibility that is a fundamental feature of its historic development into an industrial, urbanizing world. This Yorkshire is less a part of a developing story of the British nation, or of the region as a part of that nation, than an essential world embodying a set of core values that are retained despite—perhaps even because of—historical development. *Wuthering Heights* offers Yorkshire as a structure by which to comprehend a particular set of "eternal" values rather than as a setting in a broader story of the nation.

The Eternal Rocks Beneath

As *Wuthering Heights* also demonstrates all too clearly, boundaries are hard to define and histories are hard to contain. As might be suggested by Sedgwick's own view of the decline of the yeomen, of historic change in his county, geological mapmakers were also gripped with this difficulty. Mapping regions by their surface strata, geological maps, like regional novels, assigned "different temporalities to core and periphery," "consigning" the regions "to a graduated series of pasts." While, as I want to explore here, such maps could not rest, like Brontë, on the dramatic power of leaving the region or its temporal characteristics undefined, nonetheless, *Wuthering Heights* neatly encapsulates the temptation the geologist would face in attempting to draw meaningful geographic boundaries for his map: the temptation to reach for essential characteristics of the region that could link its manifold geographies and histories.

In 1853 William Smith's nephew, the Yorkshire geologist John Phillips, wrote that "geological distinctions are nowhere more boldly marked than in Yorkshire," and are nowhere "more constantly in harmony with the other leading facts of physical geography."[64] In Yorkshire underlying structures of the land were clearly legible not only to the geologist but also to the traveler or the topographer, for Yorkshire was, as Phillips had elsewhere written, "one of the few counties of England, which are, for the most part, defined by natural boundaries":

On the west it reaches, and in some places extends beyond, the great summit ridge of the island; it has the Tees as its natural limit on the north, the Dun for a great length on the south, and on the east is washed by the German ocean. Its area is divided into several obvious sections, distinguished alike by topographical features and geological structure. Along the middle of the county, from north to south, runs a wide level vale, filled with gravel, deposited on the upper red sandstone. From beneath, rises towards the west an elevated undulated tract, of carboniferous and calcareous rocks, which ascend to the summits of Micklefell, Ingleborough, and Pendle Hill; whilst above, on the east, appear the uniform ranges of the chalk and oolite.[65]

Running from west to east in Phillips's description we are able to move through a series of progressively newer secondary rocks. Starting with the oldest rocks in the series, the carboniferous mountain limestone, we travel to the upper red sandstone of the center of the county (overlain by the newest alluvial gravels), and finally to the very newest chalks and oolites lying above them on the eastern coast (refer again to plate 4 for De la Beche's somewhat variant column of these rocks). In the West Riding, the "elevated undulated tract, of carboniferous and calcareous rocks" and hills, the relation between "boldly marked" geological formations and the surface of the land was especially plain to see.[66] In the Craven district of the West Riding, north of Settle and the Aire river, Phillips continued, "romantic dales are sunk into the mountain limestone . . . whose hills are capped by the lower members of the coal series." South of the Aire the landscape was lower but newer, a land in which "sandstones and shales with coal abound."[67] The style of Phillips' prose is significant. Describing the stark transitions between two landscapes of the West Riding as "romantic" he indulges in poetic syntax, moving the verb "abound" out of usual word order to the end of the sentence, and creatively using the verb "to sink" to suggest that the dales have gently settled into the shape of hills and valleys by an almost magical process rather than by the complex and combined actions of sedimentation, lateral thrust, uplift, and erosion. This poetic dramatization of the shifting surfaces of the West Riding makes accessible to even the most casual of observers a bipartite structure, exaggerating the geological and topographical distinctiveness of the two characteristic landscapes of the region and giving them striking visual identities easily imagined by the geological observer and the general reader alike. Phillips's poetic prose performs the work of creating visualizable boundaries for complex layers of rock that, in reality, overlap and are jumbled up across the West Riding, transforming that complexity into "romantic" and conspicuous difference.

In Phillips's 1853 text *Rivers, Mountains and Seacoasts of Yorkshire*, the agricultural, architectural, and industrial uses of the land were important too. The mountain limestone of the north, he wrote, was used for walling and erupted in jagged outcrops locally known as "scars," blanketed by "a sweet green turf." The sandstones and shales of the southern portion were overlaid by uncultivated moorlands, "deeply covered by brown heath."[68] "The formations" in the West Riding, another textbook put it, "may be studied on a scale that occurs nowhere else in England."[69]

The geological structure of West Riding was first mapped in Smith's 1815 geological map of England and Wales (plate 5), and then in more detail in his county map of Yorkshire (plate 6). Attempts to produce a more detailed geological map of Yorkshire proceeded relatively slowly in the ensuing decades, but in a paper delivered before the Geological Society in 1831 it was Sedgwick who took up the challenge. In this paper Sedgwick provided both textual and visual accounts of "a series of longitudinal and tranverse sections of a portion of the Carboniferous chain between Penigent and Kirkby Stephen" (plate 7) and a "tabular view" of the rocks of the region in ascending order.[70] Any traveler moving through the landscape with Sedgwick's paper in hand could follow the longitudinal sections to see an unfolding series of rocks pass before his eyes, a chronological panorama of a portion of the history of the earth, and could crisscross those journeys with the aid of his transverse sections, accumulating knowledge through a variety of different routes across the same terrain.[71] Like the gentlemanly Lockwood, with his interest in the word "wuthering," Sedgwick felt his home county could best be understood by retaining dialect terms even in geological writing, and in pointing out the traditional local uses of the resources of the land. "Strong post limestone" affords "a beautiful material for the construction of doorposts," he wrote, and the siliceous grit of the hamlet of "Cow Gill" forms "the most beautiful roofing slate I have seen in the carboniferous chain." In the paper he argued for the efficacy of local terms such as "scar," meaning "any bare precipice on the face of a mountain," and "four-fathom" and "twelve-fathom limestone," terms "in common use in the North of England," which "on that account ought not to be changed."[72]

If Sedgwick's commitment to a rough, plain, colloquial style of writing reminds us of Gaskell's description of the inhabitants of Yorkshire, it is because it was clearly linked to his self-fashioning as a Yorkshireman in broadly stereotypical terms. In its early days, the Geological Society had set up a Committee of Nomenclature in the hope of creating a uniform and precise geological language by which to name the rocks—one that would supersede the terms used by miners and quarriers and practical men in different counties.[73] It had failed, but its existence reminds us that Sedgwick's preference for terms already in ordinary local usage is political, privileging the provincial and practical over the metropolitan. Furthermore, it reflects Sedgwick's wider suspicion of all literary forms of language as obfuscations. For him, the dialect terms of a particular region were more authentic than any scientific terms that could be imposed on them from without, having an organic relationship with the land they described and with the peoples who knew that land most intimately, if not most scientifically.

In old age, Sedgwick elaborated on his view of the importance of dialect nomenclature, especially in his 1868 protest against the renaming of Cowgill Chapel in Dent to "Kirkthwaite Chapel" by a new curate hoping to extend the boundaries of the chapel's district to include the village of Kirthwaite (which should not, Sedgwick argued, include the second "k" in its rendering).[74] In an appendix on "the provincial dialects of the North of England," Sedgwick drew the geological and linguistic boundaries that he held so dear. He argued that the dales in Dent were "cut off from the neighbouring parts of Yorkshire by a broad chain of mountains" and could therefore "be considered, physically and geographically, as a part of Westmoreland" and the Lakes, and that this geographical isolation meant that the area's dialect was much closer to that of the Lake District too.[75] Geological features isolated the dialects associated with different historic periods, so that in different locations the languages of otherwise forgotten historic epochs were spoken. It followed that these different dialects, each representing a different period, could be mapped like the rocks of geological maps.

More importantly, words were widely considered to be subject to the same natural forces as rocks. Sedgwick's principal source for this discussion was Isaac Taylor's philological study of toponyms, *Words and Places*. In that study Taylor argued that most words were subject to forces of erosion, decay, and metamorphosis as they traveled across speakers of different languages, or as grammatical inflections, misunderstood etymological roots, or standardizations led to their alteration through time.[76] The only exception to the natural erosion and metamorphosis of language came in "the names of places." Toponyms were so indestructible that Taylor considered them a more reliable source of geological evidence than those mutilated "fragments" of rocks and fossils he had read about in Charles Lyell's *Principles of Geology*.[77] Drained marshes, coastlines that had long since fallen into the rapacious sea, dried-up rivers, silted-up lakes, and islands converted into land, all long vanished from human view, were yet preserved in the "names of places," "the beacon-lights of geologic history."[78]

Furthermore, in the sense that names are given to landforms in even preliterate societies, Taylor considered them the most authentic and undiluted forms of language.[79] Sedgwick picked up Taylor's thread, celebrating the use of "plain and homely words" in Dent and deploring the loss of the "grand sonorous guttural" of the Dales and Westmoreland on the grounds that the guttural was "too vigorous for the nerves of modern ears."[80] "We may polish and soften our language by this smoothing process," he wrote, "yet in doing so we are forgetting the tongue of our fathers; and, like degenerate children, we are cutting ourselves off from true sympathy with our great northern progenitors, and depriving our spoken language of a goodly part of its variety of form and grandeur of expression."[81] The forms of the mountains and the forms of language were bound in mutual relation: ancient dialects were preserved by the isolating powers of the mountain, and those ancient words were the evidence for vanished geological forms. Rough, hard, and isolated, the words, people, and mountains of this part of Yorkshire offered keys to a series of ancient worlds on the cusp of being forever forgotten. Of course, such an image was not unique to Yorkshire: the contrast between roughness and polish, with all the value placed on "roughness," was an important part of Highland Celticism: James Macpherson's Ossian poems had articulated the same idea. The link between Scott's valorization of a rough Highland geography, outlined in chapters 1 and 2, and this Yorkshire-bound "roughness," should be clear: the ruggedness of the rocks in these locations bespeaks a kind of authenticity from which geologists and writers of fiction alike could borrow.

Sedgwick did not provide a map with his 1831 paper on Yorkshire, but in the privacy of the Geological Society and its *Transactions* he did amplify a grand historic narrative from its crisscrossing transverse and longitudinal sections. "Each limestone group," he concluded, "commenced at the beginning of a period of repose—that the marine animals which assisted its growth were at first few in number and ill-developed—that they gradually become vigorous and full-grown; and were at length destroyed only after repeated irruptions of mud and sand."[82] The valleys in the chain were "valleys of denudation," gradually eroded and weathered by ordinary causes in Lyellian fashion. Arriving at the Craven fault, the story turned more dramatic (refer again to plate 1 for Henry De la Beche's transverse sections of the Craven fault):

> The horizontal beds of the great Scar limestone lie far below the bottom of the neighbouring valley, but the broken ends of the whole mass have been torn up from the foundations of the mountain, and jammed against the edges of the upper horizontal groups. The vast force of elevation is indicated by the enormous extent and contortions of the dislocated masses; and the line of greatest stress is indicated by an anticlinal axis, on the north side of which the Scar limestone, after many breaks and undulations, gradually falls down ... and is buried under the conglomerates of the new red sandstone, where the longitudinal section ends."[83]

Sedgwick dramatizes a series of violent past participles—"torn up," "broken," "jammed against," "dislocated"—with intensifiers—"vast," "enormous," "many"—to punch home the force of change and the scale of the damage it has caused. This hyperbolic prose emphatically underscored Sedgwick's theoretical position as an advocate of Léonce Élie de Beaumont's theory of the sudden elevation of mountain chains, and no doubt gave him a compelling oral arsenal with which to fire out his case in the halls of the Geological Society. The passage moves climactically through these intensely violent, upward-thrusting forces before falling off into a more attenuated description of the gradual falling and burial of the limestone beneath the new red sandstone. Though the landscape has partly been produced by gradual geological processes across millions of years, the rhetorical force of Sedgwick's prose is all on the side of the rapid elevation.

Just as Sedgwick accepted that the "polishing" of language into a "common standard" was necessary in the modern age, he also traced the gradual effects of weathering and erosion after an immense span of time in the ancient rocks of the landscape. But too much erosion—in language or in geological explanation—was glib, urbane, and produced forms lacking in grandeur, more suited to the feeble "nerves of modern ears." This is an aesthetic preference for the heroic men of an older northern Britain and for the heroic actions of a violent early world. Sedgwick's description of the faults and folds of the West Riding are of a piece with his self-presentation as a blunt, plain-speaking Yorkshireman with an honest, unadorned vision of a rude and powerful landscape with which he had intimate personal acquaintance. As his first biographers put it, picking up this thread of metonymic associations, it is not "a mere fancy which traces a connection between his rugged nature and the crags of that wild mountain-valley" in which he had grown up. "To the end of his days he was at heart a Yorkshire Dalesman."[84] And Yorkshire, as Sedgwick's speech, the dialect of its inhabitants, and the structure of its rocks all revealed, was an honest land in which nature held no secrets. Her violent past was unarguably plain to see.

It is unlikely that the Brontës read Sedgwick's paper on the West Riding, but they probably did read his contribution to *Guide to the Lakes*, much of which was derived from the 1831 paper and which emphasized this complex mixture of violent upthrust and gradual erosion in the creation of the limestone scars of the north West Riding—by association with Wordsworth's Lake District to the north of the Yorkshire dales. Here Sedgwick described "the great precipices under the crown of Ingleborough," "made up of the rocks of this complicated group, in which are five beds of limestone, alternating with shale, sandstone, and a few thin bands of coal."[85] "On the southern limits of the country here described," in Ingleborough, Gorsdale, and Kirkby Lonsdale, where Cowan Bridge school was located, "faults and breaks" had in ancient times been created, "which were gradually opened out into wide valleys."[86] That "limestone is, like a great potsherd, broken into many fragments, and is now elevated to the tops of mountains," resting on inclined slates in valleys in some parts of this region near Settle (though "it requires little effort of imagination" to remember that these limestones, "in this part of our geological maps, were once united").[87] "In such cases," Sedgwick continued, "the jagged edges of the slates have been worn off by the continued erosion of water, and rubbed down almost to a smooth horizontal surface."[88] "Again and again," over the whole area of the Lake District, "have the mountains been shattered by faults, and swept by denuded currents." And again:

> The grand forces of elevation had left stranded and shattered pieces of limestone like smashed-up bits of Roman pots waiting to be mapped atop the hills, while below them erosion had smoothed the older rocks from beneath which they had protruded into coffer moors and valleys. This was the region the geologist would attempt to disentangle.[89]

By the late 1860s the West Riding had been conceptually separated into two distinct and readily recognizable districts, both of which will be familiar to the reader of *Wuthering Heights*: the southern lowland moorlands and the northern upland limestone. Papers of the Yorkshire Geological Society, for instance, describe the "low country" of the south West Riding as "anciently an extensive marsh," with "desolate grit moors" and "considerable tracts of sedgy ground," while "the limestone ridges and mural escarpments of the Craven Fault," further north, as at Skipton, for instance, exhibited strata that had "been denuded so as to expose a considerable surface of limestone," wearing "the bright green aspect of limestone pastures" and marked by intermittent quarries.[90] Even further north, on a clear day, could be seen the distinct slate hills of the Lake District.[91] But for this geologist, the oval-shaped anticlinals of the Craven district described by Sedgwick, of which the limestone hills were the remnant, had not been produced by the violent upthrusts of mountain elevation:

> The regular lines of elevation extending over great areas, the contortion of the beds, and the entire absence of erupted masses, teach us to look to lateral pressure for the explanation, and not to any volcanic and cataclysmal agency. . . . Even in the days when what Professor Ramsey has called the "Jack-in-the-box" hypothesis was rampant, Conybeare and Buckland regarded the anticlinals of the Mendip Carboniferous rocks as irrefragable proofs of disturbance, produced, not by vertical eruptive thrust, but by lateral pressure.[92]

The language here, describing the phenomenon described by Sedgwick as "broken ends . . . torn up from the foundations of the mountain," "jammed against" other groups of rocks by "the vast force of elevation," is much less emotive, except when it comes to the denigration of Sedgwick's view as a "jack-in-the-box" exception to the general sober rule. Some of the reasons for the argument were lithological. The mountain limestone was "so unelastic a substance," and its bending "at sharp angles without breaking, really indicates a gradual and prolonged action" of lateral compression, the truth of which the author reinforced by means of an experiment closely resembling that suggested by Charles Lyell's image of a book demonstrating lateral thrust (refer back to fig. 1.6).[93] Other reasons were structural. Mapping had shown that there were almost no volcanic rocks in the region, making it less likely that rapid elevation had produced the mountains. Nonetheless, the gradual processes at play here were considered to have produced even more "romantic" divisions than Sedgwick allowed, unjumbling the complex groups of each region and typifying them by the starkly opposed images of northern scars and southern moors. What matters here is that these stark oppositions (reflecting the starkly opposed landscapes of Wuthering Heights and Thrushcross Grange) became increasingly embedded in geological understandings of the county, and that they depended upon metonymic correlations between the historic and cultural lives of the peoples thought to have occupied those differently imagined lands and the rocks that lay beneath them. They depended, in other words, on the boundaries geologists drew across different parts of the county, boundaries often drawn on imaginative and historical grounds. For Sedgwick, Yorkshire was associated with the Lake District, with the Vikings, with rough and violent peoples and rough and violent geological processes. And as in the creation and reception of Brontë's Yorkshire, the description of the history of the rocks or the peoples of the county is elevated to something approaching "eternal" status. The Yorkshires of geological and literary imagination overlay one another in metonymic relation.

Different Boundaries

Sedgwick never produced a geological map of Yorkshire, though his work was an important spur to the map that was finally produced of the region by John Phillips, a man also intimately associated with Yorkshire in the public imagination. Phillips published two monographs on the geology of Yorkshire, a guide to railway excursions in the county and another to its rivers, mountains, and coast. He wrote frequently on geological topics for the *Philosophical Magazine*, was a prominent figure in the British Association for the Advancement of Science, lectured widely across the country, and was professor of geology at King's College London and later at Oxford.[94] He was, as I have noted, the nephew, and also the apprentice, of William Smith, and served as a surveyor for the Geological Survey from 1839.[95] He undertook consultancy work for entrepreneurs and mining companies attempting to locate coal in Yorkshire and Lancashire, and he wrote for encyclopedias and "a flurry of general works on geology."[96] Though his geological work had a much greater compass than the geology of Yorkshire, his support of its local institutions, including from 1849 the Yorkshire Naturalists' Club and the Yorkshire Antiquarian Club, meant that he was very much considered a geologist and a man of the Yorkshire terrain.[97] Travel books, guidebooks, and popular journalistic accounts of Yorkshire frequently mention Phillips or quote his writings in large chunks.[98] If visitors to Yorkshire thought they knew the county, more than likely it was Phillips's version of it that they understood.

In his 1836 book *Illustrations of the Geology of Yorkshire*, Phillips had provided geological sections and a map of the county, though Sedgwick was critical of them on the grounds that they did not acknowledge his own earlier work on the region, were on unspecified scales, and were not colored but merely hatched and stippled.[99] Nonetheless, in the accompanying monograph Phillips showed that the coal, limestone, and red sandstone were not ubiquitous in the Carboniferous system but were merely prevalent there, that the mountain limestone was not sharply separated from the old red sandstone but merged into it via alternating beds, and that there were up to five intermediate beds between the lower scar limestone and the millstone grit above it. This was an even more complex picture than Sedgwick had drawn, and reemphasized with added force that "the lines of division between major strata were arbitrary and could be merely local in their application. It followed that any attempt to make stratigraphic correlations over large distances was hazardous," and relied on "personal judgment" as much as on the available evidence.[100] In an effort to efface his own "personal judgements" from the stratigraphic work, in this book Phillips also moved toward statistical palaeontology to define the ages of the rocks, taking average samples of their fossils to characterize them instead of using the traditional method pioneered by his uncle of finding one representative fossil that could characterize each stratum. But the reliance on "personal judgements" in drawing the three-dimensional boundaries of the landscape could never be entirely effaced. Phillips, with his own aesthetic and ideological vision of the county, did not identify the dual landscapes of the West Riding with the Lake District, as Sedgwick did, but with another region representing a different historic epoch.

In 1853 Phillips produced a detailed map of Yorkshire on a scale of five inches to a mile, which he reproduced in condensed and more accessible form in *Rivers, Mountains and Seacoasts* (plate 8). For Phillips the boundaries implied by stratigraphic columns and by sharply colored maps "seemed a strait-jacket that forced the "soft shades of mother nature" into rigidly drawn limits of line and color, and he experimented with chromolithography in the 1853 map to enable soft and delicate transitions between closely related strata, creating an almost "chiaroscuro" effect to represent a chronological sequence of rocks moving from light to dark.[101] In the western regions of Yorkshire, as depicted in the original map, a very pale blue signifies the Palaeozoic strata, "a small and singular district of Silurian slates, and a very large and varied series of Mountain Limestone, Millstone Grit, Coal, and Magnesian Limestone," the latter of which "passes into the New Red Sandstone group, which commences the Mesozoic series" of the central-eastern portion of the map: oolites, gritstone and shale, ironstone and coal. The pale blue of the western portion is concentrated into turquoise for the newer limestone formations in the northwestern corner. Similarly, green weakens into yellow, which intensifies to a pale and then a darker orange, followed by a pink and a darker orange still, in the eastern half of the county, making almost insensible gradations from one into the other. In the more clumsily rendered colors of the map as it was reproduced in the monograph, Phillips splits the county into two much more distinct regions, utterly different in their geology, and within those regions he blurs the boundaries between the strata more forcefully than in the original map. The technique results in the southern and northern districts of the West Riding being more subtly differentiated in the map than they are in Phillips's text. The complications of the interlocking, overlapping, faulted, contorted strata as they spread over the entire region are more clearly visible in the map than in the bifurcated descriptions of them that were reproduced in textbooks, periodicals, and travel literature through the period.

Most importantly, the map and its accompanying text offer contradictory kinds of definition to the boundaries of the landscape. In *Rivers, Mountains and Seacoasts*, Phillips begins by invoking the county's "long line of romantic coast, its broad fertile valleys, and ranges of barren mountains," bemoaning how little attention they had received by comparison with the Lakes of Wordsworth.[102] In doing so Phillips imagined himself doing for Yorkshire what the poet had done for Cumberland. In that sense he competed not only with the recently deceased poet laureate but also with Sedgwick, whose letters in Wordsworth's *Guide to the Lakes* had described the "internal structure" of the landscape for travelers and lovers of mountain poetry and had extended the boundaries of the dales in which Dent was situated upward to the Lake District. That sense of competition contributes to Phillips's account of the region's history, for "Yorkshire, once the home of the most powerful British tribe, and now the largest of English counties, is marked by nature with boundaries befitting such distinction; the sea on the east, rivers on the north and south, mountains on the west."[103] Yorkshire's natural boundaries are thus intimately linked with its pre-Roman "powerful British tribe."[104] This tribe, the Brigantes, helps Phillips to map the region in his reader's imagination, across both time and space. Charting a series of elevations and depressions of the landscape through geological history, above and below the level of the sea, Philips notes that many of the ancient forms of life and the strata in which they are found in the borders of Wales, from the oldest fossiliferous rocks, have no trace in Yorkshire. "Had the county now extended to its old Brigantian limit,—had it even stretched a few miles westward to the Lune, at Kirkby Lonsdale, we should have had an instructive group" of such rocks.[105] Unfortunately for Phillips, the county does not extend a few miles westward. Instead, the Brigantian tribe gives him a conceptual marker by which to expand his historical description of the rocks, to fill in the gaps that are present in the particular topographical zone to which his descriptions are limited. While Sedgwick conflated the northwest of Yorkshire and the West Riding with the Lake District, Phillips links it, at least imaginatively, with Wales and the valley of the Lune just west of Sedgwick's Dent, and with the particular kind of British history he finds there. As such, his sections of the rocks and the metonymic associations he makes between them and the people, places, and culture of Yorkshire, founded on different historical and natural boundaries, are different from those made by Sedgwick.

The point is made in the differing interpretations the two men gave to the region's toponyms. Like Sedgwick and Brontë, Phillips was interested in the etymology of the Yorkshire dialect, tracing it to the Celtic language of the Brigantes in order to imaginatively associate them with the geological structure of the land. The connection was not unusual. Taylor's *Words and Places* used the stratigraphic column as a means of imagining the history of language:

> Language is stratified: the "tertiary" languages, or the newest, are forming in the New World, the development of "Teutonic" languages came before them in the "secondary" period of languages, the "primary Celtic" period analogous to the earliest fossiliferous rocks of the Silurian, the Cambrian, and the Devonian.[106]

Taylor included an ethnographic map of regions conquered by different "tribes" in Europe, as indicated by the place names that survive there. As Sedgwick described it in his *Memoir*, such maps used color to tell the history of "the successive Tribes that have peopled this Island."[107] Indicative "test" names functioned like characteristic fossils in William Smith's geological map to determine the historic epoch to which each region belonged—whether it had been conquered by Anglo-Saxons or Vikings. And as with the geological map, this historical dimension made it both geographical and chronological at the same time: near Dent, for instance, at the mouth of the Ribble and at Morecambe Bay "the shore" is "covered with Saxon names," but at the top of the bay are Norse names.[108] From the Norwegian village of Mar-thwaite (*thwaite* being Norwegian for "town") by the tributary Dee, we come to Dent, and the scars, fosses, becks, and gills that surround it have Norse names. "Hence I conclude," Sedgwick writes after deciphering these names and others in the region, "that the Saxons had settled largely on the borders of Morecambe Bay; and that the Danes followed them and made several settlements, of which Gawthrop is the highest that we have traced. Then came the Norwegians, who overcame the old settlers, and cleared away the old forests, so as to convert Kir-thwaite into pasture land. "It thus appears," he concludes, "that if we pass the Norwegian colour (as stated above) over the Lake Mountains, we must also spread it over a part of the basin of the Lune, and extend it to the head of Dent."[109] The colors on the ethnographic map, like the colors on the geological map, represent successive stages in history, represented by complex, overlapping layers.

It is important that the Vikings are a "Teutonic" tribe and therefore part of Taylor's secondary strata of languages, just as the geology of this part of Yorkshire belongs to the secondary strata of rocks. These maps assign to the *present* region a place in a relative scale of history, in which one layered timescale is analogous with the layers of another. This relative position within these layers is colored on the map and attributed moral and physical characteristics that are intrinsic features of land, its peoples, and their languages, but which also come from the mythic and legendary features of that same landscape as they were promulgated by literary texts. For Sedgwick, these are the characteristics of the "Northmen" (or Vikings): conquerors, marauders, quick converts to a plain, rough, homely Protestant faith. He was angered by the new curate's ignorance of the historic dignity of the dialect of Dent, in which Chaucerian words (Sedgwick quoted directly from Chaucer in support of his view) continued to be spoken as colloquies.[110] Adding a *k* to Kirthwaite on the mistaken supposition that "Kir" was a corruption of *kirk*, meaning church, Sedgwick argued that the curate had failed to comprehend that the word derives from the language of "the grand Norwegian Sea-Rovers or Vikings of ancient song," who lived there, and thus was not a secondhand corruption but an authentic remnant of an even-more ancient language.[111] The people of Yorkshire, moreover, shared the colonizing, marauding, conquering spirit of their Viking ancestors. "While we are adding to our stock of verbal signs," Sedgwick pleads, "and adding both to their expressiveness and polish, let us not go on with our polishing till we rub them into the quick."[112]

Fig. 4.4. "Gordale Scar," plate 20. In John Phillips, *Rivers, Mountains and Sea-Coast of Yorkshire* (1853).

For Phillips, however, the word "scar," meaning a cliff or a steep face of rock on a mountainside (see fig. 4.4 for a particularly dramatic illustration of Gordale Scar in the region), was not a "Norwegian" word but was "derived from the British or Gaelic element *sgar*; while the generic name of Crag is the unchanged British word for 'rock.'"

Even the names which are left us of Brigantian personages are explicable as of Cymraic origin. Thus the Queen of the Brigantes, Cartismandua, has a name expressive of locality—Cathair ys maen du ... perhaps her seat of sovereignty by the black druidical stones.... Finally, the Brigantes seem clearly to be named from Braighe, G., pl. Braigheacan, elevated grounds, which in Cymraic takes even the form of Brigant, a mountaineer.[113]

MAPS AND LEGENDS

These peoples, then, are not only fiercely independent, like the modern Yorkshiremen, but are intimately connected with the landscape. Like the region's modern inhabitants they are mining people, Phillips tells us, characterized by their engagement with this landscape and its now industrial products—an argument that might support Phillips's closer (or at least more explicit) engagement with industrial and practical geology than some of his gentlemanly colleagues and friends. The effect here is almost to stop time: the region represents a slab of rock in a geological column, the identities of which will continually shape the identities, habits, and economics of the people who live there, returning them to the identity of the rocks just as Heathcliff returns the world of Wuthering Heights and Thrushcross Grange to the identity of the rocks on which they sit. They represent, too, an authentic Britishness resistant to foreign invasion: offering a potted history of the Brigantes, borrowed largely from Tacitus, Phillips tells us that the Brigantes resisted the Romans for over a century, though they did so by negotiation rather than by war, and though they were headed by the treacherous female leader Cartimandua, who handed over the leader of the British resistance, Caractacus, to the Romans.[114] Once the Brigantes were overrun, the rest of Britain capitulated too. The Brigantes therefore constitute the last moment of authentic Britishness, albeit self-betraying and superseded by the invasions of the Romans, the Saxons, and the "Northmen" with whom Sedgwick identified.

Phillips's interest in the Brigantes is partly strategic. Sedgwick had named the very early, complex rocks he later unraveled in North Wales and Devon the Cambrian, after the Roman name for Wales, and his colleague Roderick Murchison famously named the rocks above them, which he first identified in Shropshire and the southern Welsh borderlands, the Silurian System, after the warlike Silures, led by Caractacus, who had lived there—or thereabouts—and were the very last tribe to be overrun by the Romans.[115] In this sense Phillips's association of Yorkshire with the Brigantes participates in a much wider, imperialist drive to define portions of deep geological time by reference to authentically British regions, each with its own distinctive topography and culture. It was also a clear reference to Murchison and Sedgwick, his elders and superiors at the Geological Society, and places his work in respectful relation to theirs, even as it also hints at a noble history of mining peoples in the Yorkshire region. The Brigantes help Phillips define his place in the elite geological community, and to define the rocks he was describing through the lens of "British" history. At the same time, however, the Brigantes help write Yorkshire into an even earlier historical layer than expected, "the still older primary Celtic strata—Silurian, Cambrian, and Devonian," as Taylor had put it, which Phillips states are not found in the secondary rocks of Yorkshire but would be if Yorkshire extended, as the land of the Brigantes once did, to the valley of the Lune.[116] This is a valley that begins, for Sedgwick, to betray the Teutonic, Norwegian "secondary" history of the Yorkshire people and that is correlated with the secondary rocks of the region. Here is that tension between the layered structure by which the past could be apprehended, as represented by geological and ethnological columns, and the geographic regions over which those layers might extend. County or political boundaries might be suitable means for isolating a coherent series of layers, as might boundaries formed by natural geological formations such as mountain ranges—or, as here, boundaries could be defined by the peoples who had lived in a place at other historic points. So for Sedgwick, wishing to link Yorkshire with the Lake District described by Wordsworth, the land was the land of the Northmen, of secondary rocks and "Teutonic" spirit; for Phillips, wishing to link it with the prestigious series of "primary" rocks unraveled by Sedgwick and Murchison in Wales and Devon, an overlapping but different region was identified with a Romano-Celtic past. Phillips was mapping the "soft" rocks, but he didn't mind pinching some of the prestige from the "hard rocks" unraveled by his geological superiors.[117] Most importantly, Phillips continued to make metonymic connections between people and places that somehow consigned them forever to a specific historical stratum. In this, he continued, like Sedgwick, to participate in a sensibility we might imagine as antithetical to the map and more suitable to the novelistic imaginings of Emily Brontë, the rendering of histories and places as expressions of an eternal identity.

Cave Geology

Phillips had archaeological justification for this identification. The preRoman past had, in fact, become associated with the region through famous archaeological discoveries in the years in between Sedgwick's traverse sections of Yorkshire and Phillips's map. Perhaps the most famous of these was used as a device in the final layer of Yorkshire maps and texts I want to look at here, a novel called *Wooers and Winners*, published in the *Bolton Weekly Journal* in 1880 by Isabella Banks, also author of the 1870 novel *The Manchester Man*. The novel is important in this story because it may have been, as Heywood has suggested, a rewriting of the story of the Sills of Dent, a story that in fact had involved Sedgwick and in fiction had previously been retold in *Wuthering*

Fig. 4.5. "Thornton Force," plate 22. In John Phillips, *Rivers, Mountains and Sea-Coast of Yorkshire* (1853).

Heights. Banks may have uncovered the true sources of the *Wuthering Heights* plot, including the fact of Sedgwick's involvement, and given her novel an explicitly geological setting: in the limestone scars of the West Riding, largely in Settle, where the Craven Fault is strongly pronounced, and in the famous limestone scars, caves, waterfalls, and "ebbing-and-flowing well" of neighboring Giggleswick—the region of "play and beauty" described by Sedgwick in *Guide to the Lakes* (see fig. 4.5). The action of the novel centers on the very juncture of the two West Riding landscapes, amid "opposing lines of moor and scar."[118] This is an explicitly stratigraphic landscape, where a great and famous fault lays bare the layers of the rock for even the least observant of visitors to detect. And it is the project of this novel to bring together the various maps, pamphlets, gossiped stories, scientific papers, and literary texts whose concrete and yet shadowy connections it has been the business of this chapter to relate.

The plot of *Wooers and Winners* hinges on the actions of a well-meaning but negligent geologist, Archibald Thorpe (whom one critic has considered an alias for Sedgwick), a man "who had more knowledge of plants and fossils than of humanity" and whose geological preoccupations distract him from acting as a good executor to a will concerning the dark-skinned Martin Pickersgill. Pickersgill, bullied by his classmates but in love with the geologist's step-daughter, Edith, is the defrauded inheritor of a colliery in Osmanthorpe and a sugar plantation in Jamaica, the first of which he reforms, the second of which he joyfully disbands once the slaves are emancipated in 1834.[119]

The unraveling of the story occurs around the joint discovery, by Thorpe and Pickersgill, of the Victoria Cave. Subsequently, "a party of scientific friends, mostly members of the Yorkshire Geological and Polytechnic Society, or of the Yorkshire Union of Mechanics' Institutes," come "to visit the curiosities of the neighbourhood, and to examine the Victoria Cave, and the coins, fibulae, bones, and other relics found therein."[120] They examine "the relics, Roman, Samian, Celtic, pottery, coins, bones, pins, beads, brooches, &c., which had been disinterred up to that time" at the museum at Giggleswick Grammar School (now, I think, in the Craven Museum), and then at "Scalebar Force" "science again grew speculative, now on the denuding power of the water on the limestone," a discussion raised in Sedgwick's papers for the Geological Society.[121] As they debate the "mysterious origin and action" of the region's ebbing and flowing well, "their discourse sent Edith's memory back to a memorable hour beside that well" when she had first learned of Martin's love for her. The discovery and its subsequent discussion provide an analogue for the unraveling of the plot's central mysteries and misunderstandings.

And yet, as the narrator notes, close to the end of the narrative, a remnant of the mysterious and even the magical remains. "It was left for geologists and antiquaries nearer our time, with Government aid, to discover the vast proportions of the cavern, chamber beyond chamber, and to lay bare the secrets of prehistoric man, buried layer below layer under their feet."[122] The story was relatively well known by 1880. One of the original discovers, Joseph Jackson, became passionate about exploring the cave: his notes were read on his behalf to the Society of Antiquaries in April 1840, and a lecture was given on the "Caves of Craven" at the Mechanics' Hall in Settle in 1862, but only in 1869 was a funded excavation of the cave organized, after the Oxford geologist William Buckland became excited by Jackson's later discovery of a spotted hyena jaw. That same year saw the founding of the Settle Cave Exploration Committee. Jackson was site superintendent, with financial support from the British Association for the Advancement of Science. The story was widely circulated in a variety of antiquarian and scientific journals, as well as in the popular press.[123]

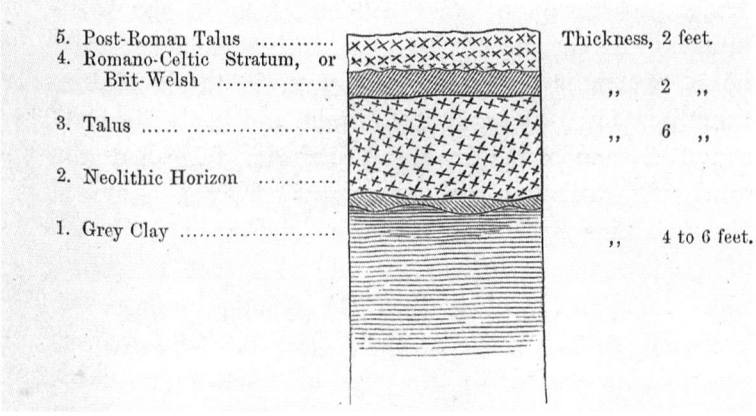

Fig. 4.6. "Vertical section at the entrance to the Victoria Cave," fig. 21. In W. Boyd-Dawkins, *Cave-Hunting* (1874), 87.

Perhaps one of the best-known accounts was in William Boyd Dawkins's 1874 book *Cave-Hunting*. The book's summary of the treasures of the Victoria Cave is particularly useful here, for in it he divided the cave into the layers of a "Vertical Section" of the strata at its entrance (see fig. 4.6) and discussed each layer in turn, from the newest to the oldest, as was by then customary. The fifth and newest stratum, two feet thick and labeled "Post-Roman Talus," was made from angular stone fragments that had fallen from the cliff, burying a "Romano-Celtic" or "Brit-Welsh" stratum, dated as such from bones, weapons, utensils, coins, and the food eaten by the cave's human inhabitants, imagined to be fleeing several waves of invaders after 449 AD.[124] A third stratum, six feet thick and labeled simply "Talus," could not be precisely dated: instead, geological methods of "relative dating" were employed, correlating its evidence with other caves and calculating the rate at which the talus had been deposited, based on current rates of weathering.[125] Below this the "Neolithic Horizon" and a "stiff grey clay of unknown depth" with fragments of limestone and blocks from the roof into which shafts had been sunk.[126] Only one of these, near the entrance, had reached the bottom, (see fig. 4.7) revealing bones and teeth of animals dating from the preglacial Pleistocene period, including spotted hyenas like those Buckland had discovered in the Kirkdale Cave, and a preglacial man.[127] Beneath that were further, unknown strata.[128]

The layered remains of the cave, Dawkins wrote, were "of almost equal value to the archaeologist, to the historian, and the geologist, and prove how

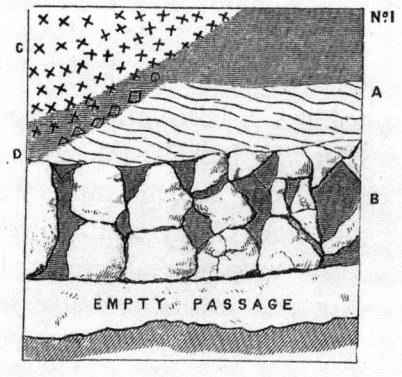

Fig. 4.7. "Section below Grey Clay," fig. 29. In W. Boyd-Dawkins, *Cave-Hunting* (1874), 117.

close is the bond of union between three branches of human thought which at first sight appear remote from each other."[129] That "bond of union" is partly chronological, but only partly so. To some extent the boundaries between history, archaeology, and geology are overlapping and chronological: the most recent layers offer archaeological evidence for the historic periods of Roman and Celtic Britain; the Neolithic layer again uses archaeological evidence but relies on the relative dates of the geologist rather than the precise methods of historical dating; and the Pleistocene stratum pushes at the limits of archaeological evidence, since only a single human bone fragment exists there, and is the gateway to preglacial epochs whose treasures can only be discovered by the geologist. The form of the rock in layers, then, offered imprecise disciplinary divisions between history, archaeology, and geology, organizing knowledge into discrete but overlapping categories. Though there is a connected history between them, the effect is to create, in Dawkins's words, "three distinct occupations—first by hyaenas, then by Neolithic man, and lately by the Brit-Welsh," three distinct moments in time, the geological, the archaeological, and the historic.[130] Nonetheless, geology encompasses the other two disciplines, for all the strata in the cave are the product of geological processes, from the "talus" created by the weathering of the cave to the "laminated clay" created as water, either from heavy rainfall or from the melting of the glaciers (a theory to which Dawkins objects), ran into the cave at various epochs. The cave emerges as an emblem of all kinds of pasts imagined through the structural and visual techniques of the geological map, and especially of the geological sections and columns such maps provided. The cave brings into physical relation the human, the natural, and the geological layers of a single location, so that each may be read in metonymic relation to the other.

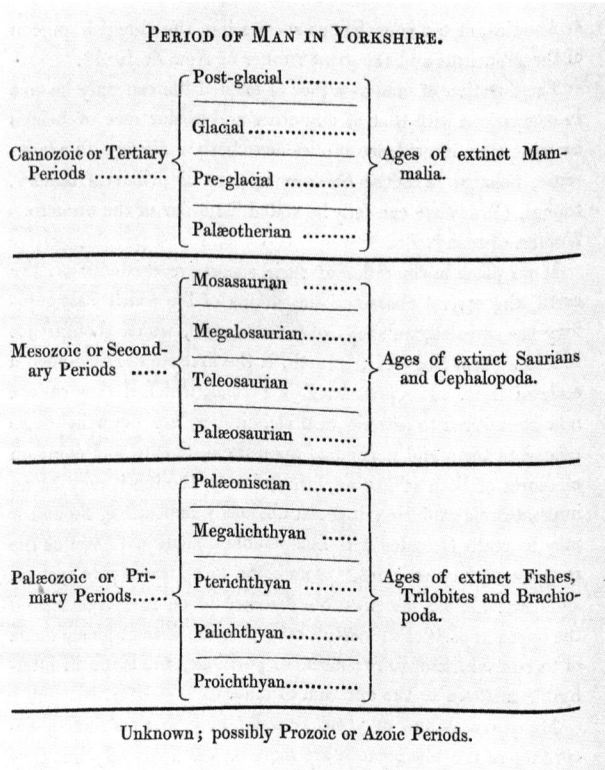

Fig. 4.8. "Period of man in Yorkshire." In John Phillips, *Rivers, Mountains and Sea-Coast of Yorkshire* (1853), 170.

As such, it avoids the problems of creating the correct kinds of boundaries, such as those exploited for imaginative purposes by Phillips, Sedgwick, and Brontë. The cave offers a closed and bounded world in which a layered past could be easily and comprehensively understood. Moreover, the cave linked the human and geological pasts in a single column (see fig. 4.8), facilitating the metonymic transfer of attributes across layers it has been the business of this chapter to describe.[131]

Wooers and Winners, whose plot centers on a fictionalized account of the discovery of the Victoria Cave, argues that a past concealing layer upon layer of pain and experience, emblematized by the landscape's many overlapping histories and "scars," is the safest ground on which to tread: "What lay under the scars was not for mortal to know," we are told about Edith, and Martin too is made nobler by his "ineffaceable scars covered up and out of sight, with which the work-day world had no concern."¹³² Elsewhere, gossip can be heard "filtering through diverse strata."¹³³ When confronting the rich railways speculator Absalom Metcalfe for disowning his daughter when she marries his employee, the geologist Thorpe "could sooner have chipped into the heart of a mountain than have moved" him. Thorpe responds, "with pointed emphasis":

> I happen to be a student of geology, and yesterday was present at the museum when Professors Buckland and Agassiz examined the fossils there collected, and the latter gentleman decided that one large fossil, hitherto regarded as a saurian, was a fish. If that fossil, your heart, could have been submitted for inspection, I wonder what verdict would have been pronounced on *that*!¹³⁴

As insults go, it is hardly a killer, but it is sufficient to say that those characters who most accurately reflect the living, breathing nature of their local environments—especially Edith and Tom Lister, who function as its poets (Edith becomes a poet and novelist named "Alkald")—are those whom the novel rewards. The Victoria Cave, with its geological, archaeological, and historical layers, offers a fitting location at which the novel's many threads and strands finally come together for Edith, enabling her to admit her love for Martin and to bring together the various strands of the plot into contiguous and meaningful relation.¹³⁵

Wooers and Winners suggests a possible triumphalist reading of the history of geological maps, at least of Yorkshire, across the nineteenth century. In the retellings of the same local legend that this chapter has related the geological map and the columns it produced were used with increasing confidence to give form to complex and changing relationships between humans and the natural world. In his 1868 memoir on Cowgill Chapel, Sedgwick noted that in recent years the lawless isolation of Dent had begun to be relieved by emigration, education, roads, railroads "touching the extremities of the valley," and reviving markets, so that "we are not now so isolated in England as we once were."¹³⁶ Twelve years later, when Banks wrote *Wooers and Winners*, those connections were plain to see. Despite its possible derivation from *Wuthering Heights*, the novel announces itself from its first pages as a realist text with a clear penchant for mapping and fits much more squarely in the tradition of the "regional novel" than its predecessor. While its opening lines state, "In all the land there is no district more romantic and picturesque than that section of the West Riding of Yorkshire where the great Pennine chain is broken up into rugged grandeur and beauty by what is known to geologists as the 'Craven fault,'" the romance of the place goes hand in hand with its isolation. "It has been so much the fashion for seekers of the picturesque to rush to the Continent in search of it that it is questionable whether, barring the geologists, the parish of Giggleswick was not a terra incognita to all but a few antiquaries, artists, and anglers, a scholar here and there, and traders in lime, yarn, or cattle, until very recent days."[137] But Settle is now, she tells us, a stop on the railway line.

Furthermore, Banks rewrites important scenes from *Wuthering Heights* and gives them much more definite, much more mappable shape. Like the narrative proper of *Wuthering Heights* Banks's novel begins in a snowstorm, which gets thicker from Leeds to Otley, then Ilkley, then Skipton, but here, as we approach the setting of the novel, the snow "appeared to grow thinner and lighter." When we reach Settle, "there was light enough to reveal the stone walls mapping out the high moorlands . . . and to demonstrate that the downfall there had not been snow."[138] Unlike the land between the Heights and the Grange in Brontë's novel, this landscape is legible, mapped out to the eye of the casual observer. The roads remain passable. The narrator is also more insistent than Brontë that we pay attention to her descriptions, exhorting her "impatient reader—you, I mean, who have skipped all this 'dry description'"—to "oblige me by turning back and reading it carefully. I have been at some pains to describe an actual habitation, so as to make events transpiring therein thoroughly intelligible. Do not defeat my good intentions."[139] "Real" personages proliferate in the text too, from the working-class poet and naturalist Tom Lister, in whose poetry "the voice of Yorkshire breathed in every cadence and inflection," and "a geological Quaker friend . . . who makes verses and sells linen, and goes rambling over the mountain moorland" collecting plants and rocks, to the geologists Louis Agassiz and Buckland, who visit the geological collections at the Yorkshire Philosophical Society in Leeds.[140] The novel documents the growth of Mechanics' Institutes, a movement founded by George Birkbeck, originally from Settle, and of the Yorkshire Geological and Polytechnic Society. While Brontë in *Wuthering Heights* traded on the illegibility of the unmapped landscape, offering only tantalizing hints about its underlying structure, *Wooers and Winners* transcribes that same landscape into legible, mappable terms. Banks reinstates a firm sense of history and of geography, offering clear causal relationships between people and the places that determine them. Perhaps the development of the geological map was one technology making this vision possible.

And yet we should, by now, be wary of such stories, compelling as they might be. That triumphalist reading comes with two significant caveats. First, the pivotal character of Archibald Thorpe is so preoccupied with geological pursuits that he is removed from the plot almost entirely; indeed, his forgetting to attend to his duties, to protect his daughter or step-daughter from dangers, or to intervene directly in almost any important matter is a recurring theme of the novel. In fact geology is insistently opposed to plot throughout the novel: Thorpe's geological obsessions repeatedly threaten to prevent the lovers from uniting. Second, *Wooers and Winners* schematizes the kinds of action that might happen within these different landscapes. Love triumphs in the cavernous scars and precipices of the West Riding limestones. The industrial centers of Leeds and Bradford give vitality to the novel and to its characters, but they are also the scenes of moral corruption, transgression, and wild desires. The caves and springs are purely natural features, while the collieries nearby are hellish, industrial, and enslaving. The rocks and the landscapes that produce them have been sorted into Romantic and industrial categories, so that no moral difficulties can occur. The novel reveals, like the maps, columns, and geological descriptions of Yorkshire by Sedgwick and Phillips, that the more transparent the connections between the metonymic layers of experience as they move backward in time, the more carefully drawn must be the geographical boundaries of the map from which those layers have been drawn. Here, a single cave provides an analogue for experience, while the novelist must carefully work to differentiate the beautiful West Riding landscape from the cities that lie within it.

In its bringing together of archaeological, mythological, natural, and historical features in a single landscape, *Wooers and Winners* reminds us that, as the layers of geological maps and columns became standard forms for understanding region metonymic slippages between literary, mythological, and emotional values, historical or geological layers helped define the boundaries of the maps themselves. As the landscapes became better known, those boundaries required ever more intensive fences. In this, the literary, the historical, and the imaginative would continue to play an important role. If the nineteenth century saw the emergence of a stratigraphic imagination, in which layered columns of blocks of rock gave shape to a view of history in which each layer of the past had a contiguous, or metonymic, association with the others, and in which particular places could be isolated and defined as indicative of particular sets of characteristics shared between rocks and peoples, *Wooers and Winners* reminds us just how important was the drawing of clear boundaries in making those associations felt. It also reminds us, I think, that the "regional" novel always included a tension between the description of a place *as* a region and the definition of that place through extrageographical and extrahistorical markers. Paradoxically, what *Wooers and Winners* reinstates to *Wuthering Heights* are precisely those markers, but it does so at significant cost to the identity of the place it describes. It is almost as if the clearer narrative line and the greater topographical and historical specificity of the novel make it less distinctly knowable as a "region" with its own peculiar, isolatable identity. *Wuthering Heights*, then, might be reconsidered as the preeminent "regional" novel of the nineteenth century, for in its collapse of time and place into an ill-defined and essential Yorkshire, it captured best the metonymic connections between landscapes and peoples that drove even the precise geological mapping of regions throughout the nineteenth century. The elaboration of the "region" as a seemingly self-evident unit of historical comprehension relied upon imaginative, historical, and cultural associations in which rocks and people could embody a particular kind of sensibility, which was seen as essentially tied to place and transcended the contingencies of historical change. Layers of meaning, knowledge, affect, experience, and identity sat atop one another in this definition of place. In this stratigraphic imagination, story, we might say, was often subordinated to structure.

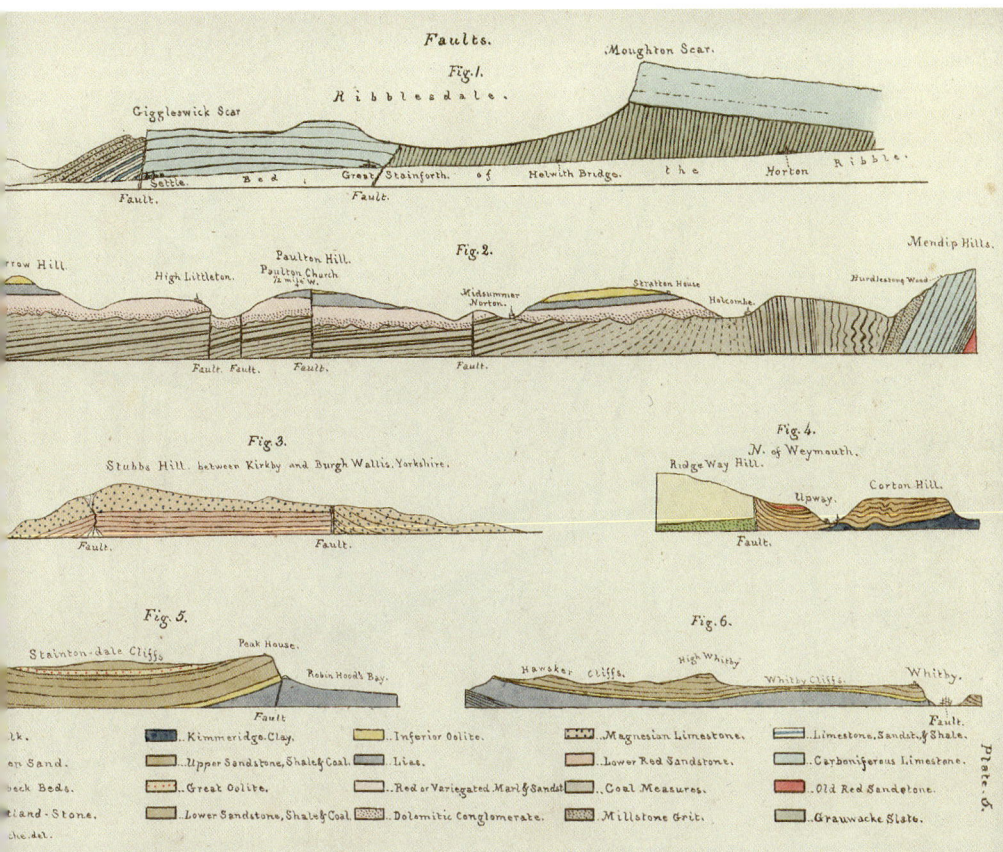

PLATE 1. "Faults," figs. 1–6. In Henry De la Beche, *Sections and Views Illustrative of Geological Phenomena* (1830).

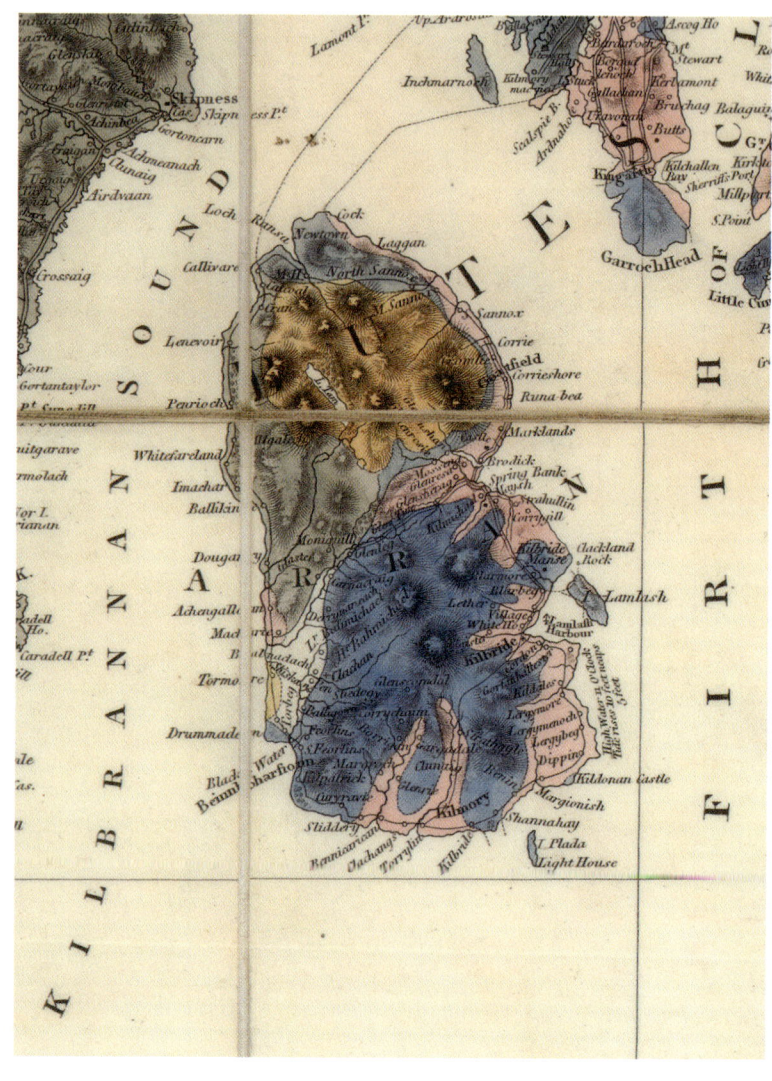

PLATE 2. John MacCulloch, map of Arran (1836). IPR/14-48CT, British Geological Survey © NERC. All rights reserved.

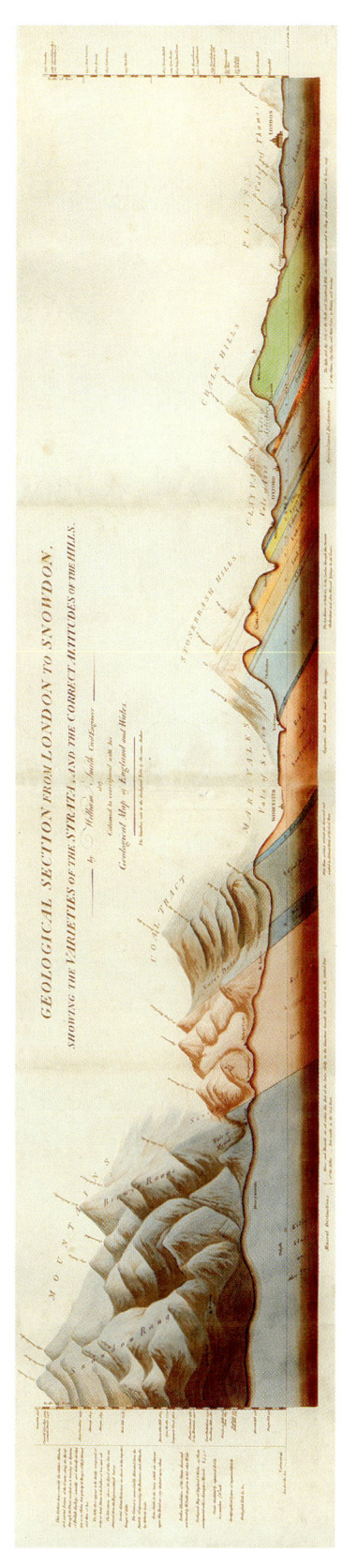

PLATE 3. William Smith, transverse section of London to Snowdon. 1PR/14-48CT, British Geological Survey © NERC. All rights reserved.

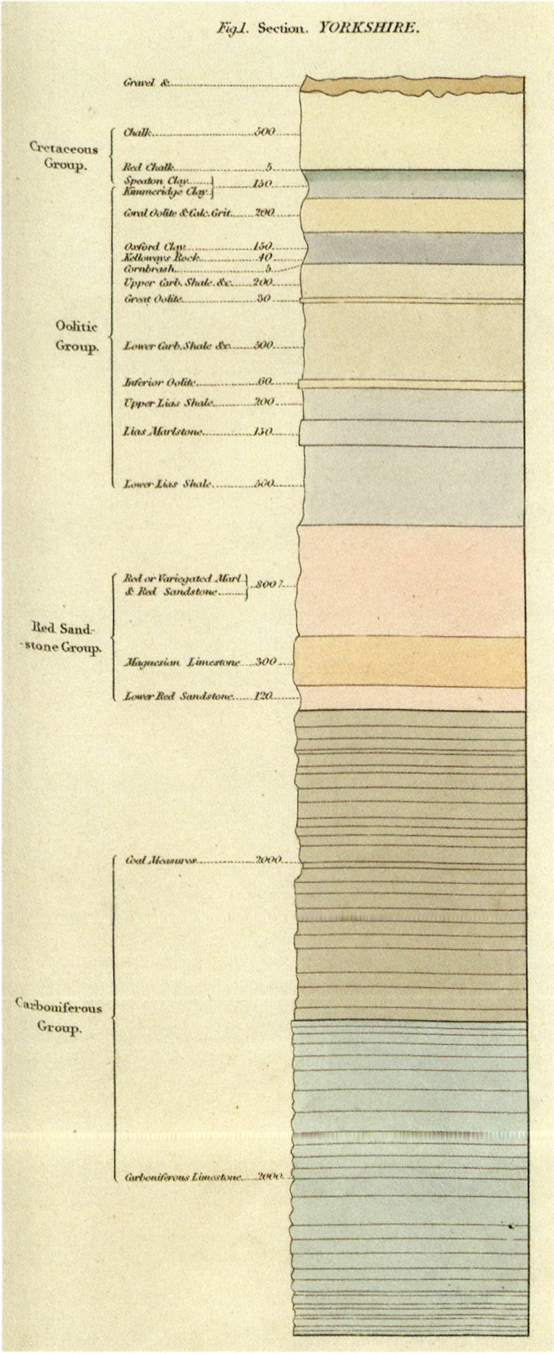

PLATE 4. "Section. Yorkshire," fig.1. In Henry De la Beche, *Sections and Views Illustrative of Geological Phenomena* (1830).

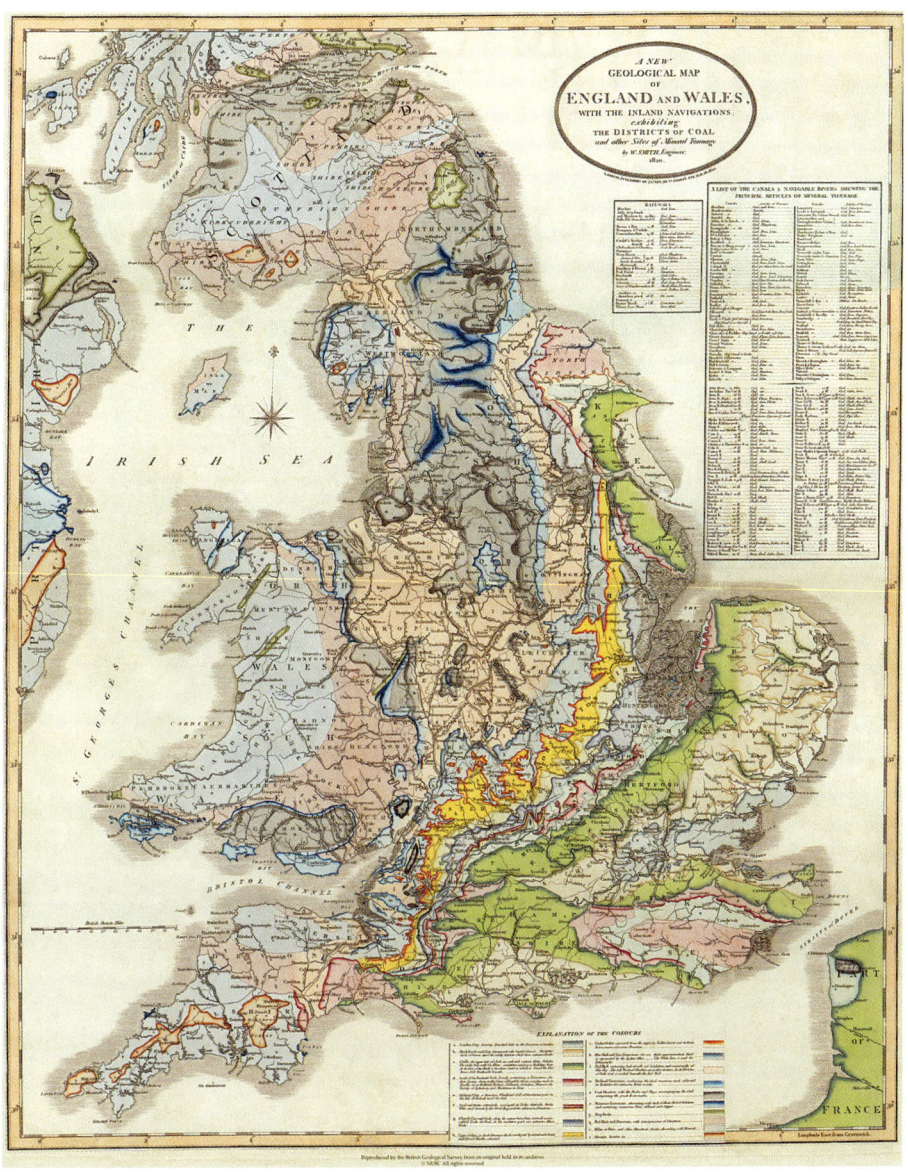

PLATE 5. William Smith, map of England and Wales (1820).
IPR/14-48CT, British Geological Survey © NERC. All rights reserved.

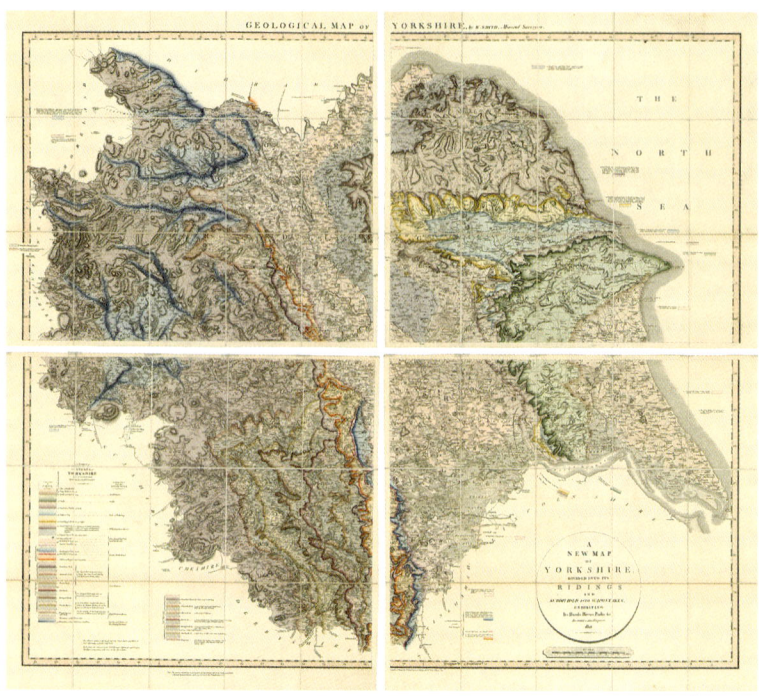

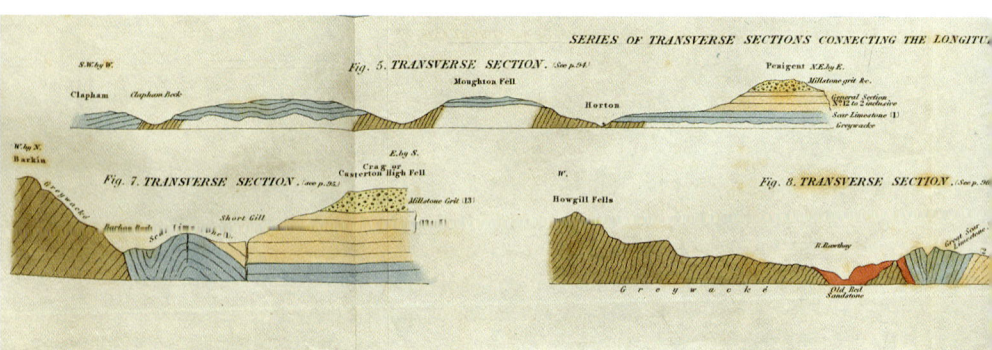

PLATE 6. William Smith, county map of Yorkshire (1820). IPR/14-48CT, British Geological Survey © NERC. All rights reserved.

PLATE 7. Adam Sedgwick, transverse sections, figs. 5–9. *Transactions of the Geological Society of London*, vol. 4, plate 6 (1831).

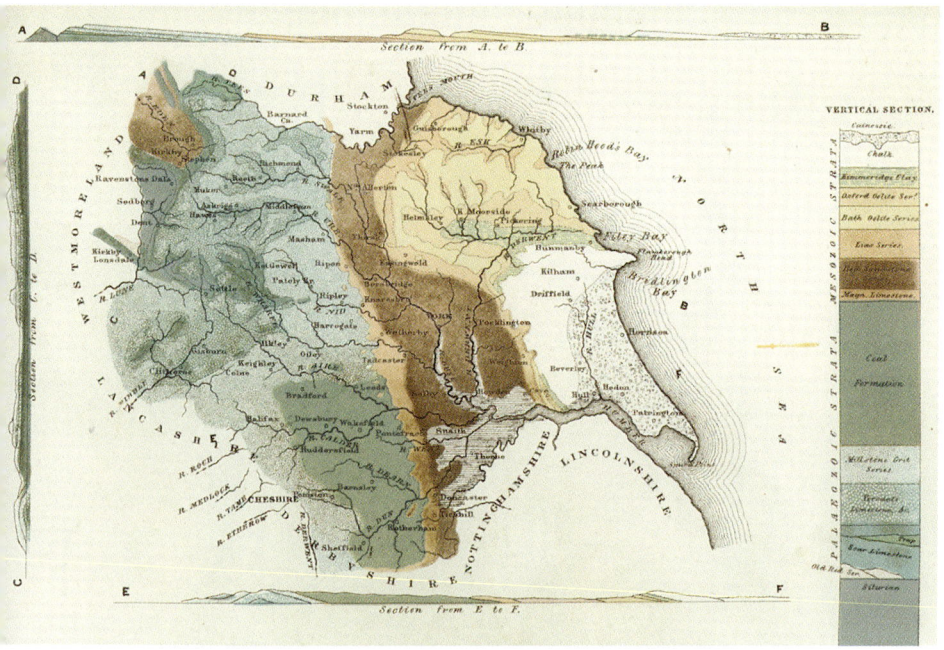

PLATE 8. "Geological map," plate 30. In John Phillips, *Rivers, Mountains and Sea-Coast of Yorkshire* (1853).

PLATE 9. George Bellas Greenough, map of Cornwall and Devon (1839). IPR/14-48CT, British Geological Survey © NERC. All rights reserved.

PART TWO *Science in Stories*

Kingsley's Cataclysmic Method { **CHAPTER FIVE**

I am exceedingly well and strong, though we did dine yesterday off raw ham, and hock at 9d. a bottle. My knapsack and plaid weigh about two stone, which is very heavy, but I go well enough under it, having got a pair of elastic cross-straps, which divide the weight over the breast-bone. . . .

. . . Kiss my darlings, and tell Rose I have got for her all sorts of curious lava-stones from the volcanoes, and shall carry them 200 miles on my back before she gets them. . . . God be thanked that I ever came here to see so much. . . .

. . . I cannot tell you what moral good this whole journey has done me. . . . Exceedingly well and strong, as lean as a lath, as one would be, who carried two stone of baggage daily increasing in weight from the minerals and fossils I find, on his back through broiling suns.[1]

Of all Victorian novelists Charles Kingsley had the most palpable interest in geological science and the most direct links to the gentlemen geologists who had pioneered its "heroic age." In 1863, having been nominated for membership in the Geological Society of London by Charles Lyell himself, Kingsley was elected to its ranks. It was an achievement he saw as the fulfilment of a "long . . . ambition"[2] cherished since boyhood, as he wrote in a letter to Lyell—a boyhood in which, he wrote in a lecture in the same year, he had "employed every half-holiday" enthusiastically engaged in the activities of "natural history and geology."[3] Once a member of the Geological Society, Kingsley's published letters reveal that he regularly corresponded with several of its leading geologists, including Lyell, Thomas Huxley, Charles Darwin, and Philip Egerton, often asking them scientific questions and questioning their research from the evidences he had collected on the Bagshot sands and gravels of the region surrounding his parish at Eversley.[4] When in 1870 he founded the Society of Natural Science, aimed at delivering scientific instruction to the middle-class men of Chester, Kingsley secured the honorary membership of many of these geologists and taught geology from the textbooks they had written. Believing it to be "my duty as an Englishman" to "train every Englishman over whom I can get influence in the same scientific habit of mind," to make him into "a rational and an able man,"[5] Kingsley substantiated his interest in geology in print in three pedagogical natural-history books: *Glaucus, or the Wonders of the Shore* (1855); the children's geology book *Madame How and Lady Why* (1870); and *Town Geology* (1872), which was derived from a series of lectures he delivered to the Chester Society in 1870.[6] Geology, Kingsley wrote, had been his "favourite study since" he "was a boy."[7]

Most work on Kingsley and science has hitherto focused on his largely positive reaction to the publication of Darwin's *Origin of Species* (1859).[8] Certainly evolutionary science was important to Kingsley. He wrote to his mentor, the Christian Socialist theologian F. D. Maurice in 1863, "Darwin is conquering everywhere and rushing in like a flood, by the mere force of truth and fact."[9] He also enthusiastically read the American Darwinist Asa Gray's *Natural Selection Not Inconsistent with Natural Theology* (1861).[10] Darwinian science "opened a new world" to Kingsley, according to his wife Fanny, "and made all that he saw around him if possible even more full of a divine significance."[11] As its first reviewers were quick to point out, the education of the naughty child-protagonist Tom in *The Water-Babies*, also written in 1863, "would have been impossible had not Mr. Darwin published his book on the *Origin of Species*."[12] For Kingsley, evolutionary biology disclosed an all-powerful God who had created a complex and evolving world at a single stroke. His support of the theory was directly acknowledged in the second edition of *Origin of Species* (1861), in which Darwin suggested that in Kingsley's wake other readers could feel comfortable about the theological implications of natural selection: if such "a celebrated author and divine" could see the "noble ... conception of the Deity" implied by evolution, then so, Darwin intimated, might they.[13]

Though *The Water-Babies* is an important document in the cultural history of evolutionary biology, and though Kingsley's voice was therefore of some importance in 1860s evolutionary debates, the significance of evolutionary ideas to Kingsley's fictional works can only be dated after the publication of the *Origin* in 1859. But before this date Kingsley not only wrote five of his seven novels, but also consistently rejected existing theories of "development," including those expressed in the anonymously published *Vestiges of the Natural History of Creation* (1844). Kingsley's participation in a much broader range of geological debates in his realist fictions, *Yeast* (1851) and *Alton Locke* (1850), have barely been addressed by literary critics, except to point out moments at which Kingsley might be said to have anticipated Darwin's theory. Clearly there is more to Kingsley's scientific fictions than that which pertains to evolution.

Most importantly, Kingsley's geological studies gave him a means by which to interrogate his practice as a novelist. His fictional heroes regularly practice geology, from the hero of *Yeast* (1851), Lancelot Smith, who we first meet attempting to finish a review of Roderick Murchison's *Silurian System* (1846), to Alton Locke, who takes lessons in "arranging and naming" the geological specimens owned by the Reverend Dean Winnstay in the 1850 novel of the same name. Tom Thurnall, the doctor-hero of *Two Years Ago* (1857), too, discusses geology with friends and family and has exploited his geological knowledge to lucrative effect on the gold fields. At the same time and in the same novels, however, Kingsley also struggled to find a fictional form through which to reveal social evils and to ignite an active sympathy for the suffering of the working classes among his readers. Literary criticism often censures Kingsley's fiction as caught hopelessly in the realist's desire to record "facts," a desire that abrogates the novelist's need to tell stories. *Yeast*, it has been said, "is more of a tract than a novel."[14] For other critics there is a "helter-skelter quality to Kingsley's works" because he is "less interested in the conventions of plot than in the need to make statements"; the novels are "ragged" and "baggy," "tugging sharply against regularities of order and form," because he could not comprehend that "powerful fictions have never had to depend upon factual originality for their power."[15]

Kingsley's novels thus exhibit the same tension exposed by geologists as they gave structural and stratigraphic form to the "ragged" and "baggy," "helter-skelter" natural world they encountered in the field. What has not been recognized hitherto is that Kingsley, deeply embroiled in geological debate, was not simply a bad novelist unable to reconcile the messy forms of nature with the exigencies of good storytelling; rather, he was continually questioning the limits of the novel as a form through which to tell truths about the world outside the text.[16] During the composition of *Yeast* he wrote in a letter that he would henceforth "write no more novels," instead turning to "the symbolism of nature and the meaning of history" as more appropriate "studies" for an intellectual clergyman.[17] Of course, Kingsley did not "write no more novels," producing six works of fiction in the fifteen years that followed *Yeast*. Yet he remained in many ways suspicious of novelistic form throughout that time. In 1857, while writing *Two Years Ago*, Kingsley wrote in a letter to the essayist and critic George Brimley, "I have copied nature most carefully" in the novel's descriptions of Wales and Devon, "having surveyed every yard of ground this summer."[18] Transcribing the natural world into text, Kingsley argues for his accurate novelistic methods and for the transparent relation between his written words and the landscapes they describe. But he is also anxious about the relation of fictional form to such an enterprise, going on to note that "if any man could write," in novelistic form, "the simple life of a circle of five miles round his own house ... no-one would believe it."[19] Kingsley's ambivalence circulates around the tension between the scientist's observation of the natural world and the expectations of the reader of fiction: accurate observation and verisimilitudinous description are not enough to convince readers of the truth of the novelistic representation. He sees, with some frustration, the novel reader's belief in a description of "reality" as paradoxically dependent upon its fictionality. Literary realism, for Kingsley, is merely a falsifying mode that succumbs to the reader's conventional expectations of how disbelief might be suspended, not a truthful transcription of an objectively verifiable reality. "Writing novels," Kingsley concludes, "is a farce and a sham."[20]

By contrast, Kingsley wished his scientific interests to be more widely known. He wrote in 1854:

> Those who fancy me a "sentimentalist" and a "fanatic" little know how thoroughly my own bent is for physical science; how I have been trained in it from earliest boyhood; how I am happier now in classifying a new polype, or solving a geognostic problem of strata, or any other bit of hard Baconian induction, than in writing all the novels in the world. ... [I] do believe more and more utterly, that the peculiar doctrines of Christianity ... coincide with the loftiest and severest science.[21]

Presenting himself as a trained physical scientist "classifying a new polype, or solving a geognostic problem of strata," Kingsley emphasizes the labor required to produce "hard Baconian induction," a reverence for fact, classification, and stratigraphy that is set against "writing all the novels in the world."

Aligned with "the peculiar doctrines of Christianity," this rhetoric of "lofty" and "severe" science staves off his anxiety about fictional form. In *Madame How and Lady Why*, Kingsley again bifurcates the idle pleasures of fiction and the honest, hard-working rigor of geological inquiry, so the "dear child" the text apostrophizes "will feel that novels and story-books are scarcely worth [his] reading, as long as [he] can read the great green book, of which every bud is a letter, and every tree a page."[22] Alluding to the Baconian image of Nature's Book, Kingsley's hierarchy exalts the "classifying," problem-solving, and inductive skills of the scientist above the mere telling and reading of stories. Could fiction ever be "realist," and was "realism" ever anything more than a set of fictional conventions designed to suspend the reader's disbelief? Was the novel a luxurious, idle, "sentimental" form opposed to the labor-intensive "hard bit of Baconian induction" undertaken by geologists? If it was, in what ways do Kingsley's geological novels transform or complicate its epistemological status and the claims to truth it might make?

A "Cataclysmic Method"

Kingsley's *Yeast* carries an epilogue parrying the "many criticisms" the new writer can "foresee" that the book will face.[23] His anxiety particularly centered on the novel's form and its "mythical and mysterious denouement" (364), and he attempted to relate this fictional shape to the shape of history itself:

> What if the method whereon things have proceeded since the creation were, as geology as well as history proclaims, a *cataclysmic* method? What then? Why should not this age, as all others like it have done, end in a cataclysm, and a prodigy, and a mystery? And why should not my little book do likewise? (365)

Kingsley's Carlylean sense of cataclysmic historical patterning offers a perspicuous example of a novelist considering issues of the form of fiction directly through the ideas and language of nineteenth-century geology. The world's "geology" and "history" is violent, cataclysmic, and mysterious, and the "realist" novel must reflect those cataclysmic shapes, absorbing them into its own structures in order to mirror the world it describes. The form of the novel is mimetic, reflecting the form of the world outside the text.

In these ways the passage supports the arguments of critics like Gillian Beer and George Levine, that Victorian narrative form is predicated on the "natural" patterns revealed by contemporary science.[24] But unlike the evolutionary "plot" of history, geology in this passage justifies formal breakdown rather than narrative organization. The novel itself abruptly terminates when the hero, Lancelot Smith, who has been searching fruitlessly for a mode through which to make sense of his place in an England that is "a chaos of noble materials ... polarised, jarring and chaotic" (345), suddenly meets a prophet, Barnakill, who instructs him to travel to the East and embrace his faith. This clearly instantiates the accusation that Kingsley was unable to produce solutions to the "facts" his novel exposes that could be manifested in narrative form. Instead of narrating the *Bildung* of his hero and his conversion to God, Kingsley simply displaces it onto an Eastern world he barely describes and that Lancelot will visit only after the novel has ceased. Geology here, instead of providing a "plot" of history, a pattern through which change and transformation can be imagined, belatedly justifies the breakdown of narrative with which Kingsley's novel concludes.

For Kingsley, in the epilogue to *Yeast*, geology provides an analogue to social and formal breakdown rather than a mode of narrative organization, and it is this powerful sense of fracture that his novels work to reconcile. In this way, geology poses a series of questions around which the novels take shape: How can the knowledge of geology, which emblematizes fracture and violence, but which is important to all of his male heroes, help us understand the real world better? What moral responsibilities does such knowledge incur? How might a quest for knowledge of the geological world transform a man into a heroic figure capable of transforming his society? If tightly constructed narrative is not the form for geology, or for truth-telling, then what is? And what role might fiction have to play if self-determining "plot" is to be excised from these practices?

"The Romance of the Field"

Both Kingsley's belief that society can be transformed and his understanding of nineteenth-century geology depend on his construction of a powerful masculine ideology. Aside from being the author of *The Water-Babies* and *Alton Locke*, Kingsley was also the Victorian "apostle of muscular manliness," advocating "muscular Christianity" (a term he rejected) or "Christian manliness," the term he preferred for its greater stress on the importance of Christianity in the ideal male.[25] For Kingsley, cold baths promoted moral virtue, and sporting prowess was an index of moral health. As one critic summarizes, "physical manliness" is both "the basis ... for a higher manliness" in Kingsley and "an index and a condition of psychological, moral and spiritual health."[26] Body, mind, and soul are interdependent, so that a broad range of cultural values coalesces in the figure of the ideal male, including violent strength, passionate faith, chivalry, virtue, and tenderness: Kingsley "insists upon direct correlations among social, spiritual, and physical forms of health, with the aggressively-poised male body . . . a central site for the enactment and resolution of conflict."[27] Reconciling physical aggression and Christlike tenderness, the healthy male body encompasses both sexual desire and moral purity. In its ability to reconcile these contradictory impulses, the "aggressively-poised male body" is uniquely placed to forge the social cohesion for which Kingsley, Christian Socialist and social reformer, spent much of his life campaigning.

Kingsley's masculine ideal, moreover, was often articulated most directly in his scientific writing. Perhaps the most important of these expressions is found in a passage in *Glaucus* in which Kingsley delineated the ideal temperament of "the perfect naturalist." He "must," Kingsley explained, "like a knight of old,"

> be first of all gentle and courteous, ready and able to ingratiate himself with the poor, the ignorant, and the savage; not only because foreign travel will be often otherwise impossible, but because he knows how much invaluable local information can be only obtained from fishermen, miners, hunters, and tillers of the soil. Next, he should be brave and enterprising, and withal patient and undaunted; not merely in travel, but in investigation; knowing (as Lord Bacon might have put it) that the kingdom of nature, like the kingdom of heaven, must be taken by violence, and that only to those who knock long and earnestly, does the great mother open the doors of her sanctuary.[28]

The "knight of old" must be physically "brave" and "undaunted," possessing stamina enough to survive the hardships of travel and rigorous scientific "investigation"; he must be socially adept, "gentle and courteous" toward "the poor," working in the service of the "ignorant, and the savage" who might most profit from scientific advance; taking the kingdoms "of nature" and "heaven" by "violence," he is both a man of action and a man of God. He is also sexually potent, penetrating the "sanctuary" of the "great mother" nature. As Mark Girouard notes, Kingsley's "knight" was opposed to the celibate male figure of "medieval Christianity," the "feminine" monks who "diseased their minds and hearts by trying to unsex" their bodies. Instead, Kingsley's concept of chivalry "aimed at producing whole men and consecrating their masculinity to God in all activities of normal life." In his integration of social, spiritual, physical, and sexual strength the "perfect naturalist" is "consecrated to God," a "whole" man capable of synchronizing contradiction and fracture into harmonious totality.²⁹

This knightly image of the ideal scientist—physically strong, socially adept, spiritually motivated—was equally central to the self-presentation of the gentlemanly geologists, those "brethren of the hammer, knights errant, a spiritual fraternity in search of a stratigraphical grail."³⁰ We may not be surprised that the ex-military foxhunter Roderick Murchison, whose imperialistic smash-and-grab on the classification of the strata was "both motivated and shaped by the . . . ideals of military honour, national prestige, and territorial expansion," was one of Kingsley's geological heroes.³¹ He is an "illustrious veteran" in *Town Geology*, and he figures as counterpoint to the suggestion that natural history is an "effeminate" pursuit in *Glaucus*.³² There Kingsley wrote:

> There are those who can sympathise with the gallant old Scotch officer . . . who, desperately wounded in the breach at Badajos, and a sharer in all the toils and triumphs of the Peninsular war, could in his old age show a rare sea-weed with as much triumph as his well-earned medals. (37)

This passionate avowal of scientific masculinity through martial rhetoric pervaded scientific thinking throughout much of the Victorian period, so that even the editors of Adam Sedgwick's *Life and Letters* characterized the first rumblings of Sedgwick's debate with Murchison over the classification of the oldest strata as a "skirmish" of two "combatants": it "did not amount to a formal declaration of war" but "was a sort of reconnaissance in force—a massing of troops on the frontiers, to be ready for active service should a need for action present itself."³³ At the same time, the rhetoric of the battlefield was closely connected to an association of geological science with field sports: Murchison himself had spent five years "escap[ing] from insufferable *ennui*" after his military career, becoming "one of the greatest fox-hunters in the north of England," a practice his biographers claimed had sparked his interest in the natural world and had proved his masculine mettle.³⁴ Victorian geology presented itself as a quixotic activity comprising travel, hard physical labor, the chase of the hunt, and the pursuit of truth. And Kingsley, whose lecture on natural history at the public school Wellington College in 1863 blamed a lack of science education for the tactical mistakes made in the Crimean War, and warned that unscientific men would be unsuccessful colonialists, "everlasting pale" "if they be out in a hot country," powerfully assimilated the gentlemen geologists' ideology of the "perfect naturalist" and of the moral, spiritual, and intellectual values he embodied into all his projected answers to social problems.[35]

The image of the "perfect naturalist" performs important cultural work for both Kingsley and the geologists. In their deployments of martial imagery, geologists characterized themselves as men who shared a patriotic faith in the expansion of empire and in the defense of British soil, men among whose number counted military officers who had risked their lives on behalf of such noble aims, and they defined the practice of geology as a masculine, adventurous scene of physical endurance. From there, it was only a small step to suggest that these men were modern-day knights pioneering a new age in which British classificatory systems would be exported across the globe and British geologists would unlock gold, coal, or diamonds from treasure chests buried in domestic and colonial soils. For Kingsley the image of the geologist-knight represented the God-given activity of the geologist, his modern crusade against unenlightened beliefs and views, his pursuit of truth and his quest to become not only the "perfect naturalist" but also an ideal Christian hero. Such an ideology stabilized any latent fears Kingsley may have had about the materialistic implications of the science, for, theological problems aside, geology produced men who were exemplars of Christian manliness and who united, rather than fractured, religious and scientific knowledge.

The geologists' exploits in the field constructed an ideology around the inductive scientific methods of "Lord Bacon," as Kingsley referred to him in *Glaucus*. Bacon was a politic choice as a figurehead for a science seeking cultural authority, for, though "the nature and value of" his "methodology... were hotly contested in nineteenth-century culture," Bacon's name connoted scientific prestige long into the second half of the century, particularly based on his advocacy of fact-collection over the construction of theories.[36] Disavowing the role of hypothesis and theory in scientific methodology, and restricting the investigator to the patient study and exhaustive collection of facts, Bacon offered those geologists who sought to claim public prestige for their work a conservative model, and one that suggested cultural and scientific consensus over geological methods and discoveries. The fictionalized images of the knight-errant geologist and his inductive work in the field worked to ground anxieties about the fragmentary nature of geological evidence and its subsequent need for interpretation, and about the potentially culturally divisive aspect of the science, in an image of a strong, healthy masculinity that could transcend such division through a devoted adherence to the inductive ideal.

Of course, geologists did speculate and theorize and hypothesize. This meant that the constant retelling of tales of physical and mental heroism in the "field" not only linked scientific fieldwork with gentlemanly sports and military exploits, then, but also acted as an ideological guarantee that geology was an inductive science and not a set of speculative and dangerous cosmogonies. Geologists who could monopolize these rugged, gentlemanly outdoor spaces were not armchair theorizers speculating on a world they had not witnessed firsthand—men sitting at the theater or club heretically dreaming up the past. They had proved their scientific mettle in the field, had submitted themselves to arduous forty-mile-a-day expeditions with little food and in all weather, even putting their health at risk, brushing defiantly with death in the pursuit of scientific "truth" and in order to examine the "facts" at close range. The point is clear if we remember Sedgwick's accusation that the *Vestiges of the Natural History of Creation* (1844) was a "fictitious narrative." Such an argument rests on the author's failure to observe "facts" for himself in the field, a failure only excusable in a woman: "The ascent up the hill of science is rugged and thorny," Sedgwick had written, "and ill-fitted for the drapery of a petticoat."[37] Only the strongest, most intellectually able men could participate in geological science, because only they possessed the strength and stamina required to follow inductive methods. The physical trials these men endured evidenced their commitment to the "truth," just as their commitment to the truth confirmed their venerably masculine status and gave them a prestigious and reputable public persona. The "romance of the field" appropriated a culturally accepted story that legitimized geology as little more than the newest object of the age-old knightly quest for truth, romanticizing its appeal. This space offered an imaginative realm in which the methodological and epistemological difficulties that racked geological science might be fully and finally reconciled.

Yeast

Of course, this "romance of the field" did not entirely dispel narrative from the realms of science. Instead, it displaced the "story" from the natural world onto the image of the geologist himself: though "truth" was merely a set of "facts" objectively observed and recorded in the field, free of any distortions or hypothetical links between them, the geologist was a knight-errant participating in an age-old quest narrative. This displacement of "fiction" from accounts of the geological world onto the sturdy image of the geologist-knight is central to

Kingsley's first novel, *Yeast*. In that novel the name of the central protagonist, Lancelot, obviously recalls Arthurian medieval romance, and he spends the novel searching for an anchor on which to make sense of an England whose poor live in execrable conditions, whose aristocracy are being displaced by the rise of democratic politics, and whose religious life is entrenched in embattled conflict between Tractarians and Anglicans. When we first meet Lancelot, then, it is unsurprising that he is "not . . . at all happy":

> First, the hounds would not find; next, he had left half-finished at home a review article on the Silurian System, which he had solemnly promised an abject and beseeching editor to send to post that night; next, he was on the windward side of the cover, and dare not light a cigar; and lastly, his mucous membrane in general was not in the happiest condition. (7)

Suffering from the same "*ennui*" that had been said to have afflicted Murchison before he became a geologist, Lancelot lies indolently in a space that conjures up every sense of the "field" as it pertains to the geologists' masculine ideology.[38] Lancelot's chivalric name and the reference, by way of the Silurian System, to the veteran soldier Murchison conjure up associations with the knightly militaristic field. Kingsley also draws on Murchison's distinguished reputation in nineteenth-century field sports, detailing Lancelot's fruitless attempt to hunt "hounds" that "would not find." Finally, Lancelot's "half-finished . . . review article" transforms the landscape in which he sits into a potential geological field in which inductive scientific research might be undertaken. The passage's confluence of gentlemanly, militaristic, and scientific pursuits casts Lancelot as the hero of a romance narrative, a narrative that embodies the potential for a knightly truth-seeking adventure.

Of course, however, having been raised in an ailing society, Lancelot falls short of the masculine promise implied in his name, just as England falls short of the heroic past Kingsley's allusion to a romanticized medievalism remembers. The bathetic contrast of his storied forename with his surname, "Smith," effectively registers the banality of his current life. Lancelot is only a potential "knight," as the spoilt tone implicit in Kingsley's use of free indirect speech to ironically voice Lancelot's excuse for his indolence, that "his mucous membrane in general was not in the happiest condition," intimates. Furthermore, his "half-finished" review, his unwielded gun, and his unlit cigar indicate more than temporary unhappiness. Phallic objects all, they signify Lancelot's castration anxiety, his fear that he cannot achieve his potential either as an individual or as a man.[39] Indeed, later in the text Lancelot is instructed that in order to become a "whole" man he must disregard the "self-pleasing passions" he has indulged in, for they have made him "inconsistent, dismembered, helpless, purposeless" (354). Surrounded by possibilities, each interpenetrating with the other (hunting, fighting, science), Lancelot is dismembered, unable to reconcile his impulses to each activity within a coherent frame of action. In this opening description of its central protagonist and of an English landscape, we therefore find the entire narrative of *Yeast* writ small: the struggle of youthful masculinity to achieve perfection in a landscape teeming with conflicting alternatives, the need to find an effective channel for masculine strength and intellect, and the sense of lack in both the masculine body and the physical environment Kingsley sees as characteristic of the English condition.[40]

It might be expected that Lancelot's "dismembered" state would arise from a lack of the stamina and strength required in an "inductive" scientist—qualities equally necessary to the geologist in the field, the knight in battle, and the gentleman on the hunt. But Lancelot, though not yet a "whole" man, is resolutely an inductive scientist: "Bacon has taught us to discover the Eternal laws under the outward phenomena," he vociferously proclaims (94), Bacon being a catchword for strict enumerative induction. If the ability to undertake "a hard bit of Baconian induction" is the guarantee of the cohesive masculine identity of a "perfect naturalist," then how is it that the central conflict driving the narrative of *Yeast* is Lancelot's inability to live up to that identity? He possesses the physical, moral, and intellectual capacity to believe in the inductive method. Why then is he so dissatisfied, and so unable to see that method through?

In fact, the Baconian method has led Lancelot to an impasse: "I still dread my own spectacles," he declares toward the novel's close. "I dare not trust myself alone to verify a theory of Murchison's or Lyell's" (361). Unable to verify geological theories by patient attention to "facts," Lancelot is paralyzed by his determination to adhere to the inductive method, to the extent that he sees even his spectacles as a barrier to unimpeded vision. In altering his picture of the world so that he may see better, he fears that his spectacles may instead be distorting his vision of "reality." Though Kingsley's "bit of hard Baconian induction" produces ideal men and is the only appropriate method by which to know the world, here induction alone is not enough, because it does not give a meaningful form to the world it should describe. The plot thus hinges on Lancelot's prevarication about his capacity to observe inductively and about the ways in which he might interpret the facts he sees around him. If even spectacles make induction impossible, how can the real world ever be securely known? And what is the best form through which to realize that knowledge and give it shape and meaning? Until Lancelot can find an answer to these problems, neither his dismembered state, nor the fractured English society in which he lives, can be resolved into cohesion.

Kingsley foregrounds a discussion of "induction" in the final scenes of the novel, the "disarray" in which "the conventions of realistic narrative are abruptly suspended and Lancelot Smith follows the strange prophet-figure Barnakill into a mysterious otherworld."[41] Lancelot's quandary, as Barnakill diagnoses it, is not that he has failed to grasp the inductive method but that he has not extended that method to the study of faith. Barnakill asks him: "Did you never read the 'Novum Organum'?" (360), the text in which Bacon articulated the principles of his inductive method, adding that he has heard that Lancelot is "a geologist." "Facts, and the reasons of them," Barnakill tells Lancelot, "will be as impalpable to you" in the study of faith as they would have been in the study of science, "unless you can again obey your Novum Organum" (361). He continues this adjuration to "see" facts through an inductive method buttressed by faith:

> obey your Novum Organum . . . [renounce] the idols of the race and of the market, of the study and of the theatre. Every national prejudice, every vulgar superstition, every remnant of pedantic system, every sentimental like or dislike, must be left behind you, for the induction of the world problem. You must empty yourself before God will fill you. (361)

This passage is the most lucid account of the particular method of "induction" and its applications that the novel provides. Barnakill here teaches Lancelot a lesson in the "scientific disinterest" necessary to the inductive scientist, a method in which Lancelot must "empty himself" of all "prejudice," "superstition," or "sentimental like or dislike," like any good inductive scientist, leaving his vision clear and unimpeded. But in emptying himself in order to undertake "the induction of the world problem," Lancelot will also render himself open to God. Though, as Levine points out, Victorian novels searching for an epistemological anchor on which to ground their discussions of the "real" world often lead to death, the only state in which "knowing" can be securely divorced from the prejudices and partial superstitions experienced in life, here Lancelot's empty self is not dead but reaffirmed in the eyes of God.[42] Barnakill tells Lancelot:

You took for granted what you read in geological books, and went to the mine and the quarry afterwards. ... According as you found fact correspond to theory, you retained or rejected. Was that implicit faith, or common sense, or common humility and sound induction? (360)

Barnakill's argument here is that "spectacles" need not get in the way of accurate vision, for vision is guided not by the material function of the human eye but by an inner faith that not only enables the "fact" to gain meaning within a spiritual order but also transforms the observer into a man capable of enacting social change. The method of "sound induction," the retention or rejection of "theory" on the examination of "facts," reciprocally makes possible the individual subject's faith in the spiritual order. Each is intrinsic to the other. But this solution itself involves an acknowledgement of the instability of facts. If this denouement is a lengthy exposition of the method required to see facts accurately and to comprehend their meaning, then both the novel's driving problem and its solution are structured around the difficulty of ever knowing the "real" world accurately and of the forms such knowledge could, or should, take. It is precisely because the fact cannot be considered true without verification by a community of independent scientists schooled in a particular method and enlightened by faith in God, that facts and their observation can produce this group of men in the first place. The instability of the fact, the need to find a way of seeing it accurately, is a call to arms to the kind of men who are capable of transforming society. Moreover, because they have learned these complex and arduous methods, have learned to see facts by the shared lights of the inductive method and of God, their status as "ideal men" is, in turn, guaranteed.

The field, as defined by Kingsley and the geologists, is the space in which the fulfilment of the masculine heroic function can take place, the space in which the full meaning of an inductively observed "fact" can be registered. It is also a space in which, like the knight on the battlefield, the geologist risks his life in the noble pursuit of truth, venturing forth to collect specimens and observe strata, and in which he abandons his self-interest in order to honestly and accurately determine the facts. In the field both the knight and the geologist subordinate the claims of the self to the needs of the community and the God they serve. This is the perilous space of self-sacrifice in which, paradoxically, the masculine self is given meaning as an agent within a nation, a community, and a spiritual order. In the obliteration or emptying of the self, the "perfect naturalist" makes his body the site at which multiple seemingly contradictory cultural values may coalesce, and thus a fractured male body, and through it a fractured society, is suddenly made whole.

This has considerable impact on the narrative form of the novel. As Kingsley explained in the 1851 epilogue to *Yeast*, "the fragmentary and unconnected form of the book" is "an integral feature of the subject itself [the condition of England], and therefore the very form the book should take" (365). And the landscapes of the novel, in England and the East, are not merely descriptive backdrops to the main action, but are symbolic of the epistemological problems around which the text is constructed. As Barnakill tells Lancelot, "Henceforth you will study, not Nature, but Him." He continues:

> Yet as for place, I do not like your English primitive formations, where earth, worn out with struggling, has fallen wearily asleep. No, you shall rather come to Asia, the oldest and yet the youngest continent,—to our volcanic mountain ranges, where her bosom still heaves with the creative energy of youth, around the primeval cradle of the most ancient race of men ... where men have learnt to tame and use alike the volcano and the human heart, where the body and the spirit, the beautiful and the useful, the human and the divine, are no longer separate. (346–47)

Narrative, or at least the narrative of conversion, which Lancelot must undergo, is no longer possible in this England. It is "asleep," lying dormant and decadent atop a much more restless and energetic past, out of connection with the violent history that produced it. There is no perilous space in which the English male can adventure, in which his physical and moral strength may be tested. The landscape is "worn out with struggling," and the culture this has produced is fragmentary and chaotic, unable to be reconciled into narrative, unable to give Lancelot a space in which he can convert to a unified perception of God and nature. In some ways, Lancelot's journey to the East into an unnarratable "otherworld" is an indictment of English culture and its inability to produce change at the particular moment of the mid-nineteenth century, for change itself—a central component of narrative—is impossible without venturing outside this landscape.

In this passage Kingsley also denies the narrativity inherent in the concept of conversion, its transformation of an individual from one state into another. For Lancelot, conversion is not a transformation into another state but the acknowledgement of what he already knows: "Turn ... to Him whom you have known all along, and disregarded," Barnakill admonishes (352). Not only that, but in the Eastern landscape to which Lancelot will travel, not only are the "volcano" and the "human heart," the natural and spiritual worlds, unified, but so too are different moments in human and earth history, for "the most ancient races of men" are gathered around a volcano that symbolizes the earliest stages of geological history, the violence and "creative energy of youth." The Eastern landscape is an expression of the "field" and its values *in extremis*, a dangerous space reachable only by arduous travel, in which masculinity may be tested, strengthened, purged, and remade whole. But it also emblematizes ahistoricity, collapsing the conflict that motivates Lancelot's journey into a static symbol of wholeness. In doing so it represents the endpoint of plot, the place in which the inductively observed "fact" is given a meaning that exists outside the materialist and unscientific stories so derided by the gentlemanly geologists, in which the "fact" and the methods used to achieve it derive their meaning from a transcendent Godhead rather than from their position within a connected and meaningful plot. The unraveling of a plot relies on conflict and transformation. The England of *Yeast* is so conflict-ridden that it has become paralyzed, unable to produce the kinds of men who can generate change at all. The East, by contrast, symbolizes a space unified, whole, and complete, in which there is no conflict, no need for transformation, and in which Lancelot will be converted only to something he already knows. As such, the lurch to the East with which the novel concludes represents a deliberate evasion of plot as a form for deciphering the truth.

A study of Kingsley's engagement with the ideologies of the gentlemanly geologists renders simplistic correlations between the methods of literary realism and the "inductive" scientific method redundant. Kingsley is not merely enumerating "factual details" or making "statements" in *Yeast* but utilizing fictional form to explore the epistemological basis on which the fact itself may stand and searching for a form for the authentic transcription of the unity he perceives in the moral, spiritual, and natural worlds. The novel can never provide him with the unity of form he seeks. This explains Kingsley's residual discomfort with novelistic form and his reluctance to assimilate the "facts" the novel yields into an all-encompassing narrative that might sugarcoat discrepancies and smoothe contradictions between those facts into the soothing pleasures of plots, instead of revealing them and laying them bare. Stabilizing factual knowledge within a spiritual schema, rather than a narrative order, Kingsley's novel stages a complex epistemological discussion about the nature of the fact and the forms in which it might be adequately known.

Alton Locke

Kingsley returned to the spiritual and geological Eastern landscape in his next novel, *Alton Locke*. In this case he does narrate the transformation of his eponymous hero, though Alton does not visit the Himalayan landscape but rather dreams of it during a fever. During this dream Alton imagines himself as "a madrepore, the lowest point of created life," and moves through various geological epochs, becoming, in turn, a remora, a crab, a flying fish, a mylodon, a monkey, and a tribesman journeying from "the valleys of Thibet," the "primeval cradle of man," bringing religion to the West.[43] As he moves through this geological past he is converted from a misguided Chartist into a Christian man able to write his autobiography as a moral exemplar to youthful working-class men bent on transforming their society. Alton effectively becomes a narrator through this experience, because it is this dream that prompts his conversion and forms the subject of his first foray into prose writing.

This "plot" of self-transformation through the geological past, moreover, has been consistently interpreted by critics working within Beer's critical framework as a recapitulation of the "plot" of evolutionary development or, as several critic have recently suggested, as a developmental narrative based on the transmutationist text *Vestiges of the Natural History of Creation* (1844).[44] Simultaneously this dream sequence, like the mystical ending of *Yeast*, is most often interpreted as nonrealist, taking us "far beyond the boundaries of . . . realistic fiction" as it seeks to make sense of the facts it has exposed in a hitherto "realist" fashion.[45] So Kingsley's *Alton Locke* is both a novel in which the most important narrative movement, the transformation of Alton from a radical Chartist into a Christian male and a narrator, has been understood as "proto-evolutionary" and a novel in which Kingsley's "use" of science has been said to lead not to realism but to fantasy.

Kingsley's voice seems to have been lost in these accounts, however. In fact, it is impossible to interpret the dream sequence of *Alton Locke* within an evolutionary framework, not only because was the novel written nine years before the publication of Darwin's *Origin of Species*, but because in the 1850s Kingsley, as an aspirant gentlemanly geologist, vehemently rebutted *Vestiges* and its theories in a series of private letters to Thomas Cooper, the autodidact shoemaker-poet on whom the figure of Alton was based.[46] Cooper was the illegitimate son of dyers, and had taught himself Latin, Greek, and French and to recite inordinate quantities of poetry. A radical novelist, poet, and newspaper editor he had also been imprisoned after his participation in a Chartist riot in the Potteries in 1842. By 1854 he was about to embark on a new career as an itinerant religious lecturer, and he wrote to Kingsley asking him to recommend appropriate scientific texts for a series of Sunday lectures he was to give on theism at the London Hall of Science. Kingsley counseled him on how to present his striking views in a form and tone that would engage, rather than alienate, a broad spectrum of listeners, and he specifically warned him against citing *Vestiges*. It was the "method" of Baconian induction "whereby . . . science has triumphed," he reminded him, rather than via the unmethodical speculations of an unknown author. Of course, Cooper's lectures were politically sensitive, and to claim any kind of authority to pronounce on religious matters he would need to avoid accusations of materialism and radicalism even more scrupulously than the gentlemen of the Geological Society. Texts like *Vestiges*, discredited by the gentlemanly geological community, would have been tactically imprudent choices for a lecturer already on politically sensitive ground, especially one whose scientific training had not been received in the hallowed institutions of the metropolis or the universities.

But Kingsley was not simply playing it safe by recommending that Cooper avoid *Vestiges*. As *Yeast, Alton Locke, Town Geology*, and *Glaucus* amply demonstrate, Kingsley fervently believed in the value of the search for "empirical truth in nature," responding "to the didactic and exhortational value of each and every fact," and not in theories based on demonstrably shaky evidence.[47] And Kingsley could draw on authoritative scientific precedent in his rejection of the *Vestiges*: "Sedgwick's 'Notes to his University Studies,' containing his refutation of the 'Vestiges of Creation,'" is recommended, as is the "invaluable 'Footprints of the Creation,' a corroboration of Paley against the 'Vestiges'" and therefore "useful beyond any modern book" for Cooper's needs.[48] Three years later, in the novel *Two Years Ago* (1857), Kingsley imagined the psychological consequences the method of *Vestiges* might produce: when the suitably masculine hero Tom Thurnall asks the Spasmodic poet Elsley Vavasour "why you poets don't take to the microscope, and tell us a little more about the wonderful things which are here already, and not about those which are not," echoing the author's own sentiments. Elsley's response is to ask, "You believe, then, in the development theory of the *Vestiges*?"[49] Tom replies, in characteristically Kingsleyan fashion, "Doctors who have their bread to earn never commit themselves to theories" (266). Tom's reticence to avow or disavow theory reflects his (and Kingsley's) commitment to induction as the only safe method by which knowledge can be collected and trusted. Elsley's preference for "the old Cosmogonies, the Eddas and the Vedas" (266), as fit subject for poetry rather than the "wonderful things of the microscope" leads to his psychological breakdown atop Mount Snowdon and culminates in opium addiction and death. *Vestiges* was to be treated with suspicion, and its enthusiasts were imagined by Kingsley as effeminate, melodramatic, and psychologically unstable, contrasted with the manly inductive ideal embodied by Tom.[50] Fiction here performs the work of scientific disputation, personifying theoretical arguments in the imagined male bodies of Kingsley's characters.

Perhaps more importantly, we might note that if citing *Vestiges* would have been a tactical mistake for Cooper, it would have been an equally foolish choice as a source for Kingsley in his fictional presentation of a working-class autodidact interested in science to the fiction-reading public. If Alton was converted to God by radical, potentially atheistic transmutationist science, then Kingsley's novel would have been much less likely to make political ground for the working classes whose plight he had attempted to document.

Instead of exploring Kingsley's evolutionary sympathies, I want to focus on his concept of the scientific "field" as it structures *Alton Locke* and its seemingly unrealistic dreamland. The episode points to a significant problem with the nature of "induction" as supported by the knightly ideal. Lancelot Smith, of course, like the gentlemanly geologists, had the time and leisure to spend days and weeks in the field verifying "a theory to spend days and weeks in the field verifying "a theory of Murchison's or Lyell's," and regularly traveled throughout the British Isles, Europe, and beyond in order to undertake fieldwork. Such travels and terrains were clearly part of the "romance of the field" and its mythic identity. But in *Alton Locke*, Kingsley asks the following question: what if the "field," the symbolic sphere of masculine action, is not accessible to all men? A London tailor did not have the time or money to travel to the Alps, for instance, or to the distant mountains of Tibet. Could only a gentleman like Lancelot become a "whole" man? Or was "perfect" masculinity also available to the poor, the disenfranchised, and men with bodies weakened by hardship and deprivation, like Alton? If men like Lancelot were to channel their energies into the "induction of the world problem," to analyze the causes of poverty and disease and to transform the society in which they lived, then how were working men like Alton Locke, often the victims of such poverty, to participate in the transformation? In Kingsley's exploration of the mythologized scientific space of the "field," and the gentlemanly ideology that underpinned it, this becomes the central formal problem of the novel itself.

The People's Science

Kingsley often promoted geology as a particularly apposite study for working men. In *Town Geology*, echoing the conventional rhetoric of many a geological textbook, he singles out geology as "specially, the poor man's science" (xi).[51] "The simplest and the easiest of all physical sciences," "it appeals more than any to mere common sense," "requires fewer difficult experiments, and expensive apparatus," "requires less previous knowledge of other sciences," and "is more free from long and puzzling Greek and Latin words" (p xi). He continues:

> The most important facts of geology ... may be studied in every bank, every grot, every quarry, every railway-cutting, by anyone who has eyes and common sense ... geology is (or ought to be), in popular parlance, the people's science—the science by studying which, the man ignorant of Latin, Greek, mathematics, scientific chemistry, can yet become—as far as his brain enables him—a truly scientific man. (3–4)

Here Kingsley redefines the spaces that constitute the "field" for men who do not have the leisure or resources to make scientific excursions. For those who have nothing but "eyes and common sense" the field may be the "bank," the "grot," the "quarry," or the "railway-cutting," spaces in which Kingsley's implied readers may even have worked. So in *Town Geology* even indoor, urban, and built spaces—wherever there is access to "the pebbles in the street," "the stones in the wall," "the slates in the roof"—become spaces in which field research may take place. In extending the concept of the field to include such spaces, Kingsley attempts to democratize it, but in doing so he transforms the "field" from the physically real landscape into a metaphor for the world that opens up to the geologist only once he has learned to see it anew. The field, then, becomes a space available to any man capable of seeing geological significances in the world around him, regardless of the material conditions that might constrain him. Furthermore, it is implied that only the "truly scientific man" can appreciate the potential for the field to exist in this way. So the field is democratic not only because it is a state of mind available to all rather than a physical reality, but also because only he who is democratic in spirit can appreciate the plethora of spaces in which geology might take place and can become, in turn, "a truly scientific man." Beginning by attempting to remove the practical barriers to egalitarian participation in the natural sciences, Kingsley ends up transforming the physical reality of the field into a utopian symbol of the scientific spirit. The irony is, of course, that in his fictions he deviates from realism solely in order to preserve his belief in inductive science as an authoritative and objective route to truth.

But this vision of the field is also envisioned as a pragmatic solution to real social problems. In *Town Geology*, Kingsley advocates the study of geology for men like Alton Locke "on social, I had almost said on political grounds," because

> I have known again and again working men who in the midst of smoky cities have kept their bodies, their minds, and their hearts healthy and pure by going out into the country at odd hours, and making collections of fossils, plants, insects, birds, or some other objects of natural history. (xxvi, xii)

Town-dwelling men are better citizens, Kingsley argues, if they are also collectors of "objects of natural history," exposed to the purifying and invigorating influence of the field. But amid the familiar correlations Kingsley makes between nature and health, and between the city, smoke, and corruption, there is a point of stress in this passage in the phrase "at odd hours." In his casual tone Kingsley figures natural history as a leisured activity for the working man, who simply pops into the country every now and again to collect specimens. But this breeziness cannot entirely conceal the irony that *only* "odd hours" are available to working men, who spend their lives toiling for little pay "in the midst of" the self-same "smoky cities" Kingsley hopes they might escape. Kingsley's implicit (though suppressed) acknowledgment of the socioeconomic barriers that fence off the "field" from working men threatens to disrupt his picture of the utopian social and political harmony natural science makes possible.

In *Alton Locke*, Kingsley's belief that geology is a democratic pursuit available to all men presents him with several formal and representational difficulties. Though early in the novel Alton tells us that "the narrowness of my sphere of observation [in the London slums] only concentrated the faculty into greater strength," still, unlike Lancelot Smith, he does not compose reviews on the Silurian System, "half-finished" or not, and he cannot voyage bodily to Asia in order to practice those observational powers.[52] For him, the East is a region it was "the longing of my life to behold," one he has only known through the mediating forms of print or painting. Throughout the novel his "fascination with nature and natural history" and the "philosophy of Bacon" are repeatedly emphasized, as Vance notes.[53] But this fascination is constantly mediated by such forms: he encountered only a "few natural objects" (7) in his childhood in the London slums, such as leaves, cabbages, clouds, and water, objects that he "pored over ... not in the spirit of a naturalist, but of a poet" (8), and he must be content with the reading of missionary tracts describing "pacific coral islands and volcanoes" (14) for contact with the world outside the city. Alton cannot be inductive as Lancelot can in the early part of his life, and the implicit reason, never fully articulated in the text, is that "the spirit of a naturalist" is an ideal only achievable when one is without financial barriers. "The spirit of a naturalist," as much as Kingsley argues for its classless, timeless status and the "inductive" method through which truth can be achieved, means an emptying of the self, as Lancelot Smith learned. How can a poor man struggling to survive renounce the claims of the self entirely? Furthermore, how can he get outside the spaces of work and necessity and into the "field"?

The problem remains even when Alton travels out from the "stagnant cesspools" of the city into the Cambridgeshire countryside and secures work as a geologist's assistant for the aristocratic Dean Winnstay, "arranging and naming" (263) the dean's "room lined with cabinets of curiosities . . . hung all over with strange horns, bones, and slabs of fossils" (240), and "writing out answers to the dean's searching questions" on a geological pamphlet. Writing and arranging, we are reminded here, were equally considered to be geological practices. Dean Winnstay, despite being an aristocratic naturalist, giving "oration[s] on the geology of Upper Egypt" (239), and serving as a man of God, sees science and faith as essentially separate:

> He seemed to ignore utterly anything like religion, or even the very notion of God, in his chains of argument. Nature was spoken of as the willer and producer of all the marvels which he describes; and every word in the book, to my astonishment, might have been written just as easily by an Atheist as by a dignitary of the Church of England. (249)

Winnstay teaches Alton science, anticipating Tom Thurnall's preference for a scientific poetics when he tells Alton that "science has made vast strides, and introduced entirely new modes of looking at nature, and poets must live up to the age" (246). As he puts it, echoing Kingsley's own sentiments, "in the study" of natural history Alton will achieve "a mental discipline superior even to that which language and mathematics give" (242). But dissociating science and religion, and urging Alton to "eschew politics, once and for all, as I have done" in his poetry (266), he cannot teach Alton his final lesson. In this poetic lesson he advocates that Alton "learn "the truth of Jarno's saying in *Wilhelm meister* when he was wandering alone in the Alps, with his geological hammer, 'These rocks, at least, tell me no lies, as men do'" (266). But as Catherine Gallagher points out, for Alton this journey is impossible. He lives surrounded by charnel houses, sweatshops, and slums.[54] Bounding over mountains with a geological hammer like Goethe's Jarno, then, remains an unlikely dream for Alton, a working man, a tailor, who cannot see nature in these Romantic terms. Furthermore, in encouraging the dissociation of the political and the poetic, the scientific and the spiritual, the dean simply perpetuates divisions created by the selfish claims made by individual parties and groups (such as Chartists) and reinvigorates barriers to a "whole" and "consecrated" masculinity, which Alton must achieve and which English society is presented as lacking. Even when Alton voyages into the countryside, then, his acquaintance with geology and with natural forms is mediated, subordinated to the social and political views of the dean on whose employ he depends. Alton is too poor to be free of this employment, too poor and uneducated to become the emptied self of the "perfect naturalist." So Kingsley, ever the "negotiator," must find a space in which material concerns are irrelevant to the knowledge produced there, in which the image of the classless "perfect naturalist" may be sustained.[55]

The paradox becomes, then, that in order for Alton to witness "nature" firsthand, for his narrative to be credible as one based on inductive methods, Kingsley must step outside the bounds of realist convention and take Alton to the field through dream. The dreamspace is simultaneously nonrealist and a space in which Alton can finally be an inductive scientist, witnessing the world firsthand rather than through books or Chartist pamphlets or in the dean's private study. Almost dead through fever, Alton can finally renounce the pressing claims of the self in this space, submitting his body to the physical trials of the questing knight. Only through a dream can the "real" finally be witnessed in its entirety by a working man, whose view is otherwise occluded by his material wants and concerns.

The dream sequence, like the mystical ending of *Yeast*, involves the hero journeying to the "volcanic mountain ranges" of the East, where the "bosom" of earth "still heaves with the creative energy of youth, around the primeval cradle of the most ancient races of men."[56] Before the sequence begins, in fact, Alton justifies his involvement in the Chartist movement through reference to the language of the volcano: "To understand the maddening allurement of that dream," he explains, "you must have lain, like us, for years in darkness and the pit . . . you must have sat in darkness and the shadow of death, till you are ready to welcome any ray of light, even though it should be the glare of a volcano."[57] Both the language of the dream and the glare of the volcano are repeated at the beginning of his fever:

> An enormous, unutterable weight seemed to lie upon me. The bed-clothes grew and grew before me, and upon me, into a vast mountain, millions of miles in height. Then it seemed all glowing red, like the cone of a volcano. I heard the roaring of the fires within, the rattling of the cinders down the heaving slope. A river ran from its summit; and up that river-bed it seemed I was doomed to climb and climb for ever, millions and millions of miles upwards, against the rushing stream. . . . A raging thirst had seised me. I tried to drink the river-water, but it was boiling hot—sulphureous—reeking of putrefaction. Suddenly I fancied that I could pass round the foot of the mountain; and jumbling, as madmen will, the sublime and the ridiculous, I sprang up to go round the foot of my bed, which was the mountain. (211–12)

Associating his fever with the volcanic glare of April 10, the day of the failed Chartist march in London, Alton's political volcano is transposed into his body through the "raging thirst" and "boiling" heat of his fever. Just as the dream of Chartism is literalized into the physical dream Alton experiences during his illness, here the metaphoric volcano of political revolution comes closer to literalization, not only because his bedclothes begin to assume a volcanic appearance, but also because the symptoms of the fever itself, with its "rushing stream" of red-hot sweats, mimic volcanic action. Alton internalizes the failed revolution into his body so that its violence is written not as a new, desired-for political order but into a physical disorder. What would have been a moral violation of the natural order through political uprising becomes a physical violation through disease. But even more importantly the movement is one from the world of politics into the body. At this moment Alton's body explicitly becomes the text on which the novel's meanings will be written. The body is the appropriate form for spiritual, natural, and psychological reintegration.

Like Lancelot, Alton must "learn what it is to be a man," and he must do so in a volcanic space that he, like Barnakill, describes as the "cradle of mankind," the space in which primitive religious impulses, uncorrupted by centuries of "civilization," are written onto a turbulent landscape. As in the correlation between the metaphoric volcano and Alton's literal fever, there is a peculiar closeness between metaphor and literalness in this passage: underpinning Alton's vision of the "cradle of mankind" is the dogmatic presence of his mother "seizing the pillars of the portico" (213) and causing earthquake and tornado, choking, blinding, and burying him (214). He must endure the physical violence of the dreamworld even as it forces him to repeat the psychologically traumatic events of his past. Physical and psychological planes of "reality" coincide in this passage and manifest themselves in his still-suffering body. Furthermore, the fact of Alton's physical suffering itself enables his access to that most important of functions of the geologic "field": it forces him, through physical trial, to subordinate the claims of the self to a higher truth. The element of physical suffering, of near self-annihilation, is crucial to the strategy of the geologist and the claims to descriptive truth he will make. That he also suffers psychologically does not undermine his ability to see the "real" world before him, but reinforces his masculine ability to survive the traumatic experience of seeing "reality" undisguised.

These moments of correlation between physical, psychological, and political suffering, signified by the multiple shifts between levels of metaphor and literalness, make possible the final reinterpretation of the volcano, which transforms it from a political and then a physical image into an icon of spiritual belief. The volcano that began the "dreamland" passage also ends it, as Alton returns to "the valleys of Thibet" and the "primeval cradle of man" (224), becoming one of its westward-moving tribesmen, bringing early religion to the world. The dreadful volcano of the beginning of the dream, with its "unutterable weight" and its "millions of miles in height," is replaced by an accommodating "snow-capped" mountain, cool and refreshing, its "snow-peaks ... seeming to beckon upwards, outwards," giving Alton the feeling of "strange unspoken aspirations, instincts which pointed to unfulfilled powers, a mighty destiny" (225). Here again there is a shift between directly referential and symbolic language: the physical world of the mountaintops bears a metonymic relationship to the spiritual "mighty destiny" and "unfulfilled powers" they seem both to point toward and to beckon Alton to reach. Instead of the glare of volcanic light, or the "boiling-hot—sulphurous ... putrefaction" of volcanic lava, we get "the tribes of the Holy Mountain" who "poured out like water to replenish the earth and subdue it—lava-streams from the crater of that great soul-volcano" (226). The violence of the first volcano and its manifestation in Alton's body is transformed into a "soul-volcano," where physical suffering has been reinterpreted as spiritual nourishment. And the imperious maternal force Alton has remembered by linking "the cradle of mankind" with his guilt over his own mother is transmuted into the paternalistic benevolence of "an All-Father, whose eyes looked down upon us from among those stars above; whose hand upheld the mountain roots below us" (225). The maternal image remains viscerally physical, dovetailing with the corruptions of the flesh that Alton experiences in fever, indicating womblike suffocation and destroying the "mountain roots below us" in earthquake and tornado. By contrast the All-Father is not physical but spiritual. The son of an All-Father, Lancelot is emancipated from the nurturing but constrictive forces of the mother and becomes part of a spiritual brotherhood of the sons of God.

The multiple linguistic shifts of the passage, and Kingsley's frequent play between figurative and referential modes, ultimately enable this constant reinterpretation of the significance of the volcano and its all-inclusive final meaning as a spiritual and physical space, just as Alton, through enormous physical suffering, comes to accept his "mighty destiny" within a higher spiritual order. In quivering between "real" and symbolic spheres of meaning, as well as between scientific and religious modes of reference, Kingsley facilitates a sense of cohesion between the spiritual, intellectual, political, physical, and emotional aspects of being that have been represented as disconnected and fragmented throughout the novel, and which have been taken as indicative of the "eclectic" and "dismembered" Babel of modern England.

The meaning of the volcano in Alton's dream shifts meaning according to the consciousness of the observer. It is in these shifts that we can see that Kingsley is not enacting a developmental transmutation of his hero based on the *Vestiges* or other preevolutionary texts and theories but, rather, resituating progressive narrative by locating it in the mind of the observer. The natural world has no story to tell, but human consciousness can shift to an ever more expansive vision of its meaning. So geological epochs spin past Alton's consciousness, each expanding his viewpoint and altering his perspective, until he finally achieves the whole picture. This is not an enactment of a geological or biological plot in literature but is in fact an instance of a well-known, often-used genre of writing frequently found in geological texts. Many geological books written for broad, nonspecialist readerships included geological "retrospects" at the end of their texts, just as this retrospect ends *Alton Locke*. The retrospective "procession" or "pageant" of a portion of the geological past, a fictional imagining of how the anatomical or palaeontological or geological "facts" elucidated in the preceding chapter might appear if the geological past was suddenly reconstructed, was a standard mode of geological writing, its authors including Hugh Miller, Gideon Mantell, and William Buckland. It was likely to have been familiar to Kingsley's readers in 1850, as it was to

Kingsley himself.[58] Such forms often presented a multitude of geological epochs, moving from past to present (and sometimes to future) without asserting any "genealogical links[s]" between the life forms found there. The species represented in each successive scene do not develop out of one another. In fact, the representation of geology through these shifting, panoramic scenes of the earth actually enabled the geological writer or artist to imagine the geological past without making connective narrative links between the different epochs and images he was presenting: the scenes project "earth history as a discontinuous succession of scenes rather than as a gradually evolving narrative."[59] Martin Rudwick writes that "the earlier scenes from deep time, with few exceptions, remained isolated from each other. They were single scenes based on various parts of the geological succession," often chosen to illuminate "particular fossil discoveries" but "not deliberately selected as representative of the successive periods of earth history."[60] The violent effects of tornadoes, earthquakes, and volcanoes, which partition the "succession of scenes" in Kingsley's text, further mark the "discontinuous" nature of its movement, as do the lines of asterisks printed between scenes (see fig. 5.1). Together these violent geological interludes and typographical demarcations imply a crashing division between each scenes and emphasize the fragmentary, discontinuous nature of the dreamworld.

Kingsley's "dream" sequence, then, is not a recapitulation of an evolutionary narrative but a geological retrospect, and it performs the same functions as the retrospect common in geological writing. It offers a glimpse of narrative after a presentation of ostensibly factual information, a carefully delineated series of scenes in which the "science" of describing geological data or of describing the events and conditions of the life of a working-class poet, may be given meaning. It is a virtuoso ending offering both the pleasures of narrative and the scientific repudiation of the geological story, a spectacular device for imagining a scene from the deepest past in which the meaning and significance of the preceding passages and chapters is revealed, but one that refuses an overarching plot that makes sense of the whole. The retrospect specifically enabled geological writers, including Kingsley, to tell tales of the geological past without subsuming them into a materialist or speculative story.

The form also functions to assuage potential readers' fears that a working-class hero like Alton might threaten the existing social order. Tacking the telling of Alton's self-transformation onto a developmental or evolutionary narrative of geological change, Kingsley would have been unable to imagine Alton as a hero: Alton could not reach God through the pleasing shapes of

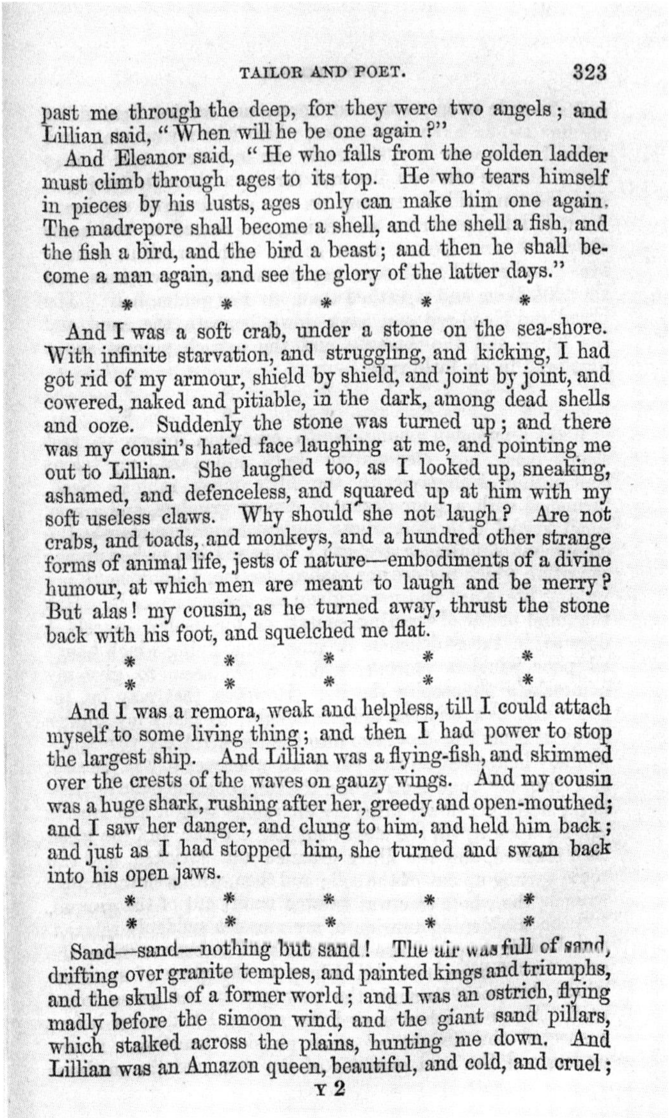

TAILOR AND POET.

past me through the deep, for they were two angels; and Lillian said, "When will he be one again?"

And Eleanor said, "He who falls from the golden ladder must climb through ages to its top. He who tears himself in pieces by his lusts, ages only can make him one again. The madrepore shall become a shell, and the shell a fish, and the fish a bird, and the bird a beast; and then he shall become a man again, and see the glory of the latter days."

* * * * *

And I was a soft crab, under a stone on the sea-shore. With infinite starvation, and struggling, and kicking, I had got rid of my armour, shield by shield, and joint by joint, and cowered, naked and pitiable, in the dark, among dead shells and ooze. Suddenly the stone was turned up; and there was my cousin's hated face laughing at me, and pointing me out to Lillian. She laughed too, as I looked up, sneaking, ashamed, and defenceless, and squared up at him with my soft useless claws. Why should she not laugh? Are not crabs, and toads, and monkeys, and a hundred other strange forms of animal life, jests of nature—embodiments of a divine humour, at which men are meant to laugh and be merry? But alas! my cousin, as he turned away, thrust the stone back with his foot, and squelched me flat.

* * * * *
* * * * *

And I was a remora, weak and helpless, till I could attach myself to some living thing; and then I had power to stop the largest ship. And Lillian was a flying-fish, and skimmed over the crests of the waves on gauzy wings. And my cousin was a huge shark, rushing after her, greedy and open-mouthed; and I saw her danger, and clung to him, and held him back; and just as I had stopped him, she turned and swam back into his open jaws.

* * * * *
* * * * *

Sand—sand—nothing but sand! The air was full of sand, drifting over granite temples, and painted kings and triumphs, and the skulls of a former world; and I was an ostrich, flying madly before the simoon wind, and the giant sand pillars, which stalked across the plains, hunting me down. And Lillian was an Amazon queen, beautiful, and cold, and cruel;

Y 2

Fig. 5.1. Charles Kingsley, *Alton Locke* (1850), 323.

evolutionary narrative, for this was exactly the form associated with atheism and the French Revolution, the notion that one class may evolve and dominate the others. If the working classes could develop into equality with their social superiors, then there was no firm basis on which rigid taxonomical social structures could be said to be based. But if, in witnessing a geological retrospect or succession of panoramic scenes, the working-class man could experience the evidences of the geological past firsthand, he could achieve an appreciation of the inductive methods that otherwise elude him, without the disquieting suggestion that the intellectual appreciation of science would eradicate social hierarchies and dissolve social distinctions.

As Ralph O'Connor notes, the retrospect was also often used by Hugh Miller, a working-class stonemason-turned-journalist who was not of the band of gentlemanly geologists but made significant contributions to geological discourse as the author of such works as *The Old Red Sandstone* (1841) and *Testimony of the Rocks* (1857).[61] Kingsley would commemorate him in *Town Geology* as "the late illustrious Hugh Miller, who made himself a great geologist out of a poor stonemason," exhorting his imagined working-class readers to follow Miller's example.[62] Kingsley moves from attempting to teach Alton the inductive methods of the geological scientist through the elite discourses and processes exemplified by Dean Winnstay and into the particular genre of the geological retrospect, and the conclusion of the novel is ultimately different for this: Alton cannot be assimilated into elite culture, and his difference from the elite cannot be leveled. For Alton the retrospect is accessible, with its restrained (though still present) impulse to narrative and spectacle, its fictionality, and its imaginative possibility of conclusion, and this, paradoxically, is the only way he can achieve the vision of the truly inductive scientist.

The retrospect that concludes *Alton Locke* thereby enables Kingsley and his readers to imagine the spiritual transformation of their hero. But it also participates in the gentlemanly geologists' prevarication about plot, narrative, and their relation to "reality" by placing far less emphasis on creating a pattern for the past than on simply visualizing it free of explicit narrative links. In this way, though the novel's recourse to a dreamworld in order to reach its explanations and conclusions is clearly not a straightforwardly "realist" technique, still it cannot be said to signify the failure of the realist novelist to find an appropriate narrative shape in which to frame his text. Instead of instantiating the idea that "realistic description could not provide material for plots," the dream sequence at the end of *Alton Locke* foregrounds the problematic of plot when used to formalize the natural world.

Devon Asleep

The final Kingsley text under discussion here is the 1857 novel *Two Years Ago*, in which the enigmatic figure Barnakill returns, and in which Kingsley's geological fictional form, the form of breakdown, is translated into something more constructive. In this, in my view Kingsley's most formally successful fiction, the layered blocks of rock and fossils known as the strata give shape to the novel's events. Here the geological map is the form Kingsley chooses for his truth-telling.

In 1884 *Murray's Handbook to Yorkshire* stated that "the only county which can at all rival" Yorkshire for geological richness was Devonshire.[63] Here again was a rich site of geological exploration, marked by dramatic outcrops and coastal cliffs, traversable by the individual geologist, and rich in fossil remains. By the time Kingsley wrote *Two Years Ago* it had also become classic ground for geological mapmakers. Its rocks, for the most part, were older than the secondary, coal-bearing rocks of Yorkshire, but as in Yorkshire, it was the precise relation to coal that made it the site of unprecedented geological scrutiny in the 1830s and 1840s.

The strata under dispute were widely known, at the beginning of the debate, as "greywacke" strata, after the most characteristic rock in the group, though they were also made up of other rocks, including different kinds of limestone and shale. These rocks lay between the secondary strata (which lay in usually conformably horizontal beds, were rich in fossils, and contained the coal measures) and the very oldest rocks (which appeared not to be stratified or to contain any fossils and had often undergone metamorphosis so that their original composition was difficult or impossible to detect). The greywacke rocks between these layers were highly folded, faulted, and out of sequence, but they were of economic significance because they lay beneath the coal that was so important to the British economy. Mapping this confusing mass of rocks might help define the base limit in the search for coal. Accordingly, the first project of the Geological Survey, founded in 1835 to map the counties of England at the detailed scale of six-inches-to-the-mile, began in Devon, the home county of its director, Henry De la Beche (see plate 9).

At the same time that De la Beche began coloring in the Ordnance Survey maps of Devon in geological colors, Sedgwick and Murchison turned their attention to these rocks as they occurred in Wales. Sedgwick began a tour of North Wales, concentrating on the older greywacke strata around Snowdon; Murchison set out from the borderlands in South Wales near Shropshire, working on the newer greywacke, and the two men met in the middle, identifying a sequence of relative ages for the greywacke that stretched from north to south through the kingdom. The general consensus in Devon was that the rocks there were consistent with the older or "lower greywacke" deposits that Sedgwick had identified in North Wales. Accordingly, De la Beche classified a deposit of anthracite, or "culm," as it was called in local dialect, at Bideford in the north of Devon, as part of the "lower greywacke" system, making coal integral to the greywackes. For Murchison, this was unacceptable, because he wanted the "upper" greywacke deposits he was now claiming as his own property to operate as the base limit in the search for coal.

By 1836 the debate had reached a stalemate, partly because the structure of the deposits in Devon could not be adequately ascertained. If the deposits sat in a trough, as De la Beche thought, it was possible that the culm was older than Murchison's upper greywacke; if they did not, the culm could be considered to be sitting atop the upper greywacke and to be younger than it. From the available outcrops, and from the order of the rocks themselves, it was very difficult to tell if there had been local overturning or what the overall structure might be, and the fossil evidence was inconclusive at best, suggesting that the whole deposit at Bideford was Carboniferous in age, so that the rocks of the whole county might be younger than anybody on any side of the debate had yet supposed. In 1839 Sedgwick and Murchison went to the Eifel region in the Rhine Province, another region well known for its greywacke rocks and fossils, to look for a correlation with the rocks in Devon. After several disagreements about the complex structure of the region and its fossils, they moved not only the culm but the entire sequence of the Devon lower greywacke to a much more recent position equivalent to the oldest secondary strata, typified by the old red sandstones best known in Scotland, so that Devon became the site of an intermediate age that linked the secondary and the older rocks, between the coal and Murchison's upper greywacke deposits (see figs. 5.2 and 5.3). This formation they called the Devonian, despite the fact that the evidence for its intermediate position in the Devon strata was actually rather weak.[64]

Part of what is interesting about this debate is the way in which, as fossils became increasingly more important as evidence for deciphering the order of the strata, topographical locations replaced rock types as names for periods of earth history or sections of the stratigraphic column. The Devonian was not just a description of the rocks and fossils of Devon, but was applied to landscapes in Germany and then across the world deemed to be of the same age. During the course of these debates Sedgwick named his lower

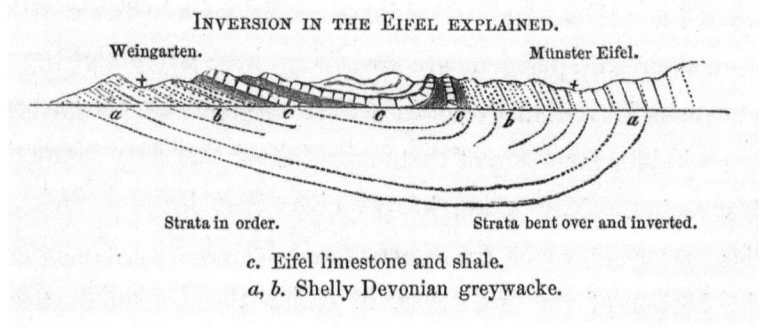

Fig. 5.2. "Inversion in the Eifel explained."
In Roderick Murchison, *Siluria* (1854), 377.

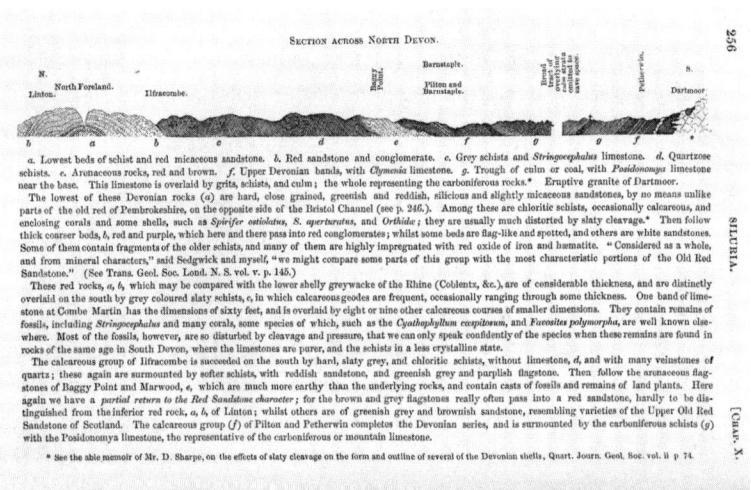

Fig. 5.3. "Section across North Devon."
In Roderick Murchison, *Siluria* (1854), 256.

greywacke of North Wales the Cambrian, after the Roman name for Wales, but saw that section of the strata obliterated in both Devon and Germany after Murchison reclassified them as newer Devonian rocks and stretched his own "Silurian" system ever-downward into Sedgwick's Cambrian. In this debate, and in Murchison's work thereafter, Devon and Wales, by correlation with the Rhenish Province of Prussia, became types on which to base geological analysis of the rest of the earth.

The "Devonian" strata, the strata that most clearly exemplify the problems and procedures of geological mapping in the nineteenth century—extrapolating world history from a distinctive local region—were part of a wider concern to think about history in spatial and structural terms. They are also the setting of *Two Years Ago*. Kingsley introduces us to this landscape thus:

> Look on fifty yards to the left. Between two ridges of high pebble bank, some twenty yards apart, comes Alva river rushing to the sea. On the opposite ridge, a low white house, with three or four white canvas-covered boats, and a flagstaff with sloping cross-yard, betokens the coast-guard station. Beyond it rise black jagged cliffs; mile after mile of ironbound wall; and here and there, at the glens' mouths, great banks and denes of shifting sand. In front of it, upon the beach, are half-a-dozen green and gray heaps of Welsh limestone; behind it, at the cliff foot, is the limekiln. . . . Above, a green down stretches up to bright yellow furze-crofts far aloft. Behind, a reedy marsh, covered with red cattle, paves the valley till it closes in. . . . in the very bosom of the valley a soft mist hangs, increasing the sense of distance, and softening back one hill and wood behind another, till the great brown moor which backs it all seems to rise out of the empty air. For a thousand feet it ranges up, in huge sheets of brown heather, and gray cairns and screes of granite, all sharp and black-edged against the pale-blue sky.[65]

The description is worth quoting at length because in it, Kingsley not only describes the what is known as the Heddon Mouth in North Devon in geological terms, demonstrating sensitivity to the geological formation of the landscape with its "shifting sand" and screes of granite, but also offers his readers a structural template by which he will give shape and meaning to the features of the natural world through the rest of the novel. First, notice the way he structures the scene as if it were on a canvas, in explicitly spatial terms. We are asked to look fifty yards to the left, and then to the right. Having established the width of the scene, Kingsley begins to explore its depth: beyond the mouth of the river "rise black jagged cliffs," in front of which is the beach with its heaps of limestone and behind which is the limekiln. Here there is both structural depth and a gesture toward narrative, as the scene implies the arrival of limestone from South Wales as it is shipped onto the Devonshire beach and its subsequent transport over the jagged cliffs for burning in the kiln. And then, continuing to move backward into the depth of this still life, Kingsley uses color to establish perspective. Mist increases the sense of distance in the ever-diminishing series of hills limned in softening colors, from the red cattle in the foreground to the neutral tones of the "brown moor," "brown heather," and "grey cairns and screes of granite" behind. This is important because *Two Years Ago* is a novel that continually argues with the Pre-Raphaelites, painters and poets widely criticized in the press for having sacrificed perspective in favor of brilliant, jewel-like colors, and for having replaced the painter's duty to paint objects from the natural world in hierarchical or otherwise meaningful relation. The narrator of *Two Years Ago* prefers, with apology to John Ruskin, the conventional Turner of the 1820s to the visionary Turner of the 1850s, prefers Thomas Creswick with his brown rocky landscapes carefully structured on the canvas to the Pre-Raphaelites with their bright blades of grass and deference to the flatter images of medieval painting. Kingsley concludes this description of North Devon by orienting us not only in the width and depth of his picture, but finally also in its height, as the moor rises upward from and into nothing, "boding rain and howling tempest."

Into this Devon scenery and out of the threatened storm washes up Tom Thurnall, a practical man with wide scientific interests, fresh from the gold diggings. He is rescued from the ocean by Grace Harvey, the local schoolmistress, who lives up to her name as the novel goes on by teaching him to seek God's grace and to add faith to his already quite lengthy list of masculine attributes. Devon is beautiful, which is important to Kingsley, but more important is the danger of this landscape, which makes not only shipwrecks, but also acts of heroism, inevitable. In his 1855 novel *Westward Ho!* Kingsley described the "delightful glens, which cut the high table land of the confines of Devon and Cornwall, and opening each through its gorge of down and rock, towards the boundless Western Ocean," each with its "wierd [sic] black cliffs which range out right and left into the deep sea, in castles, spires, and wings of jagged ironstone," its "black field[s] of jagged shark's tooth rock in strata set upright on edge or tilted towards each other at strange angles by primeval earthquakes," which belie the "richness, softness and peace" of the land and make the shores "hopeless to the shipwrecked mariner."[66] Once again, we are reminded that Kingsley's plots revolve around danger and the men whose spiritual and moral lives are enriched by confronting it.

Seeking to escape the cholera epidemic that takes up much of the novel's plot, several of its protagonists take a vacation to the "rock-girdled paradise" of Beddgelert in North Wales, where they travel to "the narrow gorge of Pont-Aberglaslyn,—pretty enough, no doubt, but much over-praised; for there are in Devon alone a dozen passes far grander, both for form and size."[67] Though the narrator thus reminds us of the Devon landscape, the place is the scene of forgetting and of self-forgetting in ways that bring the plot to its climax. At first, "the horrors which they have left behind them hang, like a black background, to all their thoughts," but "gradually, the exceeding beauty of the scenery, and the amusing bustle of the village, make them forget, perhaps, a good deal which they ought to have remembered."[68] There are two sides to this forgetting. On the one hand, the Devonshire curate Frank Headley, "a true Lowlander; born and bred among bleak Norfolk sands and fens," has his "first sight of mountains" and begins to drop his High Church Catholic notions, just as he learns to see the mountains in their true perspective.[69] Stunned to discover that the mountains "one could almost stretch out a hand and touch" are actually four miles distant, he is also made newly capable of managing the distance between himself and the woman he loves.[70] Turning to her, he says, "The first time that one sees a glorious thing, one's heart is lifted up towards it in love and awe, till it seems near to one—ground on which one may freely tread because one appreciates and admires; and so one forgets the distance between its grandeur and one's own littleness.'" "The allusion was palpable," continues the narrator, with pointed irony, "but did he intend it?"[71]

The question of scale mattered. Geology made traditional techniques for registering scale difficult for artists:

> Deprived of the traditional schemata of landscape painting—repoussoir tree, recessional perspective carefully marked out, dark foreground, light middle ground, precisely located figures, and so on,—the artist found it difficult to indicate scale. There was a feeling that the vaster the area depicted, the vaster the canvas had to be. Thus Ruskin found himself complaining that "there is no occasion that a geological study should also be a geological map" and thought that Brett's *Val d'Aosta* ... "would have been more precious if it had been only half the *Val d'Aosta*."[72]

This, in part, is what Kingsley alludes to in this novel. Almost immediately after Headley learns the rules of scale and perspective as defined by academic landscape painting, two things happen. First, the group sit down for a picnic "on one of the lower spurs of the great mountain of the Maiden's Peak, which bounds the vale of Gwynnant to the south," and the narrator embarks on another description of the landscape replete with deictic markers, or markers of time and place that can only be understood in contextual relationship to one another:

Above, a wilderness of gnarled volcanic dykes, and purple heather ledges; below, broken into glens, in which still linger pale green ashwoods, relics of that primeval forest in which, in Bess's days, great Leicester used to rouse the hart with hound and horn.... Above them, and before them, and below them, the ashes shook their green filigree in the bright sunshine; and through them glimpses were seen of the purple cliffs above, and, right in front, of the great cataract of Nant Gwynnant, a long snow-white line zigzagging down coal-black cliffs for many a hundred feet, and above it, depth beyond depth of purple shadow away into the very heart of Snowdon, up the long valley of Cwm-dyli, to the great amphitheatre of Clogwyn-y-Garnedd; while over all the cone of Snowdon rose, in perfect symmetry, between his attendant peaks of Lliwedd and Crib Coch.[73]

As in Devon, the power of the landscape is crowned by the height of a mountain or a cliff that thrusts upward to dominate the scene. One of the party is a photographer, commenting on "the breadth of light and shadow" in the scene before them, "how the purple depth of the great lap of the mountain is thrown back by the sheet of green light on Lliwedd, and the red glory on the cliffs of Crib Coch, till you seem to look away into the bosom of the hill, mile after mile."[74] Frank Headley, flushed with love, takes the opportunity to proclaim that he has "learnt to distinguish mountain distances since I have been here."[75] The point is explicit: Headley has learned to see the world accurately, with an eye sensitive to perspective and light and shading, and in doing so has learned to see the world aright in moral as much as visual terms. He has therefore become a potential husband, and a better curate.

The second thing that happens is that Elsley Vavasour, a weak, vain, and visionary poet as well as a Pre-Raphaelite sympathizer, fails to make this approach to correct perception, with devastating consequences. While Frank says that he does not write poetry because "God has written the poetry already, and there it is before me. My business is, not to re-write it clumsily, but to read it humbly, and give Him thanks for it," Elsley continues to transform nature into high-flown poetic raptures that hinder his capacity to read the natural world humbly and encourage him to start writing it in his own im age.[76] The point is again a point about form. Rewriting nature is an inherently clumsy business, and human attempts to formalize the natural world always potentially mask the correct "reading" of it. Too cowardly to collect a globe flower for his wife from "a dangerous cliff," Elsley is nonetheless angered when an old friend retrieves it for her, mistakenly suspecting them of infidelity.[77]

He runs away in cowardice, jealousy, and shame, up into the mountains. "In front of him rose the Glyder Vawr, its head shrouded in soft mist, through which the moonlight gleamed upon the checkered quarries of that enormous desolation, the dead bones of the eldest-born of time. A wild longing seized him; he would escape up thither; up into those clouds, up anywhere to be alone—alone with his miserable self."[78] As the careful reader of Kingsley's novel will remember from the opening descriptions of Devon, mist increases "the sense of distance." "Away he plunged from the high road, splashing over boggy uplands, scrambling among scattered boulders, across a stony torrent bed, and then across another and another;—when would he reach that dark marbled wall which rose into the infinite blank, looking within a stone-throw of him, and yet no nearer after he had walked a mile?"[79] Elsley, we realize, has not learned to read mountain distances as Frank has. The landscape looks nearer than it is, and he, despite being a poet, has not paid enough attention to nature to comprehend it.

This mistake soon means both that he exhausts himself attempting to get to his elusive destination, and that he becomes unable to separate subject and object in his perception of the mountain landscape. He feels "a strength which he never had felt before. Strong? How should he not be strong, while every vein felt filled with molten lead; while some unseen power seemed not so much to attract him upwards, as to drive him by magical repulsion from all that he had left below?"[80] Elsley has taken on the qualities of the iron or lead thought to intrude upward in veins through the strata, his own veins filled with such molten materials that he is propelled by a force he cannot explain, pushing upward through the rocks,

> stumbling upward along torrent beds of slippery slate, writhing himself upward through crannies where the waterfall plashed cold upon his chest and face, yet could not cool the inward fire; climbing, hand and knee, up cliffs of sharp-edged rock; striding over downs where huge rocks lay crouched in the grass, like fossil monsters of some ancient world, and seemed to stare at him with still and angry brows. Upward still, to black terraces of lava, standing out hard and black against the gray cloud, gleaming like iron in the moonlight, stair above stair, like those over which Vathek and the Princess climbed up to the walls of Eblis. Over their crumbling steps, up through their cracks and crannies, out upon a dreary slope of broken stones, and then,—before he dives upward into the cloud ten yards above his head,—one breathless look back upon the world.[81]

Become a Huttonian vein of molten rock hot with "inward fire," Elsley's sense of himself as a subject in the natural world begins to break down. He is embedded within the rocks, a part of the landscape that he cannot accurately see. That litany of present progressive clauses—"stumbling," "writhing," "climbing," "striding," "standing"—further muddles categories of perception, as past, present, and future collapse until Elsley feels himself in the midst of an eternal present. The natural world begins to appear animated, human, staring "at him with still and angry brows," and all he has for conceptual support is the Orientalist story of Vathek and the Princess, licentious and deplorable and climbing the walls of Hell ruled by the demon Eblis. Finally, with subject and object, past and present, on the brink of dissolution, Elsley attempts to dive "upward into the cloud ten yards above his head." This upward motion replicates the motion of molten lava bursting out atop the mountains, but is a physical impossibility for a man, who cannot, presumably, jump ten yards, nor comprehend that the clouds are further away than he perceives them to be. At this juncture, in "utter blackness" and "deepest shade," Elsley imagines himself "the centre of the universe, if universe there be. All created things, suns and planets, seem to revolve round him, and he a point of darkness, not of light." Caught in a thunderstorm, terrified by its lightning, scrambling to "a region where the upright lava-ledges had been split asunder into chasms, crushed together into caves, toppled over each other, hurled up into spires, in such chaotic confusion, that progress seemed impossible," Elsley's own progress in the narrative is arrested. The next morning, he flees to London where he becomes addicted to opium and dies. That's what happens in a Kingsley novel if you don't learn to read the mountains correctly.

Kingsley's reference to the Devonshire landscape in the opening passages of his Welsh description nods toward his attempt to train us not only to read landscapes in two-dimensional, or spatial terms, but also to read them in geological relation across time and through memory. To understand just how he does this we need to move to the final mountainous landscape of the novel, "upon the hills between Ems and Coblentz."[82] At the close of the novel all of its worthy male characters travel east to fight in the Crimean War. Wandering by the Rhine we find the last specimen of an unworthy man still alive, the American Stangrave, who has cruelly abandoned his lover having realized that her grandmother was a black slave. He languishes "along a road, which might be a Devon one, cut in the hill-side, through authentic 'Devonian' slate, where the deep chocolate soil is lodged on the top of the upright strata and a thick coat of moss and wood sedge clusters about the oak-scrub roots."[83] "On the right, slope up the bare slate downs, up to the foot of cliffs; but only half of those cliffs God has made. Above the gray slate ledges rise cliffs of man's handiwork, pierced with a hundred square black embrasures, and above them the long barrack-ranges of a soldiers' town."[84] This is the Ehrenbreitstein fortress on the mountain of the same name on the east bank of the Rhine. Built by the Prussians between 1817 and 1832, the fortress was never attacked. In Kingsley's words, it is "the yet maiden Troy of Europe; the greatest fortress of the world," warded by the "strange stairs and galleries of stone" that constitute the surrounding hills; "far beyond it," sleeping "high in air, the Eifel with its hundred crater peaks; blue mound behind blue mound, melting into white haze."[85] Here is a perfect Kingsleyan landscape: quiescent geology, the threat of war, the opportunity for travel. This is a site that may give form to the unity of moral, physical, and spiritual health Kingsley seeks.

This maiden fortress and the mountain behind it prompts Stangrave to ask himself,"How long has that old Eifel lain in such soft sleep? How long ere it awake again?" The narrator replies:

> It may awake, geologists confess,—why not?—and blacken all the skies with smoke of Tophet, pouring its streams of boiling mud once more to dam the Rhine, whelming the works of men in flood, and ash, and fire. Why not? The old earth seems so solid at first sight; but look a little nearer, and this is the stuff of which she is made! The wreck of past earthquakes, the leavings of old floods, the washing of cold cinder heaps—which are smouldering still below. . . . He had felt many an earthquake shock; and knew how far to trust the everlasting hills. And was old David right, he thought that day, when he held the earthquake and the volcano as the truest symbols of the history of human kind, and of the dealings of their Maker with them? All the magnificent Plutonic imagery of the Hebrew poets, had it no meaning for men now? . . . Or, had the moral world grown as sleepy as the physical one had seemed to have done? Would anything awful, unexpected, tragical, ever burst forth again from the heart of earth, or from the heart of man?"[86]

The irony is, of course, that believing that "change is a thing of the past, and tragedy a myth of our forefathers; war a bad habit of old barbarians," extinct in the enlightened age of "railroads and crystal palaces," Stangrave is unaware that "in his pocket" he is carrying "the news that three great nations were gone forth to tear each other as of yore."[87] This repetition in the grand history of the globe, as dormant volcanoes burst forth again, as the barbarian impulse to fight breaks out anew, is matched in the smaller plot of the novel as Stangrave, mirroring Elsley Vavasour, overhears Tom Thurnall talking with his lover and suspects them too of infidelity. The difference here is that Stangrave, though he is no geologist, as Elsley was not, is nonetheless geologically curious. He hides, listening to his lover pass by, "in a great block of black lava, crowned with brushwood, and supported on walls and pillars of... what should have been Dutch cheeses by all laws of shape and color, had not his fingers proved to him that they were stone." He ponders, as Elsley would not have pondered, "How they got there, and what they were," which "puzzled him, for he was no geologist."[88]

This is not the scene of self-forgetting that we saw in North Devon, but instead one of recollection. Tom challenges Stangrave to a duel on the Falkenhohe, "at the edge of" its "vast circular pit, blasted out by some explosion which has torn the slate into mere dust and shivers."[89] He is distracted from the matter in hand by the memory "of another glorious view, which he saw last summer" in Devon. "The likeness certainly exists," the narrator tells us,

> for the rock, being the same in both places, has taken the same general form, and the wanderer in Rhine-Prussia and Nassau might often fancy himself in Devon or Cornwall. True, here there is no sea; and there no Mosel-kopf raises its huge crater-cone far above the uplands, all golden in the level sun. But that brown Taunus far away, or that brown Hundsruck opposite... might well be Dartmoor, or Carcarrow moor itself, high over Aberalva town, which he will see no more. True, in Cornwall there would be no slag-cliffs of the Falkenley beneath his feet, as black and blasted at this day as when yon orchard meadow was the mouth of hell, and the south-west wind dashed the great flame against the cinder cliff behind, and forged it into walls of time-defying glass. But that might well be Alva stream, that Issbach in its green gulf far below, winding along toward the green gulf of the Moselle—he will look at it no more, ... lest he see Grace herself come to him across the down, to chide him, with sacred horror, for the dark deed which he is come to do.[90]

But the likeness is enough. The two men begin to duel, but Tom's second man runs off with his revolver. Chasing him through a crater "of 'explosion, and not of elevation,' as the geologists would say," Tom collapses into laughter. He and Stangrave narrowly escape being arrested, make friends, face up to their responsibilities, and go together to fight in the Crimea. That pause for the recollection of Grace (pun always intended), prompted by the geological similarity of the east bank of the Rhine and the West Country, preceded by the narrator's Old Testament–inflected account of the history of the world and its landscapes as a story of ongoing violence, makes all the difference. There is a deep, underlying unity, symbolized by the universality of geological structures that transcends history in Kingsley's novel, calling forth men into action at all periods and through all time. It is the men who recognize this that fare best, and the novel offers its readers the kind of training in geological map reading that might make them seek out danger in the world and help them to brave it.

Kingsley's recognition of the similarity between the two landscapes is no idiosyncrasy. *Two Years Ago* structures its story according to the shape and bent of the strata. It demonstrates the effectiveness of the stratigraphic imagination of time and process I am documenting here, in which the layers of rock and fossils as they were given visual and structural form as a column of "strata" enabled the imagination of history through the physical patterns of place. There is an underlying unity to place and history in *Two Years Ago*, forged as they both have been by volcanoes and cholera epidemics and wars, and which a man who has heeded his moral obligation to observe the world faithfully can detect. The "vulcanised" "hotch-potch" of the regions of the Eifel and Devon, which so frequently tripped up and confused their geological interpreters, overlain with the historic features of those landscapes, had nonetheless been made to yield to conceptual order. North Wales, once correlated with Devon, was still under a process of perpetual redefinition, as its slaty, contorted, and older rocks continued to elude systematic classification. Here it is in the novel that meaning, perspective, and temporal order break down. Each of these spaces and their place and status in the stratigraphic column come to define the kinds of experience that the characters in the novel may have: heroism, recollection, the restoration of order and meaning in the Devonian strata; confusion, forgetting, and heightened emotion in the Cambrian. Time and experience are given geological structure; history is made to yield to the description of disparate places.

In *Two Years Ago* Kingsley had moved far beyond his accomplishments in *Yeast*, in which he registered the epistemological problems of inductively gathered "facts" and the form that such facts might meaningfully take, and in which he substituted for the "story" of earth history a space—the field—in which reconciliations between fact and meaning might take place. Here, the field did not simply occasion narrative breakdown but gave rise to a stratigraphic form, even for fiction, by which the past could be given meaning without lurching into the speculations of excessively plotted fantasy. As such, the novel marks both a literary and a scientific achievement.

Eliot's Whispering Stones { CHAPTER SIX

Criticism of George Eliot's novels has often centered on a purported opposition between "uniformitarian" and "catastrophic" models of geological change. Eliot's "realism," most critics contend, is partly derived from her absorption of "uniformitarian" theory. As one influential critic has described it, Eliot's narrative method in her early works, *Scenes of Clerical Life* (1857) and *Adam Bede* (1859), was that of a "natural historian," undertaking "the simple labelling and classifying of physical reality," a reality that was fixed and could be mimetically represented by a narrator with a "confident authority in the power of definition."[1] By the 1870s, however, in *Middlemarch* (1872) and *Daniel Deronda* (1876), Eliot's scientific method has been considered that of an active scientific "experimenter," foregrounding her own creative role in the "reality" she presented. In this narrative, geology has a pivotal role to play: in the middle novels, *The Mill on the Floss* (1860) and *Silas Marner* (1861), Eliot exhibits "uncertain allegiance to ideas of uniformitarian development," where uniformitarianism is taken to mean organic accumulation through time and unceasing historical momentum.[2] But the cataleptic fits of Silas Marner and the "catastrophic" flood that kills Maggie and Tom Tulliver disrupt the linear continuity of the narrative and the possibility of organic harmony in the world depicted, suggesting the power of nature to operate outside of rational laws. In such instances, Eliot is said to employ catastrophic images and plots in order to "move beyond the empiricist theory of realism which sustained the natural history of *Adam Bede*," beginning the transition toward the experimental model that would characterize her last two novels.[3] Many critics have reiterated this assessment:

the flood in *Mill on the Floss* marks a significant deviation from what they conceive to be a gradualistic, quotidian principle of realist form, a "catastrophic" ending uniting religious belief and scientific reason or, in an alternative view, not vindicating "catastrophism and natural theology" but capping "a subtle and complex affirmation of uniformitarianism," creating a "Lyellian realism" that is "simultaneously realistic and romantic."[4] The persistent identification of "realism" with uniformitarianism stretches to Eliot's other works, such as *Middlemarch* (1871), in which she has been held to locate power and poetry in the minutely observed and seemingly mundane detail, just like Lyell.[5] In this understanding of Eliot's formal strategies, "catastrophism" and "uniformitarianism" have been routinely misconstrued as wholesale plots of earth history, one in which the earth moves violently and sporadically through a series of worlds punctuated by disaster, and another in which it creeps through the aeons, changing only gradually and incrementally.

But, as been amply discussed elsewhere in this book, there were never two "schools" of geologists in this sense. In fact, all geologists on all sides of the debate were engaged in a repudiation of wholesale theories of this kind as an appropriate form for earth science. The argument was methodological, about the safe scientific grounds on which to proceed, and it took place within the context of the stratigraphic mapping of the earth for which the geologists garnered their fame and prestige. In his *History of the Inductive Sciences* (1837), which George Eliot intensively read (and regarded as "dreary dryness"), the polymath William Whewell reviewed Charles Lyell's *Principles of Geology* and the intellectual debate that had surrounded it among the gentlemen geologists of the Geological Society.[6] It was here that Whewell coined the term "uniformitarian" to describe Lyell's geological "anti-narrative," and applied it specifically to Lyell's argument that only ordinary geological causes, operating at the same intensity at which they had operated since written records began, could be used to explain the workings of the earth throughout its history.[7] Whewell countered, on the side of the group he named the "catastrophists," that while he agreed in kind, he could not agree in degree. Like many of his colleagues at the Society, he concurred wholeheartedly with what already appeared to be a commonsensical view, that geologists could only invoke those causes currently operating as causes that may have operated in former worlds. But many, he suggested, believed it was a step too far to assume that those ordinary causes—such as volcanoes, floods, streams, and earthquakes—had never operated at greater or lesser intensities than they presently did. Nonetheless, Whewell was sure of one thing. "Catastrophist" or "uniformitarian," geology had entered a newly rigorous, newly prestigious era—and Lyell was far from the only hero of the age.

The language in which Whewell made this claim is telling. Earlier geologists, he wrote, had assumed that

> a few appearances hastily seen, and arbitrarily interpreted, are enough to give rise to a wondrous tale of the past, full of strange events and supernatural agencies. The mythology and early poetry of nations afford sufficient evidence of man's love of the wonderful, and of his inventive powers, in early stages of intellectual development. The scientific faculty, on the other hand, and especially that part of it which is requisite of the induction of laws from facts, emerges slowly and with difficulty from the crowd of adverse influences, even under the most favourable circumstances.[8]

Whewell may have disagreed with Lyell, but here he speaks in a language that echoes Lyell's, just as Lyell's language (see chapter 3) echoes that of other Geological Society luminaries. The two men clearly inhabit the same world. This passage clearly recalls Lyell's "early stage of advancement [in human understandings of the earth], when a great number of natural appearances are unintelligible," when "an eclipse, an earthquake, a flood, or the approach of a comet . . . are regarded as prodigies."[9] The "supernatural agencies" of Whewell's description closely approximate the "supernatural agents" refuted by Lyell. Both men invoke a Whiggish human history as a mode of explanation for earlier, discounted views of the earth and its history and structure. In fact, the only difference is that the tellers of Whewell's "wondrous tale[s] of the past" are classical, medieval, and Renaissance students of the earth rather than those eighteenth-century cosmogonists or (by Lyell's surreptitious sleight of hand) nineteenth-century geologists who continued to see evidence for catastrophic modes of change in the stratigraphic and fossil records. For both men, both of whom were appropriating a well-worn story of the history of their science, "prodigies," "romance," and "wondrous tales" were demons to be excised from a science with a new set of rigorous and inductive methods, and as their shared language suggests, whatever their differences on one particular point, they considered themselves to be fighting on the same side of the same battle.

It is worth repeating here, as I have stated elsewhere, that "catastrophism" was certainly not, as has generally been assumed by literary critics, "a geological theory discredited by the 1860s which postulated a series of world disasters and successive creations in order to perpetuate the theory of the fixity of species," in opposition to Lyell's more enlightened view, but a term restricted to a narrowly scientific argument about the possible causes of the great breaks in the geological record and about the limits of human observation and understanding of the past.[10] Or, as Eliot put it in a notebook in the 1870s:

> Is the interpretation of man's past life on earth according to the methods of Sir Charles Lyell in geology, namely, on the principle that all changes were produced by agencies still at work, thoroughly adequate and scientific? Or must we allow especially in the earlier periods, for something incalculable by us from the data of our present experience? Even within comparatively near times and in kindred communities how many conceptions and fashions of life have existed to which our understanding and sympathy has no clue![11]

It was anthropocentric, Whewell and others had claimed, to assume that the tiny portion of earth history that had been witnessed by man was adequate ground on which to reason about the history of a planet all geologists agreed was millions of years older than humankind. The question, as Eliot put it, was the simple statement of the so-called catastrophist view: not that the earth was a violent machine dispensing Old Testament-style justice at abrupt and sporadic intervals, but that there might be "something incalculable by us from the data of our present experience," that the earth may conceivably have been more volcanic a hundred million years ago than it is now. This is what was meant by "catastrophism" in the nineteenth century. Furthermore, debates about the overall shape and character of earth history were largely peripheral to the central research program of the science. Geologists were principally concerned with mapping the strata rather than with the creation of cosmological plots by which the whole of earth history could be understood. All this begs the question: has Eliot's fiction been misread by the light of an outdated geological paradigm?

Victorian Geology: Convenient for the Nonce

As acting editor of the *Westminster Review* in 1852, Eliot was understandably anxious about the quality of the articles she might receive under her stewardship of the liberal and utilitarian quarterly. And so it was equally understandable that she had "rejoice[d] greatly" when she received a "capital scientific article" from the Geological Survey palaeontologist and "pre-eminent naturalist of his era" Edward Forbes.[12] The article was to contribute to what Eliot would, with some relief, claim to be "the best number yet" published under her editorship.[13] Despite being entitled "The Future of Geology," the article devoted much of its time to reiterating the history of the science Lyell had set out in the first five chapters of the *Principles*.[14] Forbes also stressed that "geology" must be "contra-distinguished from cosmogony" (69), and distanced the work of contemporary geologists from the work of their overzealous predecessors, who

> thought nothing of submitting our planet to sudden extremes of heat and cold; shivering it into small fragments as suddenly as a Prince Rupert's drop... melting it down, stirring it up... and creating a new batch of beasts and vegetables with equal ease and rapidity; swamping the earth with no end of universal deluges. (78)

Forbes is not referring here to Geological Society colleagues like Adam Sedgwick or Roderick Murchison, the so-called catastrophists. He is referring to eighteenth-century cosmology and cosmogony, to the "sudden extremes" of fire and water represented by James Hutton and Abraham Gottlob Werner. In their day geology had described, Forbes wrote, quoting from a description of the African interior in a poem by James Montgomery, "A world of wonders, where creation seems / No more the work of Nature, but her dreams" (78). Like Lyell, Forbes adds a dash of exoticism to his account of a period of wild and wonderful geological speculation, hinting at a Whiggish narrative in which geological stories take on the glamour, and the untruthfulness, of African and Arabian storytelling and represent an earlier mental epoch in human history. Like Sedgwick in his famous review of the *Vestiges of the Natural History of Creation*, Forbes views such earth lore in gendered terms, comparing geology in this early stage of its development to "a woman of genius—and a handsome one too—she was opinionative and dogmatical; bold in assertions, and apt to let imagination get the mastery over judgment" (73).[15]

Forbes also drew attention, following Lyell, to the possibility that thousands of pages had been ripped from the "Book of Nature" in the process of time and that many of the strata and the fossils they contained had been damaged and lost. For Forbes, this meant that the "dogmatical" and "imaginative" cosmological approach rejected by the earlier generation of geologists, including Lyell, Whewell and Sedgwick, might not be entirely rebutted but instead reappropriated in a new form. Like his geological elders Forbes represented the desire to tell imaginative and dogmatical stories about the history of the earth as a natural desire for a science as young as eighteenth-century geology had been. But for him, that imaginative and youthful geology was not to be entirely forgotten but simply married off to a more rational, discerning, and ostensibly masculine lover: in the nineteenth century, Forbes writes, "paternal love endowed her with an Aegis in the shape of a winning presence, and the gift of the gab" (73), and gave geology "that vigour which has resulted in the strength of an immortal" (70). This metaphor is not of the transcendence of masculine "judgment" over feminine "imagination," but of an egalitarian marriage between the two principles. The reference to an Aegis endows that "paternal love" with the role of religious and military shield and protector, investing this new kind of geology with martial strength, spiritual authority, and classical pedigree. But importantly it protects, rather than destroys, the feminine imagination it exhibited in its youth. Forbes's metaphor implies that both storyteller and paternal inductivist are required for a healthy science to grow.

Forbes believed in repeated and successive creations carried out by an interventionist God.[16] But it is important to note also that he is clear in this article that "geology knows of no beginning" (80), and he predicts that "before twenty years are over, the world's antiquity and the partiality of the deluge will be taught to children in schools with no more hesitation than is now entertained about teaching the motion of the earth round the sun" (75). He is keenly alive to the role of imagination and interpretation in the process of scientific "judgment." Furthermore, Forbes is keen to move on from such debates to discuss the form given to the earth by the stratigraphic column. "We speak familiarly of Palaeozoic fossils" he continues, but "are we really convinced that we have as yet found them? Are they not merely provisional terms, convenient for the nonce, and only inconvenient when showy and unsubstantial theories are built upon the assumption?" (78). Similarly, the terms used to describe sections and systems of the geological strata, such as "silurian, Devonian, oolitic, cretaceous, eocene, miocene" are not "fixed, unchangeable, and cosmopolitan" (88), because "the geological scale of formations" is "an artificial scheme founded on local considerations, although an instrument and standard of comparison of great value when used judiciously" (89–90). The stratigraphic column itself, with its "sturdy stripes of demarcation" between different periods, is complicated by the many "janus-like fossils" that cannot be placed easily on either side of those "stripes," which do not demarcate boundaries but "really mark prodigious gaps in our knowledge of the sequence of formations and the procession of life" (90). Imagination would always have a role in a science whose evidences were so incomplete and whose classifications were subject to repeated revision and dispute. This, however, was an imagination put to the service of stratigraphic classification and nomenclature—forms from which fiction had been excised.

Like Forbes, Eliot's close friend Herbert Spencer enunciated a history of geology that again emphasized a dichotomy between "cosmogony" and good scientific practice in his 1859 essay "Illogical Geology," published in the *Universal Review*.[17] Eliot and Lewes owned a copy of this work in their edition of Spencer's *Essays* (1863), and as Nancy Paxton has demonstrated, Eliot read Spencer's work closely throughout her career.[18] In this essay Spencer argues that "the more advanced views now received respecting the Earth's history, have been evolved out of the crude views which preceded them," views he considers to have belonged to men like Werner and Hutton. Like Forbes, he does not dismiss them outright but sees them as essential paradigms on which future research was based: "In the progress of geological speculation," he writes,

> we have dogmas that were more than half false, passing current for a time as universal truths. We have evidence collected in proof of these dogmas; by and by a colligation of facts in antagonism with them; and eventually a consequent modification.[19]

As for Forbes, the "dogmas" of early geological science are not entirely rejected, despite the fact that they have since been proved to have been "more than half false." Scientific language and scientific classifications, for Spencer, are not representative of "universal truths" but are a convenience, enabling the scientist to give his ideas and observations a form, without which they would remain in the domain of the unthinkable: "If we waited till all the facts were accumulated before trying to formulate them, the vast unorganised mass would be unmanageable. Only by provisional grouping can they be brought into such order as to be dealt with."[20] Here is the problem of form all over again. Form constitutes science, and as scientific evidence shifts and changes, so too must the shapes and patterns by which it can be understood and given sense and meaning. Once enough "antagonistic" facts have been accumulated, Spencer continued, the theories of a science and its nomenclature must both change too.

For Spencer, geology had failed to abandon the older nomenclature and its concomitant habits of thought, so that geologists like Murchison were caught up in the language of the cosmogonies they had discarded, trapped in a language that contradicted their research. Lurking in their descriptions of the strata was a story they had not realized they were telling. Murchison, extending his Silurian domains ever downward in the hope he might reach the base of the fossil record and include the earliest life on earth as the base point of his system, had forgotten that there was no recognizable "beginning" to history. Spencer asked, and answered in the negative, "can we find the beginning of life?" If not, he contended, geologists could no longer usefully employ the term "Palaeozoic" to refer to strata containing the earliest organic remains. As "azoic" means "bearing no trace of life," the term misleadingly implied that there was an identifiable beginning of all life on earth, when it might simply be, as Lyell had suggested, that such rocks merely exhibited the beginning of physical *records* for life on earth. Who knew what had gone before? Murchison's stratigraphic column implied that nothing had gone before. In doing so, it took on an outmoded form. For both Forbes and Spencer, echoing Lyell's sense of the mutilation of the strata, the form of the natural world was characterized by "prodigious gaps" and "great blanks." The forms that geology took would need to account for them, and in doing so those forms would always be shadowy, arbitrary, and partial. This, I am arguing, was the geology Eliot knew—a geology whose primary interest lay in a methodological dispute about the nature of scientific evidence, rather than a geology that offered two "plots" of earth history in rivalry.

The "Geological Story"

These two essays reflect the more heterodox geological ideas that were circulating in the intellectual milieu that Eliot, editing the *Westminster Review,* in love with Spencer, and later partner of the zoological writer G. H. Lewes, was grappling with in the 1850s and beyond. And they may have appealed to Eliot as she experimented throughout her career with the ways in which plot might structure and distort the perceptions of "reality" that her characters experienced or, by extension, that her novels attempted to reveal to their readers. In a note entitled "The Historic Imagination," Eliot referred to the "vulgar coercion of conventional plot."[21] Conventional plots simplified a complex world, bullying a disparate and messy "reality" into preconceived and self-determining patterns of meaning. Not that this precluded Eliot's producing romantic and melodramatic plotlines herself. Cases for the prosecution include Hetty Sorrel's seduction, infanticide, rescue, and transportation in Eliot's first novel, *Adam Bede*; the tragic plot of Maggie Tulliver's spiritual struggle, seduction, and death in a violent flood in *The Mill on the Floss*; the fairy-tale schema of *Silas Marner*, with its subplot of secret marriage and opium addiction; and the sensation-fiction story of cursed diamonds and hidden parentage in *Daniel*

Deronda. If conventional plot is "coercive," then Eliot, it might be argued, was more than infrequently coerced by it. But in Eliot's fiction, plot is always complex, often self-reflexive, and fraught with her sense of the paradoxical claims of being both a "realist" writer communicating social "facts" and the purveyor of fictions to a readership waiting to be entertained. For Eliot, as for the geologists, the concern is not to excise plot altogether, but to be conscious and careful about the ways in which plot might facilitate, or hinder, rational and reasonable thought.

Eliot began to explore this problem as early as 1856 in her famous essay "Silly Novels by Lady Novelists."[22] Here Eliot ironically classified various plots as different "species." There was the "most pitiable of all silly novels by lady novelists, the oracular species," for instance (449), with its exaltation of "very ordinary events of civilised life ... into the most awful crises" (452). Or there was the "mind-and-millinery species" (442), in which the "vicious baronet" husband of the heroine "is sure to be killed in a duel, and the tedious husband dies in his bed requesting his wife, as a particular favour to him, to marry the man she loves best, and having already dispatched a note to the lover informing him of the comfortable arrangement" (443). The operative word here is "sure": the baronet "is sure to be killed," for the events and situations are predicated on a plot so conventional it writes itself. The "vicious baronet" dies, the beautiful heroine marries her lover in the end, and all of this is both conventionally arranged and morally acceptable. The deterministic coercions of "conventional plot" make truth-telling in such fictions impossible. Not only that, but Eliot's specimen-classification pits the authoritative voice of the man of science or the natural historian against the female author. "No!" she exclaims ironically, having led herself down a speculative avenue of inquiry. "This theory of ours, like many other pretty theories, has had to give way before observation" (443).[23]

In another essay of the same year, "The Natural History of German Life" (1856), reviewing W. H. Riehl's sociological works *Land und Leute* (1855) and *Die Bürgerliche Gesellschaft* (1856), Eliot further formulated her literary method.[24] In the essay she praises Riehl for his status as a "pedestrian" first and a "political author" second, his "views ... evolved entirely from his own gradually amassed observations" (68). In moving from facts to carefully circumscribed theories, rather than working from theoretical speculations to such "facts" as might fit them, Riehl's work is more truthful than the "abstract theorising of educated townsmen," the "many who theorise" (52). This repudiation of theory is essentially political rather than scientific, rejecting the theoretical abstractions that had imagined factory workers as a "deracinated and alienated" social group before the 1848 revolutionary uprisings across Europe.[25] But this supports the view of theory as inaccurate, untruthful, and a potential threat to the social order.

Furthermore, this widespread suspicion of "theory" operates for Eliot as plot. As A. S. Byatt puts it, Eliot "dislikes literary generalisations about buxom rosy-cheeked peasants and commends Riehl for avoiding them."[26] Her preference for concrete description is an aesthetic as well as a political choice. For Eliot, those "silly" narratives populated with archetypal characters and stock situations extend the observation of individual, concrete moments into generic articulations. They blot out inconsistencies between individuals and communities, obliterating the differences made obvious by detailed observational study and subsuming them into archetypal patterns that do not arise from a mass of studied facts. In so doing they obscure the reader's apprehension of the reality beneath the surface.

Not that Eliot does not also betray a "proclivity to the theoretical," as it has been described.[27] The key, however, is that in "The Natural History of German Life," she grasps for a language that "express[es] *life*, which is a great deal more than science" (69). This language, the language of "incarnate history," preserves "hoary archaisms 'familiar with forgotten years'" (69). It embodies archaic meaning, lost tradition, and forgotten history in its present articulations. This language is important particularly to the English "natural historian," for he, unlike Riehl, lives in a country in which "artisans" and "peasants" appear rapidly to be becoming lost. While Riehl can employ strict inductive methods in his observations of the German peoples who are his specimens, the English natural historian finds that many of the objects of his study have been lost in the mists of time. As Eliot writes, summarizing Riehl:

> This vital connexion with the past is more vividly felt on the Continent than in England, where we have to recall it by an effort of memory and reflection; for though our English life is in its core intensely traditional, Protestantism and commerce have modernised the face of the land and the aspects of society in a far greater degree than in any continental country. (70)

The natural historian in an England whose "connexion with the past" is under stress cannot merely observe and accumulate information but must also employ "an effort of memory and reflection." Such a project is less straightforwardly that of the natural historian and more specifically that of a geologist, observing the evidences of the past empirically, but thwarted by the fragmentary, incomplete, and often inaccessible nature of the world he seeks to describe. In the absence of such evidence, he must create imaginative modes of collection and interpretation. Desiring to represent "*life*, which is a great deal more than science," Eliot understood very early in her literary career that the natural-historical model could only take her only so far.

Eliot closely replicated her comment that "Protestantism and commerce have modernised the face of the [English] land" in her description of the provincial town of St. Ogg's in *The Mill on the Floss*:

> The Catholics, bad harvests, and the mysterious fluctuations of trade, were the three evils mankind had to fear: even the floods had not been great of late years. The mind of St. Ogg's did not look extensively before or after. It inherited a long past without thinking of it, and had no eyes for the spirits that walked the streets. (103)

St. Ogg's is locked in an eternal present, "not look[ing] extensively before or after," disconnected from consciousness of the "long past" that nonetheless shapes it. And two key culprits of this cultural amnesia are Protestantism and commerce: "Protestantism and commerce have modernised the face of the land and ... of society." Closely correlating her description of the problems besetting the English "natural historian" with her account of St. Ogg's, Eliot signals her novel's central concern to dramatize the "effort of memory and reflection" necessary to the realist historical novelist. The atemporal consciousness of the St. Ogg's inhabitants is partly produced by the landscape itself, the flat and "level plain" of the town and its environs. As Riehl had written, there was a "striking connection ... between the local geological formations in Germany and the revolutionary disposition of the people":

> Where the primeval physical revolutions of the globe have been the wildest in their effects, and the most multiform strata have been tossed or thrown upon one another, it is a very intelligible consequence that on a land surface thus broken up, the population should sooner develop itself into small communities, and that the more intense life generated in these smaller communities should become the most favourable nidus for the reception of modern culture, and with this a susceptibility for its revolutionary ideas. (279–80).[28]

The conservative culture of St. Ogg's is directly opposed to the "revolutionary disposition" of its German counterparts, just as the "level plain" on which it sits may be contrasted with the "tossed" and "broken up" "strata" of the German landscape. Both landscapes have "more than . . . metaphorical significance" to the "disposition of the people" who live there, but it is the difference between those landscapes that prompts a subtler method from Eliot's narrator in *The Mill on the Floss*, because the "wild" evidences of "primeval physical revolutions" in Germany make both "the reception of modern culture" and the consciousness of the past possible. After all, the "broken up" surface offers direct, tangible evidence of the geological history of the landscape that the "level plain" renders invisible.

Eliot's narrator continues in free indirect speech:

> Since the centuries when St. Ogg with his boat and the Virgin Mother at the prow had been seen on the wide water, so many memories had been left behind and had gradually vanished like the receding hilltops! And the present time was like the level plain where men lose their belief in volcanoes and earthquakes, thinking tomorrow will be as yesterday and the giant forces that used to shake the earth are for ever laid to sleep. (103–4)

In their inability to see beyond "the level plain" of their immediate horizons, the inhabitants of St. Ogg's are like the men "in an early stage of advancement," in Lyell's account, or "in early stages of intellectual development," in Whewell's, whose lack of historical perspective and limited geographical horizons make "natural appearances . . . unintelligible."[29] These are men and women for whom the geological vision of the earth is an unsuspected anathema.

Making narrative invisible to the provincial population she describes, Eliot makes it her job as a novelist to relocate, and renarrate, the story that makes the history of such lives worth telling. We might even suggest that she casts herself as a geologist of inordinate knowledge here, able to overcome her limited temporal and spatial position. But despite the seeming omniscience of its narrator, *The Mill on the Floss* testifies only to the incommensurability of the past. While the flood that ends the novel is often vaunted as a geological flood of either "uniformitarian" or "catastrophic" proportions, for instance, Eliot is in fact careful to leave its causes obscure. Geology is only one of a range of explanatory frameworks suggested for the flood. It may, Eliot hints, have been caused by an upstream farmer's substitution of the natural hydraulics of the eponymous mill for artificial irrigation techniques, for which Mr. Tulliver takes him to court, unsure of what the results of such techniques might be.[30]

Then there is the legend of "St. Ogg with his boat," who ferried the pregnant Virgin Mary across the river and received a promise of eternal safety, a story the novel both rejects and endorses—though Maggie is a virgin, her suspected seduction by Stephen Guest suggests that she has violated the terms of the legend and lost the safeguard of the saint. Other explanations abound: Mrs. Tulliver receives premonitions that her children will drown in the Floss, as they do, and there is a curse on the river if the mill changes hands, which it does. While these may appear to be little more than superstitious speculations, both turn out to be proleptic insights into the narrative's tragic conclusion, and the narrative space granted to these speculations hints strongly that this will be the case. The lost "belief in volcanoes and earthquakes" in St. Ogg's, its forgetting of the "giant forces that used to shake the earth," is misguided, but the superstitious views of its inhabitants ultimately offer an equally plausible account for the novel's events as the geological gloss of the narrator. Early critical responses to the novel puzzled over the sudden denouement of the flood, and sought to explain its presence in the novel.[31] But Eliot reminds us to take the "hoary archaisms" of the superstitious lexis of St. Ogg's seriously in *The Mill on the Floss*, deriving as it does from a long (if forgotten) tradition in the town, so that it has a certain uncanny ability to make predictive suggestions about the future. Describing the plot of the novel, we find that its workings and its logic are indecipherable. We do not know why the flood occurred. It erupts, from nowhere, into the novel, and it is over in a few short pages. All explanations of it, even scientific ones, are merely partial.

Eliot cannot, in this light, be seen simply to objectify and satirize the people of St. Ogg's as parochial and ignorant. This is the trap that the Riehlian "natural historian" could so easily fall into, observing, describing, and classifying a population that lacks the intellect to classify or comprehend itself. The geologist, by contrast, is always aware of the futility of a simplistic explanation of events. Refusing (or unable) to settle on an all-encompassing account of the history of the earth, geology was characterized by its evasion of such stories. In the end, geology seems in *The Mill on the Floss* to connote the difficulty of ever knowing the natural world outside one's own limited horizons, whether those horizons are the "level plain" of a provincial town or the more expansive (but still limited) horizons of a seemingly omniscient narrator. As such, Eliot deliberately evades the coercive force of plot and its preclusion of imagination and interpretation in determining the meaning of the events the novel describes.

The Red Deeps

Though Eliot stresses the importance of the search for story, for an expansive knowledge that looks "before and after," throughout *The Mill on the Floss* she also repeatedly warns against the dangers of too much plot. Against the "level plain" of St. Ogg's and its people is set the relationship of Maggie Tulliver and Philip Wakem, who meet and form an attachment in the Red Deeps. This landscape represents an antidote to the town's obliviating flatness, for it possesses "the charm for [Maggie] which any broken ground, any mimic rock and ravine have for the eyes that rest habitually on the level" (262). Maggie thinks that its Scotch firs with their "broken ends of branches were the records of past storms which had only made the red stems soar higher" (263), and finds passion and happiness in a landscape in which a

> line of bank sloped down again to the level, a by-road turned off and led to the other side of the rise, where it was broken into very capricious hollows and mounds by the working of an exhausted stone-quarry—so long exhausted that both mounds and hollows were now clothed with brambles and trees, and here and there by a stretch of grass which a few sheep kept close-nibbled. (262)

The "capricious hollows and mounds," the quarried stone, and the evidence of a past life of quarriers, now "exhausted," form for Maggie "an opening in the rocky wall which shut in the narrow valley of humiliation" (285), the extreme and ascetic form of Protestantism she adopted in her early teens, and which serves further to shut out the past than to expand her consciousness. In this way the Red Deeps is an ideal landscape, bearing the traces of its, and her, history—its broken rocks and stems record "past storms," but also it conceals Philip, whom Maggie has lost touch with but not forgotten. It thus offers a space in which Maggie can once again begin to make "the effort of memory and recall" that might enable her to fulfil her intellectual and spiritual potential.

But the Red Deeps is also the site of Maggie's transgression of her loyalty to her family and of a passionate sexual awakening. It resembles in color and in form the "red deeps" of Maggie's soul, with its "volcanic upheavings of imprisoned passions" (257). The narrative potential of this space is rendered dangerous by its passionate, and potentially devastating ("volcanic"), character, for on returning home after a secret visit, "at Philip's name she had blushed, and the blush deepened every instant from consciousness, until the mention of the Red Deeps made her feel as if the whole secret were betrayed" (298). Her cheeks burn redder and redder until she displays on her body the secret of the place she has been visiting, her blush making visible her sojourns into the equally red landscape. The Red Deeps is written on her cheeks. Maggie's body is transposed into narrative, the story of her yearning for knowledge and her desire for passionate fulfilment writ large upon her body. The tragic narrative train that such exposition sets in motion, and the association of Maggie's quest for knowledge with secretive sexual transgression, make her not only the most narratable human being in St. Ogg's, but an emblem of narrative itself, questing for story in the "level plain" that constricts her and denies her agency.

Again, Maggie's potentially romantic interaction with Philip in this setting is noted by Tom when he "remembered only lately hearing his mother scold Maggie for walking in the Red Deeps when the ground was wet, and bringing home shoes clogged with red soil" (299). Always rendering painfully visible the stories of "before and after," the Red Deeps is an objective correlate for Maggie's suppressed and internalized desires: her desire for knowledge, and the erotic desires in which her quest for knowledge is always bound up. But in the narrativity of the space comes the danger of exposure, the danger of readability, comes temptation and transgression.

The connections between fallenness, knowledge, sexuality, and narrative are made explicit through the novel's ironic references to Eden. Philip rails against the Creation as "sugared complacency" and "flattering make-believe" (323). Maggie also shares an ambiguous kind of Eden with Stephen Guest, who sings Haydn's "Creation" with Lucy just a few pages before he and Maggie begin to form romantic interests in one another:

> As it is always pleasant to improve the minds of ladies by talking to them at ease on subjects of which they know nothing, Stephen became quite brilliant in an account of Buckland's Treatise, which he had just been reading. He was rewarded by seeing Maggie let her work fall and gradually get so absorbed in his wonderful geological story that she sat looking at him, leaning forward with crossed arms and with an entire absence of self-consciousness, as if he had been the snuffiest of old professors and she a downy-lipped alumnus. (334)

Eliot had read Buckland's Bridgewater Treatise alongside Lyell's *Principles*, in 1841, and recorded in a letter her "pleasure" with it.[32] Though much of Buckland's book was concerned with palaeontology, in it he also retracted his earlier belief in geological evidences for the biblical Flood, arguing instead that, though there had been widespread deluges in the earth's past, none of these could be attributed to the Mosaic record. In this way Stephen's allusion offers another uncannily proleptic warning about the potentially devastating nature of floods, and gestures to further ambivalence about their causes, geological or otherwise.

That Stephen should even mention the book to Maggie and Lucy is a sign of his sexual daring. The book provided "slightly more risqué fare appropriate for discussion" among the fashionable elite only "after the withdrawal of the ladies," with its liability to provoke discussion on the nature of species, on anatomy, and on its potentially materialist implications. In middle-class circles the book was often read in more mixed, domestic company, and yet this scene of the novel, illustrating that mixture, testifies to the residual danger that accompanies the Bridgewater Treatise and its subject matter even in the domestic context: "This encounter... is a pivotal point in the book, leading ultimately to ... Maggie's disgrace."[33] Furthermore, it is specifically the narrative that Stephen concocts from the Treatise that produces that danger. Maggie, letting "her work fall, and gradually get[ting] so absorbed in his geological story that she sat looking at him ... with an entire absence of self-consciousness," pays no heed to the warning this choice of text should provide, for she is too busy succumbing to the pleasure of the story itself (334). Interlacing Maggie's erotic desire for Stephen and Stephen's attempt to impress Maggie with his "geological story," Eliot blends sexuality with storytelling. Though Stephen calls Philip "the fallen Adam with a soured temper" and himself and Lucy "Adam and Eve unfallen—in paradise" (323), his play at pedagogy in this passage casts him not in the role of Adam unfallen but of the serpent. It is apposite that the form of knowledge he offers is geological, affronting religious tradition, and that this book of the novel is entitled "A great temptation." Geology—however ironically—offers a clear language for a kind of fallen knowledge, a correlate for the language of Eve, hinting at Maggie's later status as a fallen woman after her elopement with Stephen in another of the novel's pointedly proleptic episodes.

It is specifically the pleasurable nature of the form in which geology is related to Maggie that makes it dangerous here. Eliot makes it ironically clear that Stephen is not offering Maggie the knowledge she quests for throughout the novel, but only something that looks like it, for Stephen cannot satiate Maggie's appetite for his "geological story," finding "the stream of his recollections running rather shallow" under the force of her excitement (334). This moment encapsulates Stephen's intellectual inadequacy as a lover for Maggie, sugarcoating his partial, dilettante knowledge of geology in the pleasing shape of a story. What he doesn't know, we can surmise, he is able to hide, at least initially, in the form of the story. Furthermore, though Maggie's resemblance to a "downy-lipped alumnus" (334) in this passage seems to permit her a momentary transgression of the barriers to knowledge imposed on her by her sex, it also renders her submissive, naïve, and dependent upon the knowledge of "the snuffiest of old professors" (334). If Stephen is "quite brilliant," as we are informed by the narrator, then this must be Maggie's view of him, contradicting very sharply the passage's numerous hints about Stephen's intellectual and moral unsuitability for the role of her lover. Maggie is too easily tempted, like Eve before her, and is imprisoned even further within her gender. As James Secord has written, geology's ambivalence about narrative was related to the sense that "a strong narrative line put scientific exposition uncomfortably close to novel-writing. The emotional engagement of readers might lead to a suspension of judgment and unlicensed speculation."[34] In making Maggie forget herself, forget the ties of duty to her cousin Lucy and to the community in which she lives, Stephen's geological story produces in her just such a "suspension of judgment," just as her sojourns in the Red Deeps had prompted dangerous and unlicensed behaviours. *The Mill on the Floss* emerges almost as a fictional dramatization of the plight of the readers of the narrative-driven *Vestiges* as Sedgwick described them: the "seductions" and "serpent coils of a false philosophy" in that work, he implied, tempted female readers to "pluck" its "forbidden fruit" with its pleasing fictional shapes. But Eliot, unlike Sedgwick, indicts a culture in which women are not educated to overcome those temptations, foregrounding the ways in which knowledge may take gendered form.

In the story-barren world of St. Ogg's, the pleasing shapes of narrative work to obfuscate knowledge. Maggie's absorption in Stephen's story, halfheartedly appropriated from the texts of the university professors who constituted the geological elite, renders her an emblem of the dangers of story itself and a perpetual threat to the "realist" strategies of representation Eliot is interrogating, for her protagonist continually seeks solace in stories that may console her from the hard world in which she lives. Maggie, one concludes, is at least in part a symbol for the emotional difficulties inherent in the realist project, or for the continual temptation to locate oneself in a predetermined plot.

Against Maggie's transgression and fall, moreover, is Eliot's refusal to provide a single narrative shape through which to definitively interpret the events of the novel and their meaning. The flood cannot be accounted for as a geological flood—"catastrophic," "uniformitarian," or otherwise—for it has religious, spiritual, agricultural, legal, and supernatural explanations that counter and contradict the scientific account. The flood is an event in the narrative of *The Mill on the Floss* that powerfully resists emplotment. And it does so in a novel that continually reminds us, by geological means, of the potential seductions of "silly" stories.

Blank, Grey Sides of Stone

The heroine of *Daniel Deronda* belongs to a very different social stratum to Maggie Tulliver, and the people who populate the canvas of *Daniel Deronda* are far removed from the "artisan," "peasant," and lower-middle-class citizens of St. Ogg's. Living in genteel poverty with her mother and her four half sisters at Offendene, a comfortable redbrick home "chosen ... simply for its nearness" to her rector uncle, and later marrying a duke, Gwendolen Harleth might at first glance be assumed to have more social, intellectual, and spiritual opportunities afforded her than Eliot's previous heroine.[35] But Gwendolen, like Maggie, is hemmed in by her sex and her desire to escape the boundaries of her limited existence. As she half jests at one point in the novel:

> We women can't go in search of adventures—to find out the North-West Passage or the source of the Nile, or to hunt tigers in the East. We must stay where we grow, or where the gardeners like to transplant us. We are brought up like the flowers, to look as pretty as we can, and be dull without complaining. That is my notion about the plants: they are often bored, and that is the reason why some of them have got poisonous. (119–20)

Despite the difference in their social backgrounds, Gwendolen's status as "a girl," not having the "highest breeding," and never being free from the "sordid need of income," means that her "passion for doing what is remarkable" is thwarted in equal measure to the thirst for self-knowledge dramatized in Maggie's story in *The Mill on the Floss* (47). Furthermore, Gwendolen's intellectual "horizon" is as restrictive as Maggie's, though not as blamelessly so, being "that of the genteel romance where the heroine's soul poured out in her journal is full of vague power, originality, and general rebellion, while her life moves strictly in the sphere of fashion; and if she wanders into a swamp, the pathos lies partly, so to speak, in her having on satin shoes" (47). Gwendolen is restricted by her choice of an inappropriate plot, the plot of the "genteel romance," with its poverty of empathy and its lack of psychological consequence.

Here again is that language of inevitability: just as the "vicious baronet" of the "mind-and-millinery" species of fiction is "sure to" die as required, here the use of the continuous present tense to describe the heroine's recordings in her journal, her fashionable movements, and her wearing of satin shoes registers discontent at the self-determining forms of conventionally plotted fictions. Both Gwendolen and Maggie, denied access to knowledge and to self-knowledge by their sex, seek self-determination and agency in plots whose patterns are conventional, unrealistic and predetermined. The conflict is obvious— they can never achieve the freedom to determine the course of their own lives when they cut their lives according to the cloth of conventional romance.

The "level plain" of St. Ogg's also finds its counterpoint in Gwendolen's natural environment. The narrator describes the house at Offendene in the following terms:

> One would have liked the house to have been lifted on a knoll, so as to look beyond its own little domain to the long thatched roofs of the distant villages, the church towers, the scattered homesteads, the gradual rise of surging woods, and the green breadths of undulating park which made the beautiful face of the earth in that part of Wessex. But though standing thus behind a screen amid flat pastures, it had on one side a glimpse of the wider world in the lofty curves of the chalk downs, grand steadfast forms played over by the changing days. (19)

The "flat pastures" surrounding the house almost entirely block out any sense of "the wider world," which can only be "glimpse[d]" in the "grand steadfast forms" of the "the chalk downs." The house is characterized not by what it is but by what it is not, isolating Gwendolen from the "undulating park" and "the face of the earth." It is clear that this "wider world" includes not only sociogeographical width, but also historical expansion, for the chalk downs possess a historic immutability which makes them "grand" and "steadfast" in juxtaposition with the evanescent play of "the changing days." Geological landforms give expression to an advanced consciousness of a wider world: the chalk downs give concrete and tangible form to an otherwise unimaginable history. Here is a hint of a stratigraphic imagination of the past, the imagining of millions of years of history as a layered sequence of strata. Furthermore, the "flat pastures" of Offendene are doubly disconnected from this wider world for Gwendolen because, as the narrator's regretful tone makes clear, it is a "pity that Offendene was not the home of Miss Harleth's childhood, or endeared to her by family memories," for "a human life ... should be well rooted in some spot of a native land" where "the definiteness of early memories may be inwrought with affection" (18). It might not be "trade" and "Protestantism" cutting Eliot's last heroine off from an expanded historical consciousness, but cut off she still is.

As in *The Mill on the Floss*, Eliot continues to equate "the wider world" with irregular landscapes that bear the marks of geological process. Like Maggie before her, Gwendolen, living in "flat pastures," is keen to escape the viewpoint of those whose "eyes rest habitually on the level": "I don't mind going up hill," she tells Herr Klesmer in her bid to become a singer, for "it will be easier than the dead level of a being a governess" (237). Throughout the novel the geological imagery of hills, mountains, and earthquakes is used to give form to Gwendolen's spiritual and emotional development: during her marriage she becomes "conscious of an uneasy, transforming process—all the old nature shaken to its depths, its hopes spoiled, its pleasures perturbed, but still showing wholeness and strength in the will to reassert itself" (394). That "wholeness and strength" must be broken down and destroyed, forced into fragments like the geological evidences mutilated and "shaken to [their] depths" by earthquakes. "After every new shock of humiliation [Gwendolen] tried to adjust herself and seize her old supports," to grasp onto what she knows rather than to embrace what she does not know (394). It is only after Grandcourt dies and Gwendolen is "forsaken," as she puts it, by Daniel, that she begins to make the arduous journey toward fulfilment that she has hitherto been unable to make. In other words, she is forced to accept breakdown and disintegration as an important part of self-knowledge. Daniel's revelation that he is to marry Mirah and travel to the East "inspired her with a dreadful presentiment of mountainous travel for her mind before it could reach Deronda's" (746), an imaginative "going up hill" that will take her, metaphorically speaking, away from the "level," the "flat pastures" over which she had too easy dominion. Most importantly, it also constitutes a decisive break from the conventional plot of the mind-and-millinery novel with which she had previously identified. The vicious baronet does indeed die in *Daniel Deronda*, but Gwendolen's assumption that she might therefore marry Daniel is sadly misguided. Truth and meaning are to be found in the breakdown of that plot, and in the breakdown of Gwendolen's assumption that she is a heroine in a conventional story.

Gwendolen's journey to self-knowledge begins in the landscape that shelters the "Whispering Stones," "two tall conical blocks that leaned towards each other like gigantic grey-mantled figures" (135), a landscape that, like the

Red Deeps in *The Mill on the Floss*, bears a visible record of its past. At first the Whispering Stones are no more than a curiosity on a romantic sojourn into a picturesque landscape, in which Gwendolen is "busy with a small social drama almost as little penetrated by a feeling of wider relations as if it had been a puppet-show" (133). But detaching herself from the rambling group, Gwendolen returns to the spot to be confronted by her romantic predecessor, Lydia Glasher, who reveals her prior claim to be Grandcourt's wife, the four illegitimate children she has borne him, and offers Gwendolen a dark warning not to marry him. "The 'geological structure of Wiltshire,'" Jane Irwin notes, "plays a role in a scene in DD ch. 14: the 'Whispering Stones' that turn 'their blank grey sides' to Gwendolen, as if to conceal Lydia Glasher, are scattered Sarsen stones."[36] These stones, as Eliot records them in her notes, have an interesting history:

> The geological structure of Wiltshire is chiefly chalk and oolite, the clays of the Wealden forming but a narrow band around the chalk, and those of the Tertiary beds of Southampton being confined to the vales below Salisbury. The London basin ends at Hungerford, but its former extension to the W. appears probable from the numerous blocks of Bagshot sandstone (Sarsen stones) scattered over the downs.[37]

The "former extension" of the London basin to Wiltshire can only be guessed at, only "appears probable," because the "Sarsen stones" that lie "scattered over the downs" are "Bagshot sandstone," a stone that constitutes much of the London basin. The blocks are testament to an earlier geological epoch, objects now out of context as they have been wrenched from the strata in which they were deposited and dragged, mysteriously, to their current location. As accidental remains of the past, the sarsen stones are icons of temporal displacement, implying a former reality of which they are the only extant evidence and whose history cannot be decisively established. The source for Eliot's note is unknown, but the sarsen stones had attracted considerable interest in the press. George Poulett Scrope's 1858 article on Wiltshire in the *Quarterly Review*, for instance, had outlined the "geological structure" of the region, the "sarsen-stones," "hewn from … masses of siliceous grit, belonging to tertiary strata of the age of the Bagshot sand," forming ancient stone temples at Avebury in Wiltshire and at Stonehenge.[38] The name itself "perhaps" means "Saracen (or Pagan), an epithet which, after their conversion to Christianity, the Saxons may be supposed to have applied to the stones composing these ancient heathen temples."[39] The function of such stones for their "ancient heathen" worshippers being unknown, the "Whispering Stones" embody narrative displacement and tell a story of heathen sacrifice and pagan worship that intensifies the sublimity of the "grey blank sides," the incommensurability of the past. The stones suggest conspiracy, collusion, fragmentation, and mystery.

In turning their "blank, grey sides" to Gwendolen, the sarsen stones shut her out of their collusive, "whispering" conspiracy with Lydia Glasher. Lydia's narrative bears an uncanny resemblance to that hidden history of the stones: like them, she has become disconnected from her past, as Grandcourt has moved away from her and their children, disinheriting her son and so cutting him off from his familial history. The parallel between Lydia and the sarsen stones makes clear the narrative implications of Grandcourt's power and control: in cutting her off, Grandcourt hopes to leave Lydia behind entirely, to deny her agency, to isolate her from her own story. It is appropriate that it is in the landscape of the Whispering Stones that Lydia emerges as a testament to the presence of isolated fragments of the past, divorced from their contexts, eerie in their present power. In their mnemonic function as physical—and yet fragmentary—reminders of a forgotten history, the Whispering Stones reveal to Gwendolen her place in a broader narrative, a wider landscape than her own; and yet in their sublimity and incommensurability, and in their symbolic relation to Lydia's plight, they suggest narrative redundancy, and the power of men like Grandcourt to isolate and prohibit women taking command of their own stories. The story of which Gwendolen hopes to become a part by marrying a vicious and glamorous baronet turns out not to be her own at all, for she is little more than a usurper of Lydia's role in a story over which she has little control. The geological history of the sarsen stones emblematizes the fragmentation, disorder, and breakdown of conventional plot that Gwendolen must learn to accept is an ordinary part of "real" life.

The Make-Believe of a Beginning

Daniel Deronda is famously preceded by the assertion that "men can do nothing without the make-believe of a beginning" (3). Of course, it was this difficulty to which Charles Darwin alluded with a measure of irony in the title of *Origin of Species* (1859), in which no such origin was detected, and it was this difficulty that had caused Spencer to decry geology, with its Palaeozoic strata, as "illogical." But beginnings are not only "make-believe." They are the first constituents of plot. And while they are not to be trusted, they are also essential preconditions of thought. "Men can do nothing," Eliot reminds us, without them.

Noah's Flood, whose geological existence had long been dismissed, is the most frequently deployed "beginning" in *Daniel Deronda*, picking up on the flood that ends *The Mill on the Floss*. There is the sinister depiction of Grandcourt's young son, for example, "bending his blond head over the animals from a Noah's ark, admonishing them separately in a voice of threatening command" (316). Here the child plays at being a tyrannical God, frighteningly echoing his father's paternal and destructive power at the exact moment of his disinheritance. As he plays at power, Grandcourt's son ironically loses it. There is one beginning undone. Or again there is Daniel's search for his origins, a central strand of the story. At thirteen he "had a passion for history, eager to know how time had been filled up since the Flood, and how things were carried on in the dull periods" (149). This adolescent belief in this mythic point of origins is dissolved, however, in Daniel's discovery that he is Jewish and in his search for an even earlier history. Hitherto "Deronda, like his neighbours, had regarded Judaism as a sort of eccentric fossilised form which an accomplished man might dispense with studying, and leave to specialists." And yet, he realizes, it "was something still throbbing in human lives, still making for them the only conceivable vesture of the world" (334). The fossilized past comes alive for him. What becomes important, as the end of the novel suggests as Daniel heads east, is the quest for a meaning-making story, rather than the fixing of it.

Daniel's lesson is not shared by the English culture of which he is a part, however. Though Daniel rejects the English university system as too "narrow," his Cambridge-educated friend Hans Meyrick writes of himself in a letter to Daniel that he had hitherto "held lightly by your account of" the Jew "Mordecai, as apologetic, and merely part of your disposition to take an antediluvian point of view, lest you should do injustice to the megatherium" (598).[40] The "antediluvian point of view" signals Daniel's interest in a history predating the Flood he marked as a starting point in his childhood. Hans's glib dismissal of this perspective prompts "causticity" and "heat" from Daniel (602), for whom the "antediluvian" has a living significance. Another character, Mr. Vandernoodt, gossips about the story of Lydia Glasher: "The affair has sunk below the surface," he tells Daniel; "people have forgotten all about it":

Such stories get packed away like old letters. They interest me. I like to know the manners of my time—contemporary gossip, not antediluvian.

These Dryasdust fellows get a reputation by raking up some small scandal about Semiramis or Nitocris, and then we have a thousand and one poems written upon it by all the warblers big and little. But I don't care a straw about the *faux pas* of the mummies. You do, though. You are one of the historical men—more interested in a lady when she's got a rag face and skeleton toes peeping out. (403–4)

Lydia, he implies, has already lost her story-worthiness. She is already little more than an antiquarian footnote, an "eccentric fossilised form" studied only by specialists like Scott's antiquarian hero "the Rev. Dr. Dryasdust," to whom he refers. Antiquarians, by popular opinion, simply collected mounds upon mounds of objects, which they cluttered up, one atop the other, and never fully reconciled into a meaningful historic narrative. Coins, medals, inscriptions, fossils, manuscripts, all could be found piling out from the antiquarian's study.[41] Vandernoodt's preference for "contemporary gossip" is unmistakably shallow, relegating all history to the pedantries of antiquarian "small scandal," but it is similar to Hans's accusation regarding Daniel's desire not to "do injustice to the megatherium" as "merely antediluvian." There is a preference here more for detail and for clutter than for story. With this attitude, Vandernoodt conceives of Daniel as "one of the historical men—more interested in a lady when she's got a rag face and skeleton toes peeping out." Historical interests are not only antiquarian, but necrophiliac, privileging the clutter of the dead over the living stories of romance.[42]

Vandernoodt's bit of "contemporary gossip" about Gwendolen does in fact "interest" Daniel, of course. Daniel's "mind," we are told, "had perhaps never been so active in weaving probabilities about any private affair as it had now begun to be about Gwendolen's marriage," at least "since the early days when he tried to construct the hidden story of his own birth" (404). In learning to connect his interest in "the hidden story of his own birth" and in the antediluvian past, a past that geology had taught the Victorians to accept, with the "contemporary gossip" of the stories, "packed away like old letters," that affect Gwendolen, Daniel is saved from being a Dryasdust. In *The Mill on the Floss*, narrative was both desirable and dangerous. The novel powerfully refused to transform the "hoary significances" of the past and its multiple meanings, partly produced through imagination, into a single overarching narrative explanation, making Maggie the victim of the cultural stories proliferating around women and incarcerating them within narrow plots. Now, in *Daniel Deronda*, Eliot dramatizes Daniel's gradual acceptance of a complex, feeling, historical consciousness as a narrative action, but one predicated on imagination and on an acknowledgement of the gaps that characterize experience. He does not ignore Lydia and cut her out of the story, as do Vandernoodt and the gossips, but rather accepts her surprising existence. It is this kind acknowledgement that allows him to come to terms with his Jewish heritage, to link the "contemporary" with the "antediluvian," to look both "before and after"—and to do so without succumbing to predetermined, heavily plotted forms of understanding.

The role of breakdown, fragmentation, and disconnection in Daniel's developing historical imagination is explicit in the following passage:

> The first shock of suggestion past, he could remember that he had no certainty how things really had been, and that he had been making conjectures about his own history, as he had often made stories about Pericles or Columbus, just to fill up the blanks before they became famous. Only there came back certain facts which had an obstinate reality,—almost like the fragments of a bridge, telling you unmistakably how the arches lay. And again there came a mood in which his conjectures seemed like a doubt of religion, to be banished as an offence, and a mean prying after what he was not meant to know; for there was hardly a delicacy of feeling this lad was not capable of. (152–53)

Storytelling is imagined here as a means by which to "fill up the blanks" of history, prompted by a lack of "certainty [as to] how things really had been." "Only," as Eliot's narrator points out, "there came back certain facts which had an obstinate reality." That "obstinate reality" is emblematized by Lydia, telling Gwendolen "unmistakably how the arches lay," in the seemingly sublime landscape of the Whispering Stones, where lie blocks of Bagshot sandstone, disconnected from the strata from which they were hewn, "like the fragments of a bridge." It is only by coming to terms with the "fits of dread" she experiences, the "remembered madness" suggesting the abyssal nature of experience, that Gwendolen can achieve psychological integrity. And it is only by encountering the "obstinate reality" of historical fragments, that Daniel can begin to understand his full and true identity.

Unlike Grandcourt, for whom cohesive knowledge in narrative form demonstrates and exercises his will-to-power, Daniel and Gwendolen learn the importance of fragmentation as a constituent of knowledge. Even more explicitly in *Daniel Deronda* than in *The Mill on the Floss*, geological discourse suggests the attractive power of plots, but real attention to geological landforms gives form to a "realist" vision in which fragmentation and breakdown are necessary constituents of self-knowledge in the real world. It is only once we do away with redundant discussions of a mythically constructed debate about "uniformitarianism" and "catastrophism" that we can see that Eliot's writing is often geological writing, haunted by the problems of evidence and by the role of imagination in giving form to a long-dead past.

Dickens and the Geological City { **CHAPTER SEVEN**

Nothing is high, because it is in a high place; and ... nothing is low, because it is in a low one. [*Loud applause.*] This is a lesson taught us in the great book of nature. This is the lesson which may be read, alike in the bright track of the stars, and in the dusty course of the poorest thing that drags its tiny length upon the ground. This is the lesson ever uppermost in the thoughts of that inspired man, who tells us that there are

    Tongues in trees, books in the running brooks,
    Sermons in stones, and good in everything. [*Cheers.*][1]

C harles Dickens may seem a less obvious choice in a book about the writing of the earth sciences than Walter Scott, George Eliot, or Charles Kingsley. Work on Dickens and science has proliferated since George Levine's work in *Darwin and the Novelists*, but its central problem has been the opinion that Dickens's scientific reading was "nugatory."[2] The most well-represented branch of science on his bookshelves was natural history, and even in this Dickens has been felt to display only the "intelligent interest that would be expected of a man of the world."[3] Similarly, though a handful of critics have explored Dickens's sometimes enthused, sometimes anxious response "to the wonder and romance of contemporary science" in his periodicals *Household Words* (1850–1859) and *All the Year Round* (1859–1881), others note that such readings are beset by "the difficulty of establishing the extent to which the journals ... may reliably be taken to represent Dickens's own views and opinions."[4] Levine's influential "one culture" model partly surmounted the problem by pointing out the similar structural patterns implicit in the worlds described by Dickens and Darwin, and one recent study has generated insightful new readings of *Little Dorrit* and *Our Mutual Friend* in which their protagonists are imagined as excavators of ruined and buried pasts shaped by the archaeological and geological excavations with which Dickens and his readers were familiar.[5] But in an attempt to develop more direct and concrete links between Dickens's work and evolutionary science, almost all subsequent studies have focused on his 1860s novels, written after the publication of the *Origin of Species* (1859), and on the recovery of textual interrelationships between Dickens's works and scientific writings or ideas.[6] While the previous chapters of *Novel Science* may seem also to have focused on primarily textual interrelationships, this final chapter will depart from that model and build on the book's earlier work on maps and illustrations, on learned societies and gentleman's clubs, and on periodical culture, to think about the range of sites in which "science" took place in the nineteenth century and with which Dickens was more fully engaged than many of his fellow novelists.

A primarily textual approach clearly misses a trick. For scientific writings and ideas did not take solely textual form. In geology, a visual and structural science, this is even more emphatically the case than in some others. As Ralph O'Connor has written, "Geological phenomena were displayed within, and in competition with, an enormous range of theatrical reconstructions of the landscape and assorted natural disasters, from pantomime stage-sets to outdoor simulations of volcanic eruptions. Most of these sites would not be considered 'scientific' today, and few were intended to promote geology; yet they all contributed to the science's public appeal."[7] Not only contributing to the science's appeal, moreover, these "theatrical reconstructions" also gave geological writers "a rich source of imagery for reconstructing exotic landscapes, the distant past, and historical change," and they provided "virtual tourism" for an "untravelled public."[8] Museum objects were accompanied by imaginative illustrations that drew heavily on the form and style of the panorama—a 360-degree painting suspended around the viewer without a visible frame, in a purpose-built rotunda building. Panoramas required spectators to observe the canvas from the inside, usually ascending ladders or stairs to a platform to view it, and verbal recreations of panoramic scenes from deep time abounded in geological writing, from textbooks to expository pamphlets. Such scenes might also employ shifts like the "double-effect" diorama, a related visual form that overlaid two scenes and, by a trick of the light, made one mutate into the other.⁹ It is important to note that the words "panorama" and "diorama" were used very flexibly in the period: other forms of vast or shifting displays were often referred to as "panoramic" or "dioramic," and the term "panorama" came to denote any painting of vast size. As O'Connor's analysis definitively shows, moreover, this visual aspect of geological science is not to be sniffed at as some derisory form of popularization in which elite knowledge is diffused to the less-educated masses. Recently historians have emphasized a more general "pictorial turn in British culture" and the range of sites and experiences that constituted Victorian science.¹⁰ "Authorship, editing, reviewing, specimen dealing, industrial consulting, instrument making, museum curating, lecturing, and showmanship" all count as important practices that go toward defining what we think of as nineteenth-century science. James A. Secord has called this "commercial science" and reminds us that, though "we are used to seeing the Victorian world through the printed word ... the visual field of contemporaries was dominated by shows, exhibitions, and pictorial representations." So the controversial *Vestiges of the Natural History of Creation*, he argues, "was read not only as a book," despite the fact that there were no illustrations within its covers, "but *seen* as a museum of creation," playing on the widespread visual literacy by which geologists and the public had come to comprehend the geological past.¹¹ Recovering Dickens's engagement with the scientific culture of his day, then, requires immersion in that visual culture. For despite the well-established connections between Dickens's novels and Victorian popular entertainment and between Victorian show business and the display and dissemination of science, critics have not yet fully explored the links between scientific shows and Dickens' fiction.¹²

It should be unsurprising that it was "commercial science" with which Dickens was involved. Dickens's geology is not the geology of the Geological Society of London. It is not the geology of the heavyweight quarterlies, or of the weighty geological tome. Though he appears to have been friends with the stratigraphic geologist Roderick Murchison, there is less evidence that he read Murchison's grand works *Silurian System* (1839) or *Siluria* (1854) than that he enjoyed the company of a scientific celebrity.¹³ Though his scientific *reading* may have been "nugatory" compared to that of George Eliot, for example, it was with "the showmen of science" and the science of show that Dickens was primarily involved.¹⁴ This should not be surprising: literary men and women were as engaged in the commercial, visual, and spectacular world of show business as scientific ones, and Dickens was both the most performed and the most performative of all Victorian novelists.

This chapter focuses on three "late" Dickens novels: *Dombey and Son* (1846), *Bleak House* (1853), and *Our Mutual Friend* (1865). Dickens's contemporaries often greeted these novels with confusion and disappointment when compared with the earlier picaresque-style fictions of such novels as *Pickwick Papers* (1837): "The problem, they agreed, with the later Dickens novel was— not enough plot."[15] This may seem counterintuitive to the modern reader, for it was precisely from *Dombey and Son* onward that Dickens made careful notes and plans for the structure of his novels. In these novels we have a Dickens carefully working out plot patterns, so that no single episode can be understood without reference to the pattern of the whole.[16] But what I want to argue is that this structure is predominantly spatial, relating parts to whole in a way that resists simple linearity. This is not to say, of course, that Dickens did not tell stories or construct plots in these novels, but rather to remind readers of the primarily spatial and pictorial organization with which he was experimenting in the 1850s and beyond. This newly assiduous concern with structure in the later novels has consistently been related to Dickens's apprehension of the changing organizational structures of the world around him. For J. Hillis Miller, there is a conflict in both *Bleak House* and *Our Mutual Friend*, tackled differently in each novel, between the "dispersed" individual units of experience and the possibility of "an ideal unity ... transcending the differences between individual lives."[17] Others identify "Dickens' contradictory investment in both order and fragmentation," reflecting "the social formation in which he wrote," in which he experienced both the "atomisation" and breakdown of community in urban life and the creation of new forms of institutional organization such as the Chancery courts or the education and prison systems.[18] Moreover, if critics have seen the structures of Dickens's late fiction mirroring a world both fragmenting into isolated parts and cohering all too rapidly into powerful modes of state organization and control, they have also located a new historical consciousness in these same works. Hilary Schor, for instance, suggests that plot in the later novels depends upon "the constant revelation and re-encryption of the past," "the quest to know and understand" it.[19] Experimenting with fictional forms that might relate part to whole, the late Dickens novel reaches ever further into the past for an authoritative ground on which those fragments might rest, and it is the argument of this chapter that Dickens organizes the history he finds spatially and pictorially as much as narratologically. This is not intended to be a binary. The presence of a spatial structure does not preclude storytelling, by any means, and a picture is not the opposite of a narrative. But there is a sense in which Dickens arranges lives and scenes in jigsaw fashion in these late texts, placing them in contiguous rather than obviously causal relation to one another, building an *image* of a world, a spectacular snapshot of it. Such novels as *Our Mutual Friend* (1865) excavate the "unattainable substratum" of experience, imagining the modern world "everywhere heavy with the debris of history."[20] In this context, as this chapter will demonstrate, Dickens's representations of a transforming London cityscape frequently draw on a particular mode of geological knowledge in order to visualize the city in the midst of history, reconciling chaos into a series of visual images that make sense only when they are contextualized in time, but in which time is represented via visual and essentially spatial forms. Furthermore, this geological vocabulary and imagery enters into Dickens's self-conscious considerations of his narrative art and its relation to the "real" world.

There are other showmen of science central to this story. Dickens also made the acquaintance of Richard Owen, the illustrious comparative anatomist who named the "dinosaur," and whose contribution to Victorian science often took public, spectacular, and visual forms. Owen had been conservator of the Hunterian Museum, was involved in the construction of the Crystal Palace prehistoric garden, and later pushed for the construction of the Museum of Natural History in South Kensington.[21] Owen was no stranger to a piece of scientific showmanship, fashioning himself as the wizardly creator of the fossil saurians, reconstructing their bodies from tiny fragments of bone at a single glance, resting a hand on creatures that towered above him as if they were old friends (fig. 7.1). Owen's involvement in the prehistoric garden project was merely as a consultant—and a second-choice consultant at that (after the fossil collector Gideon Mantell, who turned down the opportunity to work on the models)—to the sculptor Benjamin Waterhouse Hawkins. Though "there can be no doubt that Hawkins did make extensive use of Owen's published descriptions of extinct reptiles and mammals," he also relied on other authors and contributed his own interpretations of the available evidence. Owen "appears to have had his only systematic input just before the small clay sculptures were scaled up to their final dimensions; and in many cases his advice was simply ignored," so that many of the creatures he had named for the Victorians, including the iguanodon and the megalosaurus, were not recreated according to his descriptions of them.[22] Nonetheless, Owen's image was enough. The Crystal Palace Company publicized his involvement in the project in an attempt to enhance "the authority of the displays" "by drawing on heroic images of the man of science." "Publicity about the dinner" held in the belly of the

Fig. 7.1. Richard Owen posing with the moa bird, *Dinornis giganteus*. In Owen, *Memoirs on the Extinct Wingless Birds of New Zealand*, vol. 2 (1879), plate XCVII.

Fig. 7.2. "Dinner in the belly of the iguanodon." *Illustrated London News*, issue 662 (January 7, 1854), 22: Reproduced by permission of the syndics of Cambridge University Library.

Fig. 7.3. "Professor Owen: Riding his hobby." In Frederick Waddy, *Cartoon Portraits and Biographical Sketches* (1873), 37.

iguanodon on New Year's Eve 1853, in honor of Owen, who sat at the head of the table, "never ceased to emphasize Owen's character as a scientific wizard who could restore lost creatures as if by magic" (see fig. 7.2).[23] Improving on earlier reconstructions of the primeval monsters at the Hunterian Museum and the Antediluvian Room at the Egyptian Hall, the garden caused a public sensation, prompting a spate of illustrations of the creatures in the popular press. Together, Owen and Hawkins invented the image of the "dinosaur" for a whole generation of Victorians (see fig. 7.3).[24]

Owen and Dickens read each other's works, attended each other's lectures, and visited each other's homes.[25] They almost collaborated in the mid-1860s when Dickens only agreed to contribute to a lecture series (which did not go ahead) for the anti-Sabbatarian Sunday League on the condition that Owen would also contribute.[26] *Household Words* ran several articles in the late 1850s explicitly supporting Owen's science and its belief in the archetype—an "ideal form" for each species on which all variations were based—and in God's plan manifested in the work of Nature, and Owen himself contributed to the journal.[27] This is not to say that they agreed with one another on all counts, or that their friendship implies a common view of the earth and its history and structure—such an argument would be "sanguine" in the extreme.[28] It is rather to say that both men were engaged in the ways in which spectacle could give form to their views of the world and its workings, and give equally prominent form to their careers. In this, they may have been able to help one another.

The Poetry of Science

In Dickens's review of Hunt's *Poetry of Science*, he sets out enthusiastically his ideal version of science and its relatedness to artistic and creative endeavor:

> Science has gone down into the mines and coal-pits, and before the safety-lamp the Gnomes and Genii of those dark regions have disappeared. . . . Sirens, mermaids, shining cities glittering at the bottom of quiet seas and in deep lakes, exist no longer; but in their place, Science, their destroyer, shows us whole coasts of coral reef constructed by the labours of minute creatures; points to our own chalk cliffs and limestone rocks as made of the dust of myriads of generations of infinitesimal beings that have passed away; reduces the very element of water into its constituent airs, and re-creates it at her pleasure. Caverns in rocks, choked with treasures shut up

from all but the enchanted hand, Science has blown to atoms, as she can rend and rive in the rocks themselves; but in those rocks she has found, and read aloud, the great stone book which is the history of the earth, even when darkness sat upon the face of the deep. Along their craggy sides she has traced the footprints of birds and beasts, whose shapes were never seen by man. From within them she has brought the bones, and pieced together the skeletons, of monsters that would have crushed the noted dragons of the fables at a blow.[29]

This ideal science is geology. Dickens takes us "hundreds of fathoms underground," and describes the formation of coral reefs, chalk cliffs, rocks that reveal a primeval "history of the earth," and of fossilized creatures "whose shapes were never seen by man." He acknowledges the ways in which science is a "destroyer," annihilating belief in myth, legend, and fantasy with its empirical outlook. But this destruction is in fact a form of reabsorption: mermaids, for example, are replaced directly with "whole coasts of coral reef constructed by the labours of minute creatures." Science does not merely "destroy" fantastical and aesthetic elements of mythology but converts them into newer understandings, replete with that older sense of wonder and awe. Geological science is inherently performative. It has an awe-inspiring power to "rend and rive" in the rocks, to blow caverns "to atoms," and to "re-create," through performance, the natural world, which itself becomes a performative force: both the new science and the monsters it has revealed have "crushed" the fabulous dragons in a single blow.

In London representations of the natural world often formed part of exciting, spectacular performances.[30] At just one venue, the Surrey Zoological and Botanical Institution, near central London, volcano panoramas, with sound effects and fireworks, were displayed in 1837–1838 (Vesuvius), 1839 (Mount Hecla), 1846 (Vesuvius), and 1852 (Etna).[31] The 1839 Mount Hecla had been seen by 578,000 people by its hundredth show.[32] "One of the biggest hits of the whole Victorian era" was the panorama *The Ascent of Mont Blanc* at the Egyptian Hall in the mid-1850s, a venue that also staged a panorama of Naples and Vesuvius by 1857, coinciding with a series of earthquakes at the site that year.[33] Of the impending Mont Blanc show, Dickens wrote, "So many travellers have been going up Mont Blanc lately, both in fiction and in fact, that I have heard recently of a proposal for the establishment of a Company to employ Sir Joseph Paxton [architect of the Crystal Palace] to take it down.

Only one of those travellers, however, has been enabled to bring Mont Blanc to Piccadilly," the inventor of the show, Albert Smith.[34] *Household Words* covered the earthquake shows in an article of 1858.[35] The Colosseum at Leicester Square, a dome-shaped building that regularly housed popular panoramas, had another huge hit with the "Cyclorama" in the late 1840s.[36] The Cyclorama opened with a panoramic recreation of the Lisbon earthquake of 1755, including rumbling floors and violent light and sound effects. It caused a sensation in London in 1848 and shook on into the 1850s, when it was replaced by pictures of the Crystal Palace and then, lo and behold, an erupting Vesuvius.[37] In the 1840s dioramas often gained success recreating geological events—the avalanche that buried a Swiss town in 1820 or an eruption of Mount Etna, for instance.[38]

Dickens's essay "A Rapid Diorama," in his 1846 travel book *Pictures from Italy*, reveals the extent of his immersion in such shows and the ways in which the spectacular forms they gave to the earth also gave shape to the forms of his writing. In this essay Dickens describes his trip to the region of Naples and his ascent of Vesuvius.[39] Stories like these were popular in the 1840s and 1850s: following Edward Bulwer-Lytton's *The Last Days of Pompeii* (1834), London periodicals frequently ran travel pieces describing the ascent and descent of Vesuvius, alongside trips to Pompeii, including Dickens's own magazine *Household Words*.[40] Dickens's "rapid succession of delights" is replete with shifts of scenery that fade in and out of view, overlaying images and scenes in a montagelike transformation of the initial view, just like the diorama. In "the morning, just at daybreak," for example, "the prospect suddenly becoming expanded, as if by a miracle, reveals—in the far distance, across the Sea there!—Naples with its Islands, and Vesuvius spouting fire."[41] Once the image of Vesuvius has been superimposed on this scene, it fades away again: "Within a quarter of an hour, the whole is gone as if it were a vision in the clouds, and there is nothing but the sea and sky" (235). As in the diorama, the view enlarges and recedes before the viewer's static eye; the entire Naples area passes before him, sitting stationary in his carriage, in a succession of dioramic transformations.

Dickens also wrote on "the gigantic-moving panorama or diorama" in *Household Words* in 1850:

It is a delightful characteristic of these times, that new and cheap means continually being devised, for conveying the results of actual experience, to those who are unable to obtain such experiences for themselves; and to bring them within the reach of the people—emphatically of the people; for it is they at large who are addressed in these endeavours, and not exclusive audiences.... Some of the best results of actual travel are suggested by such means to those whose lot it is to stay at home. New worlds open out to them, beyond their little worlds, and widen their range of reflection, information, sympathy, and interest. The more man knows of man, the better for the common brotherhood among us all.[42]

For Dickens, these forms of visual representation derive their worth from their ability to provide the masses with an imaginative means of travel, a kind of vicarious tourism without the cumbersome material problems involved in travel. The moving images of the diorama cause other, less transient shifts in perspective: the imaginative movement they suggest could also "widen" the emotional and intellectual "range" of the travelers. The imaginative journey across geographical space involves an intellectual journey that expands the spectator's moral perspective. The panorama, with its extended field of vision, equally extends "the common brotherhood among us all." In this way, Dickens's dioramic writing can be related to the writing of the stonemason-turned-journalist-turned-geologist Hugh Miller, whom O'Connor singles out as the most evocative exponent of the pictorial, panoramic, or dioramic mode of geological writing in the Victorian period. Indeed Dickens professed himself "much interested in the geologist" when Miller committed suicide in 1856, having been correcting page proofs for his *The Testimony of the Rocks* earlier in the afternoon. For both men, whose careers as geologist and novelist were also entwined with their careers as journalists, the panorama was an integral mode of their thinking: panoramas displayed news and were associated with it so much that when *Illustrated London News* was first published, it "introduced itself with a bow to the panorama."[43] Perhaps it is not surprising, then, to discover that Dickens had much in common with Miller. Both were feeding in and out of the same urban cultures of print and display.

By the late 1840s, then, the London public had worked up an appetite for imaginative journeys through exotic, arctic, and historical landscapes, partly satiated through panoramas and dioramas and equally satiated by the textual recreation of visual scenes of the past in the pages of such periodicals as *Household Words* and such books as *Pictures from Italy* or Miller's *The Old Red Sandstone*. In 1851 the Great Exhibition gave even more direct access to worlds that the average London showgoer could not reach. Competing with London's numerous volcano shows, the Exhibition included a Pompeian Court: "the lava-buried city, the city of the dead—at the epoch of its destruction by a sudden eruption of Vesuvius."[44] Visitors could stroll through the Alhambra, through Egyptian, Greek, Roman, Byzantine, French, Italian, German, Elizabethan, and Renaissance courts, and could see animals, plants, and industrial products of Australia, America, and the Orient. There were six hundred geological exhibits at the Exhibition, including fossilized fish, trees, and forests; geological maps; marbles, limestones, and petrified eggs; models of Britain's deepest mines; and cross sections of geological strata revealed through railway cuttings.[45] Its view of the world past and present, with its claims to be all-encompassing, prompted the inventor Charles Babbage to call it "A Diorama of the Peaceful Arts."[46]

Dickens, of course, was ambivalent about the Exhibition:

> There is a range of imagination in most of us, which no amount of steam-engines will satisfy; and which The-great-exhibition-of-the-works-of-industry-of-all-nations, itself, will probably leave unappeased.... The Polytechnic Institution in Regent Street ... is a great public benefit and a wonderful place, but we think a people formed *entirely* in their hours of leisure by Polytechnic Institutions would be an uncomfortable community.[47]

His difficulties here are a by-product of his fears about utilitarian science, denounced in *Hard Times*, and particularly of his fears about the state control of mass entertainment. Not only did natural history, and particularly geology, frequently offer Dickens an alternative to utilitarianism by giving scientific credibility to myth, superstition, and spectacle, but when the Crystal Palace was reconstructed at Sydenham in 1854, this time with a prehistoric garden full of primeval monsters—and became a private, commercial enterprise—Dickens and *Household Words* were both far more enthusiastic: one article described the garden's recreation of "those antediluvian days when there were giants in the land" as a "Fairyland" and "pleasure ground for those whose lot it is to labour."[48] All over London, then, visual and material displays pictured a prehistoric world characterized by enormous and lifelike monsters, and the natural world as intrinsically spectacular: immense size and the pleasure of fear in the face of calamitous disasters and gigantic creatures turned nature into performance. As Secord puts it, "the fundamental criterion for contemporary sculpture was 'truth to life,'" a perception of verisimilitude that went hand in hand with "an age of the gargantuan in science and technology."[49]

In 1846, the year of the Surrey Gardens' Vesuvius panorama, the real Mount Vesuvius was active all summer. Dickens had climbed the volcano the previous year, burned his feet on its lava and ashes, and described his experiences in *Pictures from Italy*. This was also the year in which he began writing and publishing *Dombey and Son* (1846–1848), his novel of "transition," as G. K. Chesterton described it.[50] The earthquakes and volcanoes he had witnessed in Italy become one of the central motifs of that novel as he describes the building of the railroad in a working-class London district:

> The first shock of a great earthquake had, just at that period, rent the whole neighbourhood to its centre. Traces of its course were visible on every side. Houses were knocked down; streets broken through and stopped. . . . Everywhere . . . carcases of ragged tenements, and fragments of unfinished walls and arches, and piles of scaffolding, and wildernesses of bricks, and giant forms of cranes, and tripods straddling above nothing. There were a hundred thousand shapes and substances of incompleteness, wildly mingled out of their places, upside down, burrowing in the earth, aspiring in the air, mouldering in the water, and unintelligible as any dream. Hot springs and fiery eruptions, the usual attendants upon earthquakes, lent their contributions of confusion to the scene. Boiling water hissed and heaved within dilapidated walls; whence, also, the glare and roar of flames came issuing forth; and mounds of ashes blocked up rights of way, and wholly changed the law and custom of the neighbourhood.[51]

Just like "the earthquake which preceded the eruption" of Vesuvius that buried Pompeii in *Pictures from Italy*, Stagg's Gardens has been hit by both earthquake and volcano (the "hot springs and fiery eruptions" that are "the usual attendants upon earthquakes").[52] The "conical-shaped hill" and "great sheets of fire . . . streaming forth" in Italy have become the "unnatural hill" and the "glare and roar of flames . . . issuing forth" in London.[53] Dickens's Italian town has "great walls of monstrous thickness . . . obtruding their shapeless forms in absurd places, confusing the whole plan, and making it a disordered dream."[54] In his description of London in *Dombey and Son* there are, similarly, "a hundred thousand shapes and substances of incompleteness, wildly mingled out of their places," making the scene "unintelligible as any dream." As Wolfgang Schivelbush has noted, the building of the railways in London transformed the city in the 1840s, "intrud[ing] deeply into working-class neighbourhoods," so that Dickens's description closely mirrors a changing social reality.[55] But as Schivelbush also notes, the "successive scenes" witnessed through the windows of a railway carriage also transformed the landscape into panorama.[56] It is this panoramic (or, more accurately, given its shifting scenes, dioramic) mode of perception that can be felt in the texture of Dickens's prose here, so that the dioramic presentation of the disaster in *Pictures from Italy* informs its re-presentation in the novel.[57] In *Pictures from Italy* Dickens travels via railroad to a volcanic landscape. In *Dombey and Son* the situation is the reverse: the observer in the center of London remains static—and the railroad comes to him. Like the observer at a diorama, the reader of Dickens's description of London watches the transformation of a scene through the action of a devastating volcano.

This scene of "dire disorder" is implicated in a broader narrative in which the inhabitants of Stagg's Gardens perceive their neighborhood to be under threat. Though it is "regarded by its inhabitants as a sacred grove not to be withered by railroads" and other "such ridiculous inventions," the prerailroad Stagg's Gardens is in fact, the site only of "frowzy fields, and cow-houses, and dunghills, and dust-heaps," prefiguring the dunghills and dustheaps of the wasteland London of *Our Mutual Friend*.[58] With its "rotten summer-houses," this "unhallowed spot" is hardly the Edenic "sacred grove" its inhabitants claim it to be. Dickens extends this gentle parody of the Stagg's Gardens population by mocking its naïve natural history of its landscape. Some hold it to have been constructed by "a deceased capitalist"; "others, who had a natural taste for the country, held that it dated from those rural times when the antlered herd, under the familiar denomination of Staggses, had resorted to its shady precincts" (69). Dickens presents a pseudo-aristocratic register in the noun phrases "familiar denomination," "antlered herd," and "shady precincts," with their connotations of hunting and hereditary lineage. But the cacophonic "Staggses" sounds like a Dickensian-Cockney mispronunciation of the plural for the hunted "stags" and punctuates this flow of descriptive language, revealing it to be mere hyperbole. Dickens uses the comic intrusion of dialect to prove the partiality of the inhabitants of the gardens; he reminds us that this is not Eden but a shabby, working-class London district.

These old-fashioned natural historians use anecdotal evidence and inherited aristocratic and biblical traditions to depict Stagg's Gardens as an Eden (a "sacred grove," a "shady precinct") in the process of being lost (69). In this essentially linear account, the narrative of the Fall reconfigures the coming of the railway as an irreversible event that emphatically destroys the landscape. But Dickens's narrator searches for "traces of [the earthquake's] course," using the present, chaotic scene to understand the prior actions of the event like any mid-Victorian geologist (68). The geological language here is developed in direct opposition to the biblical earth history offered by the London inhabitants. Dickens's narrator in *Dombey and Son* closely observes "traces," "fragments," and "treasures" in disarray, "wildly mingled out of their places," like fossils scattered in a moment of geological upheaval. The "giant forms" and "carcases of ragged tenements" resemble the giant remains of extinct creatures famously discovered in the 1840s as navvies cut the railways themselves—an iguanodon was found at Bletchingley and an ichthyosaurus in the Oxford, Worcester, and Wolverhampton line.[59] In this passage Dickens's language reveals that the railroad connects the city with geographically distant places and the temporally distant monsters that lurk beneath its soils.

When we return to Stagg's Gardens we find that there is, in fact, "no such place as Stagg's Gardens" (233). Its "giant forms" have actually become the "tame dragons," the "monster train[s]" that "swallowed up" the old wasteground; "mountains of goods" have grown and stand "bubbling and trembling there, making the walls quake, as if they were dilating with the secret knowledge of great powers yet unsuspected in them" (234). The disastrous and the spectacular, unearthed in the railway's construction, have become a feature of its ongoing progress. Progress and geological disaster are inseparable in the endlessly shifting urban world Dickens depicts.

London in the mid- to late 1840s was full of these images of spectacular geological processes, to which the pages of Dickens's *Pictures from Italy* and *Dombey and Son* contributed, and out of which they were fashioned. In 1848 the Cyclorama was recreating the Lisbon earthquake four times daily, and it seems more than coincidental that Captain Sol's rooms in *Dombey and Son* are described as being "very small ... but snug enough; everything being stowed away, as if there were an earthquake regularly every half-hour" (126). Dickens appeals to the visual imagination of the popular audiences who attended dioramas, panoramas, and the Great Exhibition and encourages in his readers the same kind of expansive, all-encompassing view that those entertainments offered. Dickens taps into the realms of popular mythology and the pleasures of the scientific spectacle familiar to his readers through periodical articles, the "dioramic" writings of geological writers, and several interlocking forms of urban display. He does so while stressing, moreover, accurate observation and dramatizing its broadening of social vision for the inhabitants of Stagg's Gardens. Geology allows Dickens access to popular myth, fantasy, and spectacle in *Dombey and Son*, but also allows his readers to travel imaginatively, suggesting his broader social purpose. Dickens finds in geology a form for the chaos of modernization and urbanization: like the fragmented strata it too needs reorganization. But this geology is also a performed natural world, a world connected to the visual forms of urban display, one that parodies the narratives of modernization and progress built in to the construction of the railway. Geology brings order, a sense of history, and an expansive temporal vision that, for Stagg's Gardens and for Dickens's readers, promotes social harmony— "the common brotherhood among us all"—but also converts that history into spectacle, uniting readers in a shared community of affect. Such readers respond with awe, wonder, and fear to historical processes that are always experienced at one remove, abstracted into a series of images for perusal and contemplation, made objects of hermeneutic activity through the performative nature of their representation. That imagistic imagination, furthermore, converts geology into a pictorial spectacle. It is put to the use of all kinds of stories. But it does not tell one kind of plot or another about the workings of earth history. Important are scale, motion, transformation, and surprise—not the particular pattern by which geological change can be understood.

The Elephantine Lizard

Bleak House (1852–1853) was written on the other side of the Great Exhibition, when the Crystal Palace prehistoric garden was under construction. The central image of Stagg's Gardens in *Dombey and Son* was a visceral imagination of an earth rent open by the processes of history, the primeval strata laid bare in an act of industrial construction, "dragon forms" of a long-buried past reanimated by the violence of historical momentum. In *Bleak House* that paradoxical relation between the violence of urban modernization and the feral history it exposes is the opening image of the novel:

> London. Michaelmas Term lately over, and the Lord Chancellor sitting in Lincoln's Inn Hall. Implacable November weather. As much mud in the streets, as if the waters had but newly retired from the face of the earth, and it would not be wonderful to meet a Megalosaurus, forty feet long or so, waddling like an elephantine lizard up Holborn-hill.... Foot passengers, jostling one another's umbrellas, in a general infection of ill-temper, and losing their foot-hold at street-corners, where tens of thousands of other foot passengers have been slipping and sliding since the day broke (if the day ever broke), adding new deposits to the crust upon crust of mud, stick-

ing at those points tenaciously to the pavement, and accumulating at compound interest.[60]

The megalosaurus on Holborn Hill "waddles" out of the receding tide of the biblical Flood ("the waters [that have] but newly retired from the earth"). Like Stagg's Gardens, recovering from "the first shock of a great earthquake," this landscape is recuperating from a vast geological event. On returning to Stagg's Gardens, we find that new "mountains" have formed in our absence: it has been entirely remade. The London streets in *Bleak House* are in the middle of this process, accumulating ever upward, "crust upon crust of mud" piling up like geological strata, offering yet another earth history, an alternative to the biblical Flood. Holborn Hill was a scandalized area of the city in the 1850s, the subject of much agitation, because here the highway dipped to the level of the Fleet sewer, forming a barrier between the commercial center of the City and the Inns of Court, two spaces integral to the London of the professional classes.[61] Here Dickens's cynical poke at "accumulating at compound interest" gathers meaning: not only would he "and his contemporary readers ... have seen Holborn Hill as metonymic of pervasive English inefficiency," but the accumulation of sewage and mud in the streets of London both hinders and parodies the workings of finance.[62] It forms a literal barrier to those workings by physically impeding travel to the city, at the same time mocking the sheer bestiality of unfettered accumulation. Moreover, like the railway cuttings that revealed slices of the earth's strata and which Dickens visualized in *Dombey and Son*, excavations for sewers offered Victorians visual access to the layers of the past: here a broken and rent landscape opens to reveal its hideously corporeal secrets, the sewers and the monstrous megalosaurus threatening to overspill their buried realm and emerge into the contemporary world.

Several critics have noted the novel's deployment of geological language.[63] But mostly they have been concerned with what they see as the debate between two geological schools, "uniformitarian" and "catastrophist," a debate this book argues has been profoundly misinterpreted in literary criticism and should now be all but excised from analysis. George Levine, for instance, negates the impact of "catastrophism" on the text in order to assert that Dickens was a "uniformitarian": "His Neptunism and Vulcanism are a literary convenience that required no belief," he writes, referencing (and dismissing) early nineteenth-century debates about the respective roles of flood and fire in geological history.[64] For Lawrence Frank the novel is indebted to the evolutionary arguments of *Vestiges of the Natural History of Creation*. Dickens's anonymous narrator, with his present-tense opening, "will enact the excavation of . . . Victorian London" in a godless world. He is a geologist after the manner of *Vestiges*. By contrast, the narrative offered by the novel's heroine, Esther Summerson—providential, conservative, and teleological—gives voice to the "catastrophist" point of view.⁶⁵ It should not need reiterating here that "catastrophists" were not a distinct school of geologists, and those who opposed Lyell's work on the grounds that events of greater intensity than those recorded in human history may have taken place in the depths of geological time were not defending a providential view against a scientific one. Nonetheless, another critic has argued that Dickens sets uniformitarianism (represented by Lyell, gradualism, Darwin, reform, and the realism of George Eliot) against catastrophism (Carlyle, Chartism, revolution, and the sensation novel) in order to avoid "false oppositions between realists and a 'countertradition of fantasy.'"⁶⁶

In countering such claims it should be noted that the event recounted in this passage from *Bleak House* does not constitute geological "catastrophism," for there is no hint as to whether such events are part of the regular economy of nature. To the extent that they are, the occurrence could as readily be considered "uniformitarian" as catastrophic. Lyell's *Principles of Geology* is a veritable catalog of natural disasters. And this feeds into a more important point about the nature of natural and geological spectacles in Victorian show culture. At best they made only implied arguments about the overarching pattern of geological change. The panorama, as a static illustration of a large-scale natural event, makes no claim about the underlying plot of earth history. It does not argue that the earth was once more violent than it is now. Nor does it argue that volcanoes have always operated at the same intensity. It simply describes an event at the moment of its happening, and while this may imply a story, it is a localized story and not part of any wholesale plot that explains all such events and the mode and regularity of their occurrence. And while the diorama's shifting scenes tell a very limited story—of the eruption of a volcano from a quiescent state, for instance—neither do they feed into the debate about "uniformitarian" theory. Indeed, the shifting scenes of the diorama physically enact the kind of story that most geologists were, by the 1830s, prepared to tell: one scene succeeds another, much as the stratigraphic column suggested that geological ages succeeded one another. But the underlying plot of that succession—the causal principle that connects the various scenes of earth history—remains unarticulated. The Crystal Palace monsters, by contrast, did make an argument, allowing spectators to conduct a three-dimensional walk through geological time, moving from more primitive to more advanced life forms. Even here, however, the precise "plot" that underlay this movement was left largely unarticulated or implicit. Consequently, we should be wary of reading theoretical geological positions into Dickens's work. Dickens's work is part of a much broader tradition of geological spectacle, overlapping with the successive scenes of the geological retrospect (discussed in relation to Kingsley in chapter 5), than such readings are able to suggest.

Similarly, many critics suggest that the megalosaurus was a disorienting figure with which to open a novel, or an "incommensurable image."[67] But London inhabitants who attended the panorama, visited the Great Exhibition, or read *Household Words* might have known better: the megalosaurus was just one of many images of primeval monsters available to consumers of popular entertainments in 1852–1853, at the Hunterian Museum and in the Egyptian Rooms. It was also found on the pages of popular periodicals.[68] In an article in *Household Words* from 1851, for instance, the megalosaurus was a central character. The article begins:

> Now that we can visit any portion of the globe by taking a cab or an omnibus to Leicester Square, who wants a Phantom Ship to travel in? The world, as it is, has taken a house in London, and receives visitors daily. Nothing remains now for the Phantom, but sail into the world, as it was, or as it will be.... We mean to sail quite out of human recollection, to the confines of human existence, and remain in dock among the Graptolites.
>
> So we walk down Cheapside, bustle aboard at London Bridge, and sail out, leaving man behind us. Leaving man behind us; for a thousand years roll back upon themselves with every syllable we utter; years, by millions and millions, will return about us, and restore their dead.[69]

The "Phantom Ship" articles, like the panoramas to which they are indebted, were a series of descriptions of other parts of the world, unreachable to the average *Household Words* reader, including China and the Arctic. But alongside the throng of the Great Exhibition, which has brought "the world" to London, which "receives visitors daily," and which the writer, Henry Morley, desires to escape, the rival attraction of Wyld's Globe in "Leicester Square" pictured "any portion of the globe" its spectators might wish to see. Morley competes with the Exhibition and the Globe in offering this alternative journey, but in doing so he also attempts to escape from the busy London world they had created. The ability of the panorama to provide vicarious tourism, then, takes on a political function: it forms an explicit defense of the integrity of the individual against the homogenizing influence of the crowds descending upon London.

Morley's journey into the prehistoric past allows his reader the comfort of remaining in "Cheapside," in the position of panoramic or dioramic spectator, while he spins geological epochs, each more distant and strange than the last, past his or her vision. Morley relies on panoramic and dioramic mechanisms of vision to shape the reality of the other worlds that he describes, allowing his reader to visit other worlds in space and time without, as Dickens would put it, the inconvenience of traveling there oneself. In doing so, Morley anticipates what would be the next master stroke of the Exhibition itself: the recreation of the megalosaurus and its fellow creatures for the pleasure of the London throng. As he does so, he does not invoke a narrative of the geological past but a grand visualization of all its components, recreated textually to form a single view.

Furthermore, just a week before *Bleak House* was republished in single-volume format, Francis Trevelyan Buckland's article "Old Bones" reported on the discovery of dinosaur bones beneath the ground outside St. John's College in Oxford and contrasted the modern world of 1853 with "the apparition of that great leviathan [the megalosaurus] on the top of Heddington [*sic*] Hill."[70] Buckland's alliterative recall of Dickens's "Holborn Hill" and their semicomic juxtapositions of dinosaurs with modern streets sets up a mutual allusion between *Bleak House* and *Household Words*. Each advertises the other. This network of references and exhibitions of the dinosaur reveals that Dickens was, in fact, participating in a culture in which the image of the dinosaur and the geological past was made commensurable to a paying public. At the very least he advertises the work of Richard Owen, popularly known as "Old Bones" for his dramatic skeletal reconstructions of extinct monsters. In this context, the megalosaurus can be read as a popularizing image, a dramatic effect targeting the paying publics currently discovering the megalosaurus for themselves. For the 1850s reader this is not a wholly disorienting passage but a specifically orienting one: with its dioramic vision, the scene's present-tense language reads like a set of stage directions ushering in the image that overlays the Holborn district, holding modern London and a prehistoric swamp in a single view.

The novel repeatedly returns to the language of geology, from the references to Noah's Flood to the image of Professor Dingo "disfigur[ing] houses and other buildings, by chipping off fragments of those edifices with his little geological hammer," a model of professional tenacity in contrast to Richard Carstone's frequent changes of vocation (246–47).[71] Judy Smallweed attains to "a perfectly geological age" (313), and images of volcanoes and earthquakes are regularly used to describe her family (398, 698). The slum at Tom-All-Alone's is characterized by "stagnant channels of mud" "blasted by volcanic fires," recalling the "deposits" of "mud" in the opening scene. These images reinforce a sense of the violent nature of urban poverty (657, 654), the poor abandoned and left to almost apocalyptic ruination.[72]

They also provide a telling contrast with the world of aristocratic privilege surrounding the Dedlocks. In his geological description of Stagg's Gardens in *Dombey and Son*, Dickens revealed (and expanded) the narrow perspective of its inhabitants; "open[ing] out" "new worlds" to them, as he suggested the diorama did. In *Bleak House*, Sir Leicester Dedlock, more fatally than those London residents in *Dombey and Son*, fails to attain an expansive vision of his world:

> Sir Leicester Dedlock is only a baronet, but there is no mightier baronet than he. His family is as old as the hills, and infinitely more respectable. He has a general opinion that the world might get on without hills, but would be done up without Dedlocks. He would on the whole admit Nature to be a good idea (a little low, perhaps, when not enclosed with a park-fence), but an idea dependent for its execution on your great county families. (18–19)

Dedlock's faith in his uncontested privilege is similar to those images of the "antlered herd" in the London of *Dombey and Son*: both suggest an aristocratic, traditional model of history. The phrase "His family is as old as the hills" articulates this sense of privilege through proverbial language that, like aristocratic history itself, is inherited and naturalized. Of course, in a geological world, hills erode and are remade continually. Just as the people of Stagg's Gardens had to accept the panoramic broadening of vision that geological science and the railroad combined to create, Dedlock too must learn, like Dickens's readers before him, to accept his place in a much grander history, a natural world that extends back much farther than "your great county families." Only then will he accept the links between himself and the volcanic slums at Tom-All-alone's, and only then will "the common brotherhood of man" be fully affirmed. In this novel geological events form one of the structural pleasures, and denote cultural warning: when Krook arrested the flow of information, he exploded; the lack of awareness of the "connections" between all of the characters has led directly to the volcanic slum at Tom-All-Alone's and to Judy Smallweed's "geologic age." The remaking of worlds that characterizes the narrative of *Dombey and Son* is reforged in the popular-scientific world of *Bleak House* as an almost mythic form of retributive punishment for the excesses of urban culture.

Mountains of Dust

By the time Dickens came to write his final completed novel, *Our Mutual Friend* (1864–1865), many dioramas had closed down for good.[73] Dickens reflected on the decline in the popularity specifically of the moving diorama (which involves, not the shifting of scenes through light effects, but a large canvas mounted on rollers) when, in 1868, he wrote in *All the Year Round* that "I systematically shun pictorial entertainment on rollers."[74] Concurrently, the public enthusiasm for the 1851 Great Exhibition, and the 1854 recreation of the Crystal Palace in Sydenham (with different exhibits), had been replaced by the pessimistic response to the 1867 Paris exhibition, which consolidated growing fears that Britain was losing its technological edge. By 1864 the optimistic heyday of the 1850s was over, and this pessimism was reflected in London show culture.

The scientific climate had shifted too. *Our Mutual Friend* was written six years after the publication of Darwin's *Origin of Species*; five years after the controversial *Essays and Reviews* (1860), which, at least to some readers, seemed to contain views that endorsed a godless Darwinistic universe; and two years after Charles Lyell had analyzed and supported the validity of the evidence for prehistoric man in *The Antiquity of Man* (1863).[75] In the midst of this ferment, *All the Year Round* had carried favorable reviews of Darwin and Lyell's works.[76] It has been suggested that *Our Mutual Friend*'s vision of London as "a dismal swamp" drew on Lyell's prehistoric England in *The Antiquity of Man*.[77] While the book focused on a discussion of early humans, the setting Lyell had conjured up was colorfully described in *All the Year Round* as "a steaming morass . . . its awful silence only broken by the hum of the shardy beetle, the rush of hideous flying-lizards through lofty woods of ferns and reeds."[78] Similarly, primeval beetles and pterodactyls ("flying-lizards") crawl and soar through the streets of *Our Mutual Friend*, alongside alligators, prehistoric fish, and "amphibious human creatures" scavenging survival from the refuse of the river (259, 352, 356, 69). London is described as a "mountain range," thrown up "like an old volcano, and its geological formation was Dust" (13). As one critic puts it, "From this geological perspective, the mounds" are "a demonstration of the manner in which life quite literally lives upon death. What lies beneath our feet is the history of the Earth, a history made up . . . of the accumulated waste of generations" (179–80).

That emphasis on waste in *Our Mutual Friend*, and on the ways in which everything, as waste, can be recycled into further use, has meant that both the London described in the text and the formal characteristics of the novel itself have been seen as radically different from Dickens's previous work. "There are no real secrets in *Our Mutual Friend*," Hillis Miller has written, for the entire world is thrown off from a past which has no authentic origin and from a world with "no unifying center."[79] The heart of the novel is in recycling, the transformation of goods and services in the city's dust mounds, as opposed to the vanity of speculation associated with the Veneerings.[80] The city, predicated on an endless circulation of money and wasted goods, is a heady simulacrum of authentic life with no firm ground on which meaning can rest. The novel is a bleak and endless substitution of signs against which only the domestic "enclave" created for Lizzie and Bella offers an alternative. Where in *Dombey and Son* and *Bleak House* the intimidating "dragon forms" and primeval creatures parodied the processes of industrialization and commercialization, always threatening to overrun the systems and signs of the present with the visceral materiality of the past, in this post-Darwinian, hypercommercial London there is no longer any sense that the world can be exploded and made anew, that the past holds any secrets that may erupt materially into the present and reveal themselves as the hidden basis on which modernity rests.

In *Bleak House* and *Dombey and Son*, Dickens's readers were encouraged to read the scene of the city as a geological fragment of a much broader spatial and temporal vision, but also to respond to the city as a performance of nature, to be set at one remove and enjoyed at a distance—an act of communal spectatorship that enables the city to produce a "brotherhood" of men equally participating in entertainment. But the act of reading "nature's stone book," as Dickens put it in his review of *The Poetry of Science*, is no longer capable of broadening the vision of either Dickens's characters or his readers in *Our Mutual Friend*. In this world, there are no more "sermons in stones." Bradley Headstone first meets Lizzie in a square "in the centre of which . . . is a very hideous church with four towers at the four corners, generally resembling some petrified monster, frightful and gigantic, on its back with its legs in the air" (221). And Charley Hexam, attempting to "better" himself through education, crushes his sister's "romantic ideas," the fantasies of the future that she sees in the flames of the hearth, through a phallic display of his educational superiority that draws directly on the language of geology:

That's gas, that is . . . coming out of a bit of a forest that's been under the mud that was under the water in the days of Noah's Ark. Look here! When I take the poker—so—and give it a dig—. (28)

When Charley "give[s] it a dig," this violent disturbance of Lizzie's dream relates him to all the other diggers and delvers sifting through the "geological formation[s]" for economic success in this dark portrait of London. Boffin's illiteracy in this world becomes a token of his virtue: "It's too late for me to begin shovelling and sifting at alphabeds and grammar-books," he says, in a grim parody of geological "shovelling and sifting" (50). Books are "stones" that teach miserliness and petrify their readers (49): one of Boffin's books on misers contains "a remarkable petrefaction," and Bella accuses Boffin of being "changed . . . to marble" by his miserly reading (485, 599). The text, then, marks Dickens's disillusionment with a geology whose spectacles were now old and unexciting.

The Veneerings, the Podsnaps, Silas Wegg, the Lammles, Bradley Headstone, Charley Hexam—all seek to inhabit and appropriate the dust mounds for economic gain, and the novel charts their struggles for survival in what has been seen as an essentially Darwinian landscape, a world filled with "individuals seeking their own advantage and acting without either a superintending intelligence or a common end."[81] But there is an alternative moral order within the novel that partly counters the materialism of Darwin's vision. "You could draw me to fire, you could draw me to water," Headstone tells Lizzie (397). This language, in its oppositional and violent nature forces Lizzie through exactly those same processes: "With the breaking up of her immobility came the breaking up of the waters that the old heart of the selfish boy had frozen." Lizzie's flood of compassion for the brother who has betrayed her connects her to her past on the banks of the river, and with this disaster comes "the end of our pictures in the fire!" (404).

After this experience, Lizzie absorbs the metaphor of "flood and fire" into her own vocabulary: attempting to convince Bella that she has learned her lesson, she reminds her of a vision she had of her in the fire: "A heart well worth winning, and well won. A heart that, once won, goes through fire and water for the winner, and never changes, and is never daunted" (529). Bella later thinks "how right she was when she pretended to read in the live coals that I would go through fire and water for him" (686). Lizzie's images "in the live coals," set against her brother's scientific explanation of them, are fanciful and imaginative, like Dickens's geological performance of "the poetry of science" almost twenty years earlier.

The repeated images of flood and fire in *Our Mutual Friend* again recall the scriptural apocalyptic narrative and the floods and fires of spectacular geology. Flood and fire return a sense of millennialism to the text—a sense of the possibility of rebirth or regeneration. But, unlike in *Bleak House*, that possibility is not imagined through a set of controlling metaphors that dominate the landscape of the text and enter into the plot itself, as it did with Krook's spontaneous combustion or the almost-flooded landscapes of London and Lincolnshire. References to flood and fire are only found in the speech acts of the characters themselves in *Our Mutual Friend*: the remaking of worlds is a purely rhetorical alternative to the Darwinian universe.

This language of the geological spectacular acts as a purifying agent for Lizzie and Bella, allowing both women regeneration at the same time as it proves their ability to remain unchanged. Both transcend, rather than adapt to, their environments. Bella, learning this lesson, preserves her essential goodness through the disasters the volcanic "dust-heaps" have thrown in her way, and produces the novel's only offspring. "It was charming to see Bella contemplating this baby," we are told, "finding out her own dimples in that tiny reflection, as if she were looking in the glass without personal vanity" (755). This image of Bella and her child is emblematic of non-Darwinian reproduction: it is an essential image that stresses continuity rather than change. It also stresses the novel's central theme of mutuality—a dissolution of the self into an acknowledgement of what is the same in the other. Bella's baby reproduces Bella's form so that it becomes the true mirror of her essential nature: an archetypal vision of reality that transcends the mirrored, polished realities of the Veneerings set.

Bella's non-Darwinian reproduction makes her resemble something like Owen's notion of the archetype—a kind of ideal form of her species. This is supported by the multitude of direct allusions to Owen throughout the text. When Mr. Podsnap is spied "prosperously feeding," his wife is "a specimen for Professor Owen," now vehemently opposing himself to Darwin's particular brand of evolutionary development (16). Owen is mentioned several times in the course of the novel: the text pokes gentle fun at him for liking "a quantity of bone" and again by reference to "elderly osteologists."[82] Fulweiler even suggests that the skeleton outside Venus's shop might be a version of Owen, and that his shop might have reminded readers of Owen's Hunterian Museum

(63). As it was described in *Household Words*, the Hunterian contained "skulls from all parts of the globe . . . brains of various creatures, beautifully preserved . . . and stomachs sufficient to startle any number of aldermen," along with skeletal reconstructions of the mylodon and glyptodon.[83] Venus describes his shop's wares similarly in *Our Mutual Friend*:

> Bones, warious. Skulls, warious. Preserved Indian baby. African ditto. Bottled preparations, warious. Everything within reach of your hand, in good preservation. . . . Say, human warious. Cats. Articulated English baby. Dogs. Ducks. Glass eyes, warious. Mummied bird. Dried cuticle, warious. Oh dear me! That's the general panoramic view. (81)

By the 1860s the term "panoramic" denoted comprehensiveness in general, and did not necessarily refer to the panorama as a technology of display. But it is still important to note that the panorama was out of fashion in 1864–1865, since the contents of Venus's shop also hark back to the old-fashioned Hunterian Museum, once presided over by Owen, now an "elderly osteologist." The text repeatedly alludes to Owen directly, in a period in which Dickens was actively supporting Owen's campaign for the construction of the Museum of Natural History.[84] Dickens repeats his earlier allusion to Owen through the megalosaurus in *Bleak House* to present him as a benevolent, if now somewhat obsolete, figure of scientific authority in *Our Mutual Friend*.[85]

More important than that he had particular theoretical views in one direction or another, moreover, was that Owen had made a spectacle of comparative anatomy. He was a self-appointed wizard, divining the species of the moa bird, and a popular lecturer. He had created displays at the Crystal Palace, on the pages of periodicals including *Household Words*, and at the British Museum, and he would be instrumental in creating the spectacular cathedral of science that constitutes the Museum of Natural History in South Kensington. Owen lived in Sheen Lodge in Richmond Park, a home on display to all those who traversed its walkways, and he tutored the children of the queen. Indeed, his careful self-presentation and his cultivation of a wizardly persona opened him up for pillory among his colleagues and scientific opponents. Owen brought fossil forms to life for the Victorian public. Under his anatomical direction, geology *was* a spectacle. It was not simply that geology took spectacular form, but the creatures whose skeletons he reconstructed were intrinsically spectacular. He styled himself as the creator of the "dinosauria" as "elephantine lizards," like the elephantine megalosaurus of *Bleak House*, as "the crown of reptilian creation," and in the process crowned himself king of the fossilized realms.[86] They were the perfect type of the reptile class, revealing that ancient species could be more advanced than their modern counterparts. If "Old Bones" Owen created the dinosauria, we might conclude, then they created him, too, as a man of science.

Dickens, no less than Owen, was a maker of geological spectacle. Owen participated in the creation of geology as a spectacular science, a science of magically shifting scenes and transforming fossilized creatures, and he cultivated a friendship with Dickens in order to find further occasions for displaying those spectacles on the pages of *Household Words* and in league with the age's greatest performer.[87] Believing in science-as-entertainment throughout their careers, in its capacity to emphasize "the common brotherhood in us all," Dickens and Owen, whatever their differences, responded to, and created, a geological show. And the geological show contributed to their own successes and popularity. A precise analysis of the visual and material cultures through which Dickens engaged with nineteenth-century science reveals that his interest in its developments was far from "nugatory." Instead, these contexts reveal Dickens's ongoing participation in the creation of geological ideas and images for public consumption. Geology was not only a science of remote landscapes. It was also a science of the streets. "If Dickens is an urban novelist," as one critic has put it, "his prose style . . . alive with the swarming energies of his surroundings," then those swarming energies must include the scientific shows and spectacles of which his prose is one example.[88] The streets of London, in Dickens's day, sometimes were streets where ancient lizards roamed and volcanoes and quakes shook the earth.

Dickens, then, is a somewhat different case than Eliot and Kingsley. Far less engaged with debates at the Geological Society or the particular brand of geological science that emerged from them, nonetheless he lived in and participated in a culture in which geological imagery, the picture of the past in spectacular and stratigraphic dimensions, was firmly entrenched. Dickens's fictions remind us that geological science often offered the Victorians less a specific plot of earth history than a series of spectacular visions of a world gone by, whose overarching story was left implicit, unarticulated, or unknown. That, perhaps, was partly responsible for its power.

Losing the Plot { **CONCLUSION**

ovel Science, as even the most casual of readers will discern, has told a story. It has, I hope, told several, interweaving stories about the practice of geology in the Regency and Victorian periods. It has told the story of the "heroic age of geology" through a literary lens, documenting in its first half the development of a variety of methodological and practical positions taken by the gentlemen geologists of the Geological Society as they eschewed (or at least professed to eschew) the self-determining forms of fictional "plots" in their contemplations of earth history. That story started with Walter Scott and Edinburgh and with James Hutton's hypothetico-deductive methods as they were put into elegant scientific prose by John Playfair, and as they were resoundingly rejected at the Geological Society by its earliest founders. From the very strict, perhaps restrictive, methodology of enumerative induction, the next generation of geologists—Sedgwick, Murchison, Buckland, Lyell, De la Beche—moved to theorize on the strata in largely spatial and structural terms. Yet each of their methodologies and practices was distinct. It has been the argument of this book that, from Sedgwick's antiquarian "mountain geometry" to the "romantic aestheticism" of Murchison and Geikie, from Buckland's epic promise to Lyell's mock-epic prose with its *vera causa* methods, those methods and practices were intrinsically related to epistemological and ontological problems surrounding the possible uses and abuses of literary form that gained their most acute expression in the nineteenth century in criticisms of the novel, both from within and without. Even the geological map, that icon of the anticosmological stratigraphic project, I hope to have shown, was susceptible to influence from this discourse.

I have told another story, too, picking up from that story about geological method, practice, and literary form. This story is about the uses of geological forms by novelists, as they too sought to establish the intellectual authority of their writings and to claim the ability to speak "truth" in a rapidly changing society. From Emily Brontë's untamable Yorkshire, through the geological fictions of Charles Kingsley, to George Eliot's provincial Midlands and Dickens's streets, the description of places and objects was often placed in tension with the recounting of the plot. Running through their novels is the possibility that history is not a story but a structure, the possibility that the past could be imagined as a series of blocks of layered rock, as discrete worlds and places whose relation to one another might be unknown or left unarticulated. For if, as I have put it, plots were problems for geologists, not proofs, then they could hardly be used as uncontroversial frameworks for novelistic stories. More to the point, novelists and geologists were not two distinct groups who occasionally met to swap stories. No less than their geological counterparts, novelists with pretensions to a place within the intellectual firmament could hardly consider themselves mere storytellers. Mixing at the same clubs and societies, and jostling for space in the same heavyweight periodicals, literary intellectuals—whether they wrote scientific papers, novels, or both—distanced themselves from too-intoxicating stories as a matter of course. For novelists like Charles Kingsley and George Eliot, who moved in scientific circles, geology did not offer an authoritative plot pattern on which to hang a fiction. It offered instead a useful form for narrative breakdown. Tell stories they might, but the gaps and blanks of the stratigraphic column could usefully suggest dark and disturbing ruptures in the continuous flow of the plot. This formal vision of the earth shaped the nineteenth-century imagination of places and communities, created scientific institutions such as the museum and the spectacle, and determined the course and pattern of individual lives and careers. And it gave rise to a peculiarly modern imagination of history—not always as a plot, but equally often as a spatial sequence of strata whose connections were at best implicit.

Of course, neither the resistance to heavily plotted forms I have described here, nor the concomitant rise of a stratigraphic imagination, were wholly determinitive of the directions taken by geologists in the nineteenth century. There are other stories of nineteenth-century geology—both literary and resolutely otherwise—that have been told, and will be told, which overlap with mine in different ways. This study has drawn extensively on all of those stories and in many places seeks, not to challenge or overturn them, but to add another strand of thought to the ways in which geology created and was created by the worlds—natural, cultural, and social—that it encountered. As all of the geologists and novelists who are my principal subjects also remind us, plot is as intellectually useful as it is distracting, and as powerful on the side of truth as it can be on the side of confusion. All I ask my readers to do here is to handle my story, as they should do all stories, with a measure of caution.

ACKNOWLEDGEMENTS

This book was primarily researched and written in two stages, first between 2004 and 2007 at St. Hugh's College, Oxford, and for the next three years as a member of the Cambridge Victorian Studies Group, funded by the Leverhulme Foundation, and a Bye-Fellow of Newnham College, University of Cambridge. I was subsequently appointed to teach at the University of East Anglia, and I am grateful to that university for contributing toward the cost of the illustrations for this book and for giving me a year's research leave in order that I could finish the bulk of the writing. Generous grants from the Arts and Humanities Research Council and, for illustrations, from the Royal Society and the British Academy also made the completion of this book possible. Thanks also to the syndics of Cambridge University Library for their permission to reproduce the bulk of the images in this book, and to the Mary Evans Picture Library for several images. The geological maps are reproduced by permission of the Geological Survey. © NERC. All rights reserved. IPR/140-58CT.

There are several people I am extremely lucky to have worked with. My work is immeasurably richer for having been read by Josephine McDonagh, whose critical insight, unstinting support and kindness, and brilliant suggestions have been invaluable. Without her this book would never have got off the ground. At Oxford, Helen Small, Beth Palmer, Muireann O'Cinneide, and Anthony Cummins all offered important insights at a critical stage in the book's production. So too did Ankhi Mukherjee and Simon Dentith, both of whom provided the perfect blend of support and criticism just when it was required. I reserve special thanks for Beth Palmer for many sanity-saving lunches during my time at Oxford, for Aileen Fyfe for reading and discussing my work at an early stage, and for Pietro Corsi for allowing me to attend his excellent seminars on evolution before Darwin.

I barely have words to express my gratitude to my thirteen colleagues and fellow members of the Cambridge Victorian Studies Group, whose brilliance and generosity, passionate engagement with all things nineteenth-century, and copious wine consumption transformed my work beyond recognition and gave me three very happy and productive years of work on the book. I am very grateful to Mary Beard for lending me her support for a Bye-Fellowship at Newnham College, which provided teaching opportunities and sharp and lively discussion over numerous lunches and dinners. A single discussion with Jim Secord in 2007 was enough to give my work a much sharper focus and a more sensible argument. Jim read several drafts of my work, introduced me to the History and Philosophy of Science Department at Cambridge (where discussions with Anne Secord, Sadiah Qureshi, Melanie Keene, and Martin Rudwick helped hone and sharpen the manuscript), and saved me from more embarrassing errors than I dare recount. This is a very different, and much better, book than it could ever have been without his support. It is all the better, too, for Clare Pettitt's insightful suggestions and moral support, and for the challenging and stimulating intellectual environment provided by every member of the CVSG, each of whom read drafts of my work and suggested new and fruitful avenues for research. The comments, advice and inspiration of Peter Mandler, Mary Beard, Simon Goldhill, Helen Brookman, Sadiah Qureshi (whose support, technical and otherwise, requires special mention), Michael Ledger-Lomas, David Gange, Anna Vaninskaya, Astrid Swenson, Jos Betts, and Rachel Bryant-Davies are therefore to be found on almost every page of this book.

In addition, I am very grateful to Melanie Keene, Ralph O'Connor, Gowan Dawson, and Alice Jenkins, all outstanding scholars in the fields of history of science and literature and science, for their time, encouragement, support, and insight. Three anonymous readers for the University of Chicago Press thoroughly read the manuscript and gave more detailed and generous suggestions than ever would have been expected of them. I have gratefully attempted to incorporate all of their suggestions into the finished work. I have also been very lucky to have such an excellent editor as Karen Darling. It goes without saying that all errors are mine and mine alone.

Last but by no means least, I have personal acknowledgements to make: to my family—especially to my parents, Karrie and Cary Wood and Vince and Jo Garvin, to my grandmother Jan, and to the entire Buckland family—for their support, love, and encouragement and for making the time it took to write this book a lot of fun. Most of all I need to thank my amazing husband,

Patrick, who can explain what I have been doing for the last seven years much better than I, who has nurtured me through innumerable tantrums, crises, and triumphs, and who gave me the best deadline for finishing the book I could ever have hoped for—our beautiful, beautiful daughter Isobel Edith Buckland. This book is for her.

APPENDIX

Lines on Staffa

CHARLES LYELL

Ere yet the glowing bards of Eastern tale
Had peopled fairy worlds with beings bright,
Roamed o'er the palace and enchanted vale,
And dreamed a heavenly vision of Delight,
And told of realms rich with unborrowed Light,
On which the needless sunbeams never fell,
Whose noon of splendour never knew the night,
Illumed by lamps that burnt unquenchable,
And dazzling hung in air, upheld by magic spell:

All these and more, with which their wizard strain
Led far away deluded Fancy's child,
Till he would turn on Nature's self again,
And deem her charms a desert bleak and wild,
Himself from visionary heavens exiled;
While yet unheard that strain, the Time had been
When Nature's hand as if in sport she toiled
To build e'en more than could the thoughts of man,
Amid the Ocean vast, had framed a fairy scene.

For she had found a lone and rocky isle,
And at her voice a thousand pillars tall,
She bade uprising lift the massy pile,
And far within she carved a stately hall
Against whose sides the entering waves did fall,

While to their roar the roof gave echo loud—
And she had hid each column's pedestal
Beneath the depths unseen of Ocean's flood,
While towered their heads on high, amid the passing cloud.

And she had fashioned with an artist's pride
The dark black rock where hung the sparkling foam,
And many a step along its sculptured side
Had hewn, as if to tempt some foot to roam,
Some favoured foot of Mortal yet to come.
She bade no shapes of Terror there abound,
That pillar'd hall no guardian dragon's home,
But Ocean rolled his mighty waves around,
To guard from vulgar gaze her fair enchanted ground.[1]

NOTES

Introduction

1. See O'Connor, *Earth on Show* (2007), for a rigorous and evocative account of the broad cultural formations of geology in the first half of the nineteenth century.

2. See, for instance, Knell, *Culture of English Geology* (2000); Alberti, "Placing Nature" (2002) and "Objects and the Museum" (2005); and Rudwick, *Scenes from Deep Time* (1992). This list barely scratches the surface but covers collecting, museums, and visual culture.

3. As will be demonstrated more fully later in this introduction, and throughout the book, the importance of literary activities to the new science is central to key works in the field, notably O'Connor, *Earth on Show* (2007), and Secord, *Victorian Sensation* (2000). What is new here is the emphasis on literature as a form of scientific *practice* in the nineteenth century.

4. For this early use of the word "geology," see Rudwick, *Bursting the Limits of Time* (2005), 133–35, and Dean, "Word Geology" (1979).

5. Ibid., 460–63.

6. Ibid., 464–65. Also see Kolbl-Ebert, "Bellas Greenough's 'Theory of the Earth'" (2009), 115–28.

7. This explanation of the studies that preceded "geology" is taken from Rudwick, *Bursting the Limits of Time*, chap. 2 (esp. 59–115). For the important context of geognosy/geology in Germany see Guntau, "Rise of Geology" (2009), 163–77, and for the suggestion of some links between geology and mining see Shortland, "Darkness Visible" (1994), 28–39.

8. Geological Society, *Geological Inquiries* (1808), 450.

9. Ibid.

10. Ibid., 451.

11. Ibid., 450.

12. See O'Connor, "Facts and Fancies" (2009) for a strong example.

13. Laudan, "Ideas and Organisations" (1977), 537.

14. See Rudwick, *Great Devonian Controversy* (1985); Secord, *Controversy in Victorian Geology* (1986); Oldroyd, *Highlands Controversy* (1990).

15. Secord, *Controversy in Victorian Geology* (1986), 313–14.

16. Phillips, *Illustrations of the Geology of Yorkshire* (1829), 14.

17. Lyell, "Apparent Horizontality of Inclined Strata," in *Elements of Geology* (1838), 108.

18. See Rudwick, *Meaning of Fossils* (1976), chaps. 3 and 4. Rudwick, *Great Devonian Controversy* (1986), tells the story of the classification and elaboration of the "Devonian" system, which was in part a battle over the use of fossils in stratigraphy.

19. Weindling, "Geological Controversy" (1995), 247.

20. The classic statement of the gentlemanly constitution of the Geological Society is Porter, "Gentlemen and Geology" (1978), and of their lack of interest in economic or practical geology, Porter, "Industrial Revolution" (1973). Veneer qualifies the argument in "Practical Geology" (2009), suggesting that in its earliest years the Society included many with mineralogical and economic interests, but that by the late 1810s and 1820s the focus had (with minor exceptions) shifted to stratigraphy.

21. Porter, *Making of Geology* (1977), 140, 141.

22. See ibid., 142, and Rudwick, *Bursting the Limits of Time* (2007), 467.

23. In England, unlike on the continent, the state did not run or operate mines. For the importance of the mining context, nonetheless, see Guntau, "Emergence of Geology" (1978), 280–82, and Weindling, "Geological Controversy" (1995). See Shortland, "Darkness Visible" (1994), for an account of the conceptual boon mining may have given to geology in nineteenth-century Britain. Most important, perhaps, is Lucier, "Plea for Applied Geology" (1998), which argues that "the disjunction between mining and geology in Britain . . . is still open to debate" (3) and reminds readers that the state-funded Geological Survey of Britain was founded by a leading Society gentleman, Henry De la Beche, not merely with a commonplace rhetoric of economic utility but that "practical concerns were, for the most part, determining factors in its design and prosecution" (6) and that it studied mining districts and coalfields even under its second director, another Society gentleman, Roderick Murchison. Private surveys were also of significance within this geological culture.

24. See Rudwick, *Great Devonian Controversy* (1985). Hugh Torrens has done most to reveal the search for coal as a motivating factor in geology around the turn of the nineteenth century and later: see, for instance, his "William Smith" (2003). For geological maps and the search for coal, and an explanation of "base limit," see plate 9 and chapter 5 below.

25. See Knell, *Culture of English Geology* (2000), for a powerful account of the ways in which a "social understratum" of women, children, and laborers, for instance, "formed, if not the bedrock, then the infiltrating roots of an infrastructure which fed the loftiest and most esoteric pronouncements on geology" (3), providing information as often as fossils.

26. See Rose, "Military Men" (2009). Klonk's *Science and the Perception of Nature* (1996) studies domestic tourism to the Isle of Staffa during the Napoleonic years (chap. 3).

27. See, for instance, the proud presentation of one of the century's most famous

geologists, Roderick Murchison, as a famous huntsman turned geologist in Geikie, *Life of Sir Roderick Murchison* (1875), 1:89, 91.

28. The best studies of these gentlemanly and masculine aspects of nineteenth-century geology are still Porter, "Gentlemen and Geology" (1978), 809–36, and Rudwick, "Foundation of the Geological Society of London" (1963).

29. Allen, "Lost Limb" (1995), 204, 205. Allen's overarching argument that geologists and botanists had lost touch with one another by midcentury suggests that geology was a specialized discipline and not a natural-historical one, at least by 1859 (see 208), but that collecting and hammering continued to be of importance to geology up to and even beyond this point.

30. Knell, *Culture of English Geology* (2000), 7.

31. Page, *Advanced Text-Book of Geology* (1856), 289.

32. Alberti, "Placing Nature" (2002), 292; Knell, *Culture of English Geology* (2000). See also Torrens, "Geological Communication in the Bath Area" (1997), for a case study of local geological activities, and particularly of collecting, in the late eighteenth century. Torrens's collection, *The Practice of British Geology* (2002), also explores mineral surveying, the importance of landowning patrons in fostering surveys of their estates, and mining, as important strands of geological history in this period.

33. See Alberti, "Placing Nature" (2002) and "Objects and the Museum" (2005).

34. Dickens, *Bleak House* (1998), 247.

35. [Anon.], "Our Honeymoon" (1853), 51.

36. [**Morley**], "Hammering It In" (1857); [White], "Two College Friends" (1856).

37. [Owen], "Leaf from the Oldest of Books" (1856).

38. Banks, *Wooers and Winners* (1880), 1:301.

39. For a study of the representations of geology in written texts of the period, see Zimmerman, *Excavating Victorians* (2008). Though Zimmerman offers excellent readings of Alfred Lord Tennyson's *The Princess* and Charles Dickens's *Our Mutual Friend*, she pays only scant attention to the historiography on nineteenth-century geology. I seek to combine the kind of scholarship she has undertaken with much closer historical work.

40. O'Connor, *Earth on Show* (2007), 8.

41. In particular see ibid. and Secord, *Victorian Sensation* (2000).

42. O'Connor, *Earth on Show* (2007); Secord, *Victorian Sensation* (2000); Sommer, "'Amusing Account'" (2004) and "Romantic Cave?" (2003).

43. For studies of the visual cultures of geology see the following works by Rudwick: *Scenes from Deep Time* (1992), "Emergence of a Visual Language" (1976), "Caricature as a Source" (1975), and "Encounters with Adam" (2004).

44. O'Connor, *Earth on Show* (2007).

45. See O'Connor, "Epic of Earth History" (2009).

46. See Zimmerman, *Excavating Victorians* (2008).

47. O'Connor, *Earth on Show* (2007), 11.
48. Porter, "Gentlemen and Geology" (1978), 819. Shortland also discusses the prevalence of such imagery in "Darkness Visible" (1994), 41–43.
49. Heringman, *Romantic Rocks* (2004), 3, 4. For his case study centering on Dovedale in Derbyshire as demonstrating the specifically "English" nature of this kind of "romance," with analogues in Gothic architecture and medieval chivalry, see 4–5, 259–64.
50. O'Connor, *Earth on Show* (2007), 158, 161.
51. See O'Connor, "Epic of Earth History," 209–11.
52. The analogy I pursue here between writing and the museum is suggestively explored by Fyfe in "Reading Natural History" (2007), 196–230, in which she argues that the texts and images of the periodical *Pictorial Museum* offered its consumers a better experience than the British Museum, for it used similar organizational techniques but was able to draw clearer connections between exhibits and to engrave them on a more convenient scale than museum objects.
53. Lyell, *Principles of Geology* (1830–1833), 1:78–79.
54. McKeon, "Generic Transformation" (2000), 385.
55. Ibid. See also Porter, *Making of Geology* (1977), 204.
56. Turner, "Lessons from Literature" (2010), 583.
57. See, for obvious examples, Beer, *Darwin's Plots* (1983); Levine, *Darwin and the Novelists* (1988).
58. See O'Connor, *Earth on Show* (2007) and "Epic of Earth History" (2009), for examples of the first two of these. For Britain's place at the apex of history see, for instance, Buckland, "'Pictures in the Fire'" (2009).
59. Turner, "Lessons from Literature" (2010), 584.
60. See the entirety of Turner, "Lessons from Literature" (2010).
61. Quotations from ibid., 582, 581. For Turner's work on this subject, see his *English Renaissance Stage* (2006). This question about form is a problem I have with the current call for a "new formalism" that would reject historicist readings of literary texts and renew focus on the aesthetic structures and values of literary form. Its definitions are, in my view, far too narrow and restrictive, reducing form, once again, to a purely literary and linguistic question. For my longer discussion of the essentially historical nature of both "form" and definitions of form, see Buckland and Vaninskaya, "Epic's Historic Form" (2009), 163–72.
62. O'Connor, "Facts and Fancies" (2009), 333.
63. For O'Connor's discussions of the avoidance of narratives of earth history, see *Earth on Show* (2007), 16, 56, 57, 59, 61, 116, 140, 159, 254–55.
64. As Tucker shows in *Epic* (2008), subsequent criticism has frequently "depict[ed] prose fiction as the genre in which modernity stands forth over epic's dead body" (4); he cites Watt, *Rise of the Novel* (1957), Maresca, *Epic to Novel* (1974), and Hunter, *Before Novels* (1990), 4n8. Much of this stems from the influential criticisms of Lukacs, in *The Historical Novel* (1983) (see, for instance, 32–36), and Bakhtin, in "Epic and Novel," in *The Dialogic Imagination* (1981), 3–40. Though the

distinction is rhetorical rather than necessarily real (as Tucker's book amply demonstrates), Tucker also notes that "Already at the turn of 19th cent. the supersession of epic by novel was an undisputed fact" to some (5n8).

65. See McKeon, "Prose fiction" (2005), 238–63. See also McKeon, *Origins of the English Novel* (1987), chaps. 1, 3, 9, 11. Levine, in *Realistic Imagination* (1981), 12, calls this the "anti-literary thrust" of much novel writing.

66. McKeon, "Generic Transformation" (2000), 383.

67. Genette, *Narrative Discourse Revisited* (1988). For the story/discourse distinction, see Genette, but also Monica Fludernik's useful summary (1993, 60).

68. Forster, *Aspects of the Novel* (2002), 61.

69. Chatman, *Story and Discourse* (1978), 45–46. See also Genette, *Narrative Discourse Revisited* (1988), 18–20. Here Genette argues that the difference between "story" and "plot" is one of degree, not of kind.

70. Jahn, http://www.uni-koeln.de/~ame02/pppn.htm (accessed 2 February 2011).

71. O'Connor, *Earth on Show* (2007), 126, 26, 397, 396, 399, 403.

72. It is important to note that in *Earth on Show* (2007), O'Connor does argue that in the writings of his chief exponent of the geological pageant, Hugh Miller, all this *is* pressed in the service of a plot. O'Connor rightly says that Miller was "concerned . . . to yoke all histories—sacred, secular, geological—to a single Christian narrative of progress," in which early scenes from deep time are "discreetly demonized" (421), populated by creatures "whose pride and aggression earned them extinction in the Deluge," prefiguring later extinctions that themselves prefigure humanity's fallen state and future damnation. "The implication," he writes, "is that civilized humans have legitimately inherited the land once tyrannized by primeval monsters" (425), and the overall plot reveals a "progressionist" momentum "from hell to heaven." Nonetheless, O'Connor also says that evolutionary theory later "reinforce[d] the sense of rightful inheritance by providing genealogical links for the pageant's separate scenes" (425). This reveals two things: first, that the "progressionist" plot, while often implied, remains merely implicit because there are no "links" between the "separate scenes" of the narrative and, second, that those links could be filled in by an entirely different plot—the evolutionary plot (or any other). It should also be noted that Miller was opposed to any notion of a simplistic *geological* progression of species and argued against this in many works.

73. See Levine, "Why Science Isn't Literature" (2008), for an important discussion of these issues. Levine is responding, in part, to critics like Carroll, whose *Literary Darwinism* (2004) and introduction to Darwin's *Origin of Species* asserts that literature offers evidence to the evolutionary scientist hoping to study human behavior and imagination, transforming literature into the object of scientific study, explicated by scientific analysis. In these works Carroll accuses Levine and other critics, such as Beer, of reckless poststructuralism ("Introduction," 65; *Literary Darwinism*, 27, 45), wading in on the "science wars" of the mid-1990s, in which Sokal and Bricmont accused humanistic scholars of considering science as mere representation. Bricmont

and Sokal, *Intellectual Impostures* (1998). See also Levine, "Introduction," in *Realism and Representation* (1993).

74. See Dawson, *Darwin, Literature and Victorian Respectability* (2008).

75. See Dawson, "Literary Megatheriums" (2011).

76. See Menke, "Fiction as Vivisection" (2000), 617–53; Dames, *Physiology of the Novel* (2007).

77. Dawson, *Darwin, Literature and Victorian Respectability* (2008), 7. See also Dawson, "Science and Literature" (2006). It is the more genuinely historiographical approach that Dawson calls for in this latter article that I hope to achieve here.

78. Beer, *Darwin's Plots* (1983), 160.

79. Ibid.; Levine, *Darwin and the Novelists* (1988).

80. On natural history and Elizabeth Gaskell, see King, "Taxonomical Cures" (2003). On literary suspense and scientific hypothesis, see Levine, *Serious Pleasures of Suspense* (2003). On pathology, sickness, and narration, see Small, *Love's Madness* (1996); Dever, *Death and the Mother* (1998); Wood, *Passion and Pathology* (2001).

81. For a similar view on the historical rather than literary-critical literature, see Rudwick, *Bursting the Limits of Time* (2007), 1–7 (esp. bottom of 7).

82. Beer, *Darwin's Plots* (1983); Levine, *Darwin and the Novelists* (1988); Meckier, *Hidden Rivalries* (1987); Shuttleworth, *George Eliot and Nineteenth-Century Science* (1984).

83. See, for examples of critics explicitly aligning uniformitarianism with literary realism, Shuttleworth, *George Eliot and Nineteenth-Century Science* (1984), 52, 81; Smith, *Fact and Feeling* (1994), 257, 120–21; Beer, *Darwin's Plots* (1983), 181; Arac, "Rhetoric and Realism" (1979), 681; Cosslett, *"Scientific Movement"* (1982), 4–5; and Meckier, *Hidden Rivalries* (1987), esp. 243–76.

84. Shuttleworth, *George Eliot and Nineteenth-Century Science* (1984); Smith, *Fact and Feeling* (1994).

85. For a recent example of this approach, taking geology as the basis of fictional and poetic narrative structure, see Zimmerman, *Excavating Victorians* (2008).

86. For examples appertaining to the flood at the end of George Eliot's *Mill on the Floss*, see Arac, "Rhetoric and Realism" (1979), 681, and Klaver, "'I Will Ferry Thee Across'" (2004), in which "catastrophism" is said to represent "an alluring worldview" that "makes compatible scientific rationality and the moral imperative of religious belief." Though Smith, in *Fact and Feeling* (1994), argues that the flood in fact "caps a subtle and complex affirmation of uniformitarianism," he leaves the binary intact (120). Meckier, *Hidden Rivalries* (1987), takes the binary to its fullest conclusion, aligning Lyell with gradualism, Darwin, the Reform Bills, George Eliot, and G. H. Lewes, and Carlyle with catastrophism, Chartism, revolution, Dickens, and Collins (249).

87. See Dawson, "Literature and Science" (2006). For challenges to the primacy of Darwin in evolutionary studies, cited by Dawson, see Desmond, *Politics of Evolution* (1989); Secord, *Victorian Sensation* (2000); Bowler, *Non-Darwinian Revolution* (1988); and essays by Bowler, Michael Ruse, Vassiliki Smocovitis, Sandra

Herbert, Pietro Corsi, James Lennox, Jonathan Hodge, Michael Ghiselin, and David Hull, comprising a special issue of *Journal of the History of Biology* 38 (March 2005).

88. For an important rejoinder to the notion that many leading geologists (including both Adam Sedgwick and Charles Lyell) were arguing in order to defend a particular biblical view, see Brooke, "Natural Theology" (1997), 53–74. As he puts it, "The affirmation of design in nature was clearly compatible with a spectrum of deistic and theistic positions, from an extreme deistic model, which subsumed divine activity under natural laws, to a full biblical theism. Moreover, deists and theists alike could find themselves sharing virtually the same model for divine activity in the physical world" (59). He further notes that "it was ambiguities associated with design arguments that helped to sustain them" for all parties seeking to avoid cultural suspicion (67) and that they "outlived their rigour" because they were tactically useful more than they were plausible (69). He also details how "almost every geologist of note" was beset by one antigeological cleric or another, particularly over the issue of Noah's Flood (which all agreed, by the mid-1830s, was not an event for which there was geological evidence), making untenable the notion that those geologists who disagreed with Lyell were simply defending religious views (62). See also the important essay by O'Connor, "Young Earth Creationists" (2007), for an excellent summary of scholarship on this field and directions for new research.

89. Secord, "How Scientific Conversation Became Shop Talk" (2007). The entire collection containing this essay (Fyfe and Lightman, *Science in the Marketplace*) calls for attendance to the creation of new sites and audiences for science and demonstrates the fruitfulness of such an approach. For the importance of place as a category through which to think about science, see Livingstone, *Putting Science in Its Place* (2003), and Morus, "Replacing Victoria's Scientific Culture" (2006), each of which argues for the importance of different sites, practices, and performances as constituent elements of nineteenth-century science. That this includes writing in a variety of contexts and guises (the different forms and formats of which are also of great significance) is evident in Cooter and Pumfrey, "Separate Spheres" (1994).

90. This argument builds partly on Noah Heringman's assertion throughout *Romantic Rocks* (2004) that geology offered a vocabulary for "rocks of ruin."

91. This is not to say that nobody has ever suggested this before: the epilogue to O'Connor's *Earth on Show* (2007) emphasizes, for instance, just how unrecoverable and unknowable much of the geological past was (439–42). For O'Connor, as for me, there was always a tension between the need to tell stories and the need to resist them in geological science: while O'Connor's study largely explores the complex telling of stories, mine focuses on the points of most resistance.

Chapter One

1. Sedgwick to Canon Wodehouse, 15 June 1837, in *Life and Letters of Adam Sedgwick* (1890), I: 483.

2. Caroline Owen, in *Life of Richard Owen* (1894), 1:113.

3. Ibid., 1:113. Caroline repeats verbatim the description of the game in Scott, *Guy Mannering* (1999), 204. Oxford World's Classics editions have been generally used throughout this book. There is not one for *Guy Mannering*, so the Penguin edition, based on the authoritative Edinburgh edition of the Waverley novels, is used here.

4. Scott, *Guy Mannering* (1999), 204.

5. Ibid., chap. 36.

6. Caroline Owen, in *Life of Richard Owen* (1894), 1:113.

7. Scott, *Guy Mannering* (1999), 204.

8. Caroline Owen, in *Life of Richard Owen* (1894), 1:113.

9. Ibid.

10. Scott, *Guy Mannering* (1999), 204.

11. Ibid., 209.

12. Certainly drinking games were a frequent feature of gentlemen's clubs in the Regency period (see Murray, *High society* [1998]). Though there has been little scholarship on the topic it seems more than likely that other clubs imitated Scott, given the enormous popularity of his novels. This only reinforces the point that the Geological Society must be considered as a social, literary, and cultural space as much as a scientific one.

13. Yeo, *Defining Science* (2003), 35.

14. [Hutton], *System of the Earth* (1785); Hutton, *Theory of the Earth* (1788); Hutton, *Theory of the Earth* (1795).

15. See Rudwick, *Bursting the Limits of Time* (2008), 158–72.

16. Laudan, *From Mineralogy to Geology* (1987), 89.

17. Ibid., 94. See also Rudwick, *Bursting the Limits of Time* (2008), 90–94.

18. See Rudwick, "Hutton and Werner Compared" (1962); Dean, *James Hutton and the History of Geology* (1992).

19. Hutton, *Theory of the Earth* (1788). See also Rudwick, *Bursting the Limits of Time* (2005), 164, for a recent discussion of some of the difficulties of Hutton's text, including both its textual arrangement and its lack of "pictorial evidence."

20. See Rudwick, *Bursting the Limits of Time* (2005), 465–66 (quotation 465).

21. Playfair, *Illustrations* (1802), point 133.

22. Ibid.

23. Ibid.

24. Scott, "To Joanna Baillie, April 4th 1812," in *Letters* (1932), 3:101. For another description of this episode, including some excerpts from the play, see Wawn, *Vikings and the Victorians* (2000), 48–53, and Wawn, "*Gunnlaugs saga ormstungu*" (1982), 139–51. The manuscript of the play is now at the Henry E. Huntington Library, California, MS Larpent 1751. Some of these examples of Scott's imbrication in Edinburgh geological debates have also been pointed out, along with several examples of the geological nature of his landscape descriptions, by Dean in *Romantic Landscapes* (2007), 188–97.

25. Scott, "To John B. S. Morritt, 5th November 1818," in *Letters* (1832), 5:212.

26. Cockburn, "Account of the Friday Club" (1910), 184.

27. Ibid., 187.
28. Ibid.
29. Ibid., 186.
30. Surviving records of the club are at NLS, MS.15943.
31. Dean, *James Hutton and the History of Geology* (1992), 125.
32. Scott, "To Joanna Baillie, April 4th 1812," in *Letters* (1932), 3:101.
33. Lockhart, *Peter's Letters* (1819), vol. 3, letter 7. Whether or not he had his tongue in his cheek, this is a much more sympathetic portrait of Playfair than that offered in the *Blackwood*'s piece.
34. John Playfair, Francis Jeffrey, and James George Playfair, "Biographical Memoir" (1822), lxvii-lxviii.
35. For a recent, excellent discussion of this in relation to literature see Alice Jenkins, *Space and the "March of Mind"* (2007).
36. [Playfair], "Leslie's *Elements of Geometry*" (1812), 85.
37. [Playfair], "Leslie's *Elements of Geometry*" (1812), 97.
38. Horace, *Ars Poetica*, trans. C. Smart. http://www.perseus.tufts.edu/hopper/text;jsessionid=E1122B4D36D05B73B1ED75D42CD5C31E?doc=Perseus%3atext%3a1999.02.0063 (accessed, 13 June 2012). Thank you to Rachel Bryant-Davies for her help with this passage.
39. Bacon, *Two Books* (1808).
40. For Playfair's promotion of French mathematics and mathematicians, see "Mésure d'un arc du méridien" (1807), "Laplace's System of the World" (1810), "Buee, sur les quantités imaginaires" (1808), "Essai philosophiques sur les probabilités" (1814), and specifically on the synthetic versus analytical methods, "Dealtry's *Principles of Fluxions*" (1816), "Mascheroni, Geometrie du compas" (1806), 166, and "Donna Agnesi's *Analytical Institutions*" (1804). See also Ackerberg-Hastings, "Analysis and Synthesis" (2002), and I. B. Cohen, *Revolution in Science* (1985), 523–26.
41. Maseres, in Frend, *Principles of Algebra* (1796–1799), 1:254. This is discussed by Pycior, "Internalism, Externalism, and Beyond" (1984), 429.
42. [Playfair], "Buëe, sur les quantités imaginaires" (1808), 313–14. See Pycior, "Internalism, Externalism, and Beyond" (1984).
43. [Playfair], "Buëe, sur les quantités imaginaires" (1808), 306.
44. [Playfair], "Geographie mineralogique" (1812), 383.
45. [Playfair], "Werner on the Formation of Veins" (1811), 97.
46. Ibid., 95–96.
47. Ibid., 93.
48. Of course, Playfair's characterization of Werner serves his own rhetorical purpose. It is well established among historians of science that Werner's work was of long-lasting significance in the history of geology, despite the power of this rhetorical position (later reaffirmed by the prominent early historian of geology Archibald Geikie in his *Founders of Geology* [1897]). See Ospovat, "Distortion of Werner" (1976), and Seddon, "Abraham Gottlob Werner" (1973).

49. Playfair, *Illustrations* (1802), point 459.
50. Playfair, "Biographical account" (1822), 64–65.
51. The plausibility and persuasiveness of Playfair's prose did not meet with unanimous approbation. John Murray, for instance, anticipated later geologists' line of attack, believing that "because it is so mellifluous and reasonable," it "is simply more subtle and dangerous than that of the fabulists" (Heringman, *Romantic Rocks* [2004], 270). These views are expressed in Murray, *Comparative View* (1878), iii–iv.
52. Scott, "To William Laidlaw, November-December 1820," in *Letters* (1932), 6:304.
53. Scott, "To Lord Melville, 1st December 1821," in *Letters* (1932), 7:35.
54. Scott, "St. Ronan's Well" (1824).
55. Scott, "To John B. S. Morritt, 8th December 1820," in *Letters* (1932), 7:307.
56. Goldsmith (2006), *Vicar of Wakefield*, 62.
57. I would like to thank Simon Goldhill for helping me to understand this passage.
58. [Scott, Jeffrey, et al.], "Tales of My Landlord" (1817), 193.
59. Ferris, *Achievement of Literary Authority* (1991), chaps. 1-2. For a discussion of the early *Edinburgh Review* as "predicated wholly on reviews written by men about men and for men," though later developing an uneasy acceptance of women's writing, see Curran, "Women and the *Edinburgh Review*" (2002), 195–208 (198).
60. See Ferris, *Achievement of Literary Authority* (1991), esp. chaps. 1 and 2.
61. The mid-twentieth-century impulse to reevaluate Scott as a "realist" novelist is most obvious in Daiches, *Literary Essays* (1956), *Sir Walter Scott and His World* (1971), and "Sir Walter Scott and History" (1971); Lukacs, *Historical Novel* (1976); Johnson, *Sir Walter Scott* (1970); and Hart, *Scott's Novels* (1966). Hart later changed his mind and argued, in *The Scottish Novel* (1978), that, to the contrary, Scott was "inclined more to use history in the service of romance than to assimilate romance to the interpretation of history" (19).
62. Nairn, *Breakup of Britain* (1977); Craig, *Out of History* (1996); Makdisi, *Romantic Imperialism* (1998); Pittock, *Myth of the Jacobite Clans* (1995). See McNeil, *Scotland, Britain, Empire* (2007), and McCracken-Flesher's "arguably postcolonial" account of Scott's "narration ... of nation" as an open-ended process in *Possible Scotlands* (2007) for recent challenges to this view (26, 20). For perhaps the most important argument affirming Scott as "attempting to ensure and articulate Scotland's distinctive place in Britain" and concerned "with Britishness" in order to envision "Britain as a full cultural amalgam" including Scotland "rather than ... as a synonym for England," see Crawford, *Devolving English literature* (1992; quotations on 15, 115).
63. Scott, *Waverley* (1986), 24.
64. Of course Scott is the inheritor, rather than the primary innovator, of much eighteenth-century novelistic writing that drew attention to its own practices in this way, most notably that of Sterne. But as Crawford puts it, "the intrusive, self-mocking, and elaborately self-conscious tone of Scott's editorial voice is a new one.

Here is an author who delights in the synthetic nature of his text," in its being built up of antiquarian artifacts, "and wishes its eclecticism to be part of the reading experience" (*Devolving English Literature*, 1992, 125). The antiquarianism of Scott's writing, his "delight in self-conscious anthologising" (133), is essential to its appeal to the geologists. Crawford argues that this tradition united literary and scientific discourse well into the nineteenth century, through Scott.

65. Manning, "Antiquarianism" (2004), 57–76. See also Sweet, *Antiquaries* (2004).
66. [Scott], "Prefatory letter to *Peveril of the Peak*," in Weinstein, *Prefaces* (1978), 66.
67. Ibid., 67.
68. [Scott], "Introductory epistle to the *Fortunes of Nigel*," in Weinstein, *Prefaces* (1978), 45.
69. Ibid., 45.
70. [Scott et. al.], "Review of *Tales of My Landlord*," in *Quarterly Review* (1817), reprinted in Weinstein, *Prefaces* (1978), 14.
71. Scott, *Antiquary* (2002), 14. Watson's introduction to this edition is especially useful (2002), vii–xxvii.
72. Scott, *Antiquary* (2002), 18.
73. Ibid.
74. Ibid., 3.
75. Ferris, "Melancholy" (2004), 77–93 (88).
76. [Scott], "Introductory epistle to *Fortunes of Nigel*," in Weinstein, *Prefaces* (1978), 43.
77. Anon., "Rob Roy" (1818).
78. [Scott, Jeffrey, et. al.], "Tales of My Landlord" (1817), 195.

Chapter Two

1. William Whewell to Richard Jones, Trinity/Add.Ms.c/51/222. This letter is held in the Additional Manuscripts C collection at Trinity Library, Cambridge, box 51 (letters between Jones and Whewell). See also Geikie, *Life of Roderick Murchison* (1875), 1:91; Clark and Hughes, *Life and Letters*, 1:397–98.
2. Charles Darwin, in *Life and Letters of Charles Darwin* (1887), 1:40; Francis Darwin, in *Life and Letters of Charles Darwin* (1887), 1:125. As Adrian Desmond and James Moore have recently noted, Darwin's acquaintance with Scott and his works shaped his expectations of Edinburgh before he visited it. Desmond and Moore, *Darwin's Sacred Cause* (2009), 30.
3. Clark and Hughes, *Life and Letters*, 1:397–98.
4. For examples from the *Quarterly Review*, see [Murchison], "Tours in the Russian Provinces" (March 1841), 344–75, and "Siberia and California" (September 1850), 395–434; [Lyell], "Scientific Institutions" (June 1826), 153–79, "Transactions of the Geological Society" (September 1826), 507–40, "State of the Universities"

(June 1827), 216–68, and "Scrope's Geology of Central France" (October 1827), 437–83. For Lyell's early maneuverings with Coleridge, Lockhart, and the *Quarterly Review*, see Lyell's letters to his sister (December 4, 1825), to Mantell (June 22, 1826), to his father (November 16, 1826), and again to his sister (January 5, 1827), in *Life, Letters and Journal of Charles Lyell* (1881) 1:163, 164–65, 165, 167. In particular, Lyell wrote to his father "that there is most real independence in that class of society who, possessing moderate means, are engaged in literary and scientific hobbies," and that he hoped "to make friends among those that a literary reputation will procure me who may assist me" (April 10, 1827; in *Life, Letters and Journals*, 1:171).

5. Lyell to his father (April 10, 1827), in *Life, Letters and Journals*, 1:170. See chapter 3 for greater discussion of Lyell and the *Quarterly Review*.

6. For an extended discussion of the value of the quarterly reviews to the gentlemanly men of science, see Yeo, *Defining Science* (2003), esp. 77–115.

7. [Fitton], "Transactions of the Geological Society" (1817a), 177.

8. [Fitton], 'Transactions of the Geological Society' (1817b), 72.

9. [Fitton], "Transactions of the Geological Society" (1817a), 177.

10. Ibid., 176. The extent to which this was merely rhetorical has been much debated by historians. See Porter, *Making of Geology* (1977), 204–8. There he argues that, while similar rhetoric had been deployed at the founding of the Royal Society and again after "the plethora of late seventeenth-century theories" to advocate fieldwork, in the nineteenth century, when fieldwork had already long been flourishing, the rhetoric was simply "the geological community's attempt, in the face of enormous external pressures towards intellectual conformity in an age of counter-revolutionary turmoil, to convince society—and itself—of the untainted loyalty of its science" (205). But, he adds, it also gave them a liberating "research programme" of "immediacy and clarity" (208), articulating "an unprecedented certainty about the direction of future research" (207). Laudan actually argues, in "Ideas and Organisations" (1977), that the rigid inductivist program of the Geological Society was the least sophisticated methodology open to geologists at that period, and that it hindered the progress of geological science because "the members actually acted on their principles."(530) In the 1820s and 1830s, when the Society admitted younger members like Sedgwick, Murchison, and Charles Lyell, this methodology was relaxed. Nonetheless, its earlier commitment to producing a geological map lived on in the work of this new generation, many of whom were, as we will see, devoted to mapmaking as perhaps the most important exercise in the new science. It is this side of the science that I am discussing here—the places where we may take that rhetoric seriously.

11. [Fitton], "Transactions of the Geological Society" (1817b), 74.

12. See O'Connor, *Earth on Show* (2007), chap. 1

13. Murchison, in Geikie, *Life of Roderick Murchison*, 1: 214.

14. Sedgwick, in *Life and Letters*, 1: 397.

15. The story of this controversy is the subject of Secord, *Controversy in Victorian Geology* (1986).

16. Geikie, *Life of Roderick Murchison*, 2:206.

17. See Secord, *Controversy in Victorian Geology* (1986) and "King of Siluria" (1982). For further work on Murchison's imperialist classifications of the strata see, Stafford, *Scientist of Empire* (1989).

18. Lapworth, "On the Tripartite Classification" (1879), 7.

19. See Kolbl-Ebert, "Charlotte Murchison" (2004).

20. Lapworth wrote, for instance, that Murchison's *Silurian System* was met with "profound relief and satisfaction" by his peers: "His clear and brilliant presentation of the physical and palaeontological proofs of an orderly sequence among the Paleozoic Rocks below the Old Red Sandstone, as originally set forth in all their force and harmony in his magnificent volumes, naturally astonished and dazzled the majority of his scientific contemporaries, and secured for his nomenclature of these ancient deposits an almost universal acceptance" ("On the Tripartite Classification" [1879], 1). Lapworth overpraises Murchison here only as a prelude to some severe criticisms of his subsequent conduct, but the point remains that the form of Murchison's *Silurian System* was accredited with attaining "almost universal acceptance" for his classification of the rocks, at least to begin with.

21. Sedgwick to Murchison (January 10, 1839), in *Life and Letters*, 1:521.

22. Sedgwick to Lyell (April 9, 1845), in *Life and Letters* (1890), 2:84.

23. Sedgwick to Murchison (January 10, 1839), in *Life and Letters* (1890), 1:521. The "three-decker" refers, of course, to the heavy three-volume, or "triple-decker," novels of the mid-Victorian period.

24. Sedgwick to Canon Wodehouse, October 12 1837, in *Life and Letters* (1890), 1: 499.

25. Sedgwick to Murchison, 1837, in *Life and Letters* (1890), 1: 500–501.

26. Sedgwick, "Anniversary Address" (1834a), 207.

27. Ibid., 209.

28. Ibid. See O'Connor, "Young Earth Creationists" (2007), for important reminders that scriptural geologists such as Ure were not "pantomime villains, baleful but ineffectual opponents of intellectual emancipation" (despite the rhetoric of Sedgwick and others before and since) (357), and, surveying the kind of rhetoric in which Sedgwick participates here, that "caustic comments by participants on one side of the controversy such as ... Sedgwick hardly count as neutral sources" (363), See also Brooke and Cantor, *Reconstructing Nature* (1998), 57–62, and studies on evangelicalism and science such as Fyfe, *Science and Salvation* (2004), and Livingstone et al., *Evangelicals and Science* (1999). Sedgwick, of course, was evangelical himself.

29. See Secord, *Controversy in Victorian Geology* (1986), 63–68; Smith, "Geologists and Mathematicians" (1985), 49–59.

30. In Secord, *Controversy in Victorian Geology* (1986), 66.

31. Ibid.

32. Becher, "William Whewell and Cambridge Mathematics" (1980), 6; Smith, "Geologists and Mathematicians" (1985).

33. Playfair reviewed Robert Woodhouse's *Elements of Trigonometry*. [Playfair], "Woodhouse's *Trigonometry*" (1810). Woodhouse was the first to espouse the use

of analytical methods at Cambridge, and it was in his wake that the Cambridge Analytical Society was formed by the undergraduates George Peacock, George Airy, Charles Babbage, and John Herschel.

34. Ackerberg-Hastings, "Analysis and Synthesis" (2002); I. B. Cohen, *Revolution in Science* (1985). See also Schaffer, "History and Geography" (1991), 201–33; Whewell, *Liberal Education* (ref. 84), 45, as quoted in Becher (1980), 30
35. Sedgwick, *Discourse* (1850), cccxxi–cccxxxii.
36. Newton, as quoted in Sedgwick, *Discourse* (1850), 99–100.
37. Sedgwick, *Discourse* (1850), 100.
38. Ibid., cccxxxvi, 100.
39. Sedgwick, "Anniversary Address" (1834b), 298.
40. Secord, *Controversy in Victorian Geology* (1986), 58.
41. Sedgwick, "Anniversary Address," (1834a), 212.
42. "Adam Sedgwick to Dr. Livingstone, March 16th 1865," *Life and Letters* (1890), 2:412.
43. See Schaffer, "William Whewell," (1991).
44. Secord, *Victorian Sensation* (2000), 89.
45. Ibid., 90.
46. See ibid., 35.
47. Richard Altick, *English Common Reader* (1957), 287. For discussion of the Newgate novel controversy, see Hollingsworth, *Newgate Novel* (1963); Kalikoff, *Murder and Moral Decay* (1986); Altick, *Victorian Studies in Scarlet* (1972).
48. See [Thackeray], "Horae catnachianae" (1839), "Going to See a Man Hanged" (1840), and "Review of *Jack Sheppard*" (1840). See also Thackeray's novel *Catherine* (1999).
49. P. T. Murphy, *Toward a Working-Class Canon* (1994), chap. 3; Altick, *English Common Reader* (1957), 273–80.
50. [Playfair], "Geographie mineralogue" (1812), 381. Also see [Playfair], "Cuvier on Fossil Bones" (1811), 230.
51. Lyell to Mantell (March 2, 1827), in *Life, Letters and Journals*, 1:168.
52. [Playfair], "Geographie mineralogue" (1812), 381.
53. Sedgwick to Napier (April 17, 1845), in Napier, *Selection from the Correspondence* (1879), 492–93.
54. Ibid., 493.
55. Ibid., 494.
56. Sedgwick, referred to and partly quoted in *Life and Letters* (1890), 2:88.
57. [Sedgwick], "Vestiges of the Natural History of Creation" (1845), 44.
58. Sedgwick, *Discourse on the Studies* (1850), clviii.
59. Ibid., clxi.
60. Darwin, in Clark and Hughes, *Life and Letters* (1890), 1:344.
61. Ferris, *Achievement of Literary Authority* (1991), 33.
62. Secord, *Victorian Sensation* (2000), 240, wherein is quoted Sedgwick, 4 May 1845, BL Add. Mss. 34625, ff. 175–76.

63. [Sedgwick], "Review of *Vestiges*" (1845), 2.
64. Ibid., 4.
65. Sedgwick to Charlotte Murchison (August 11, 1833), in Clark and Hughes, *Life and Letters* (1890), 1:409.
66. See Cockburn, "Account of the Friday Club" (1910).
67. Secord, *Victorian Sensation* (2000), 236.
68. The significance of Sedgwick's possession of *Uncle Tom's Cabin* (and other antislavery texts) is discussed in chapter 4.
69. For Sedgwick's library, see Swan, *Valuable Library* (1873).
70. Clark and Hughes, *Life and Letters* (1890), 1:3. Recent scholarship has challenged the widespread view that Scott "rescued the genre" of the novel, as Garside notes in "Walter Scott and the 'Common' Novel" (1999), 1, mainly by suggesting that Scott did not innovate a new kind of historical novel, insofar as female novelists were producing the "national" novel at the same time. See Garside, "Popular Fiction" (1991); K. Trumpener, "National Character" (1993); Ferris, "Translation from the Borders" (1997); I. Dennis, *Nationalism and Desire* (1997). Nonetheless, the reception of Scott in the nineteenth century certainly created and promoted the view that he had rescued the genre. My analysis here supports the view that wider reception studies of literary authors (beyond the parameters of their intertextual relations with other novelists or poets) among readers and communities including men and women of science can help give a broader picture of the significance of those authors at particular historical moments.
71. Sedgwick to Ingle (March 24, 1837), in Clark and Hughes, *Life and Letters* (1890), 1:481.
72. Sedgwick to Joseph Beete Jukes (November 2, 1855), CUL Add. 7652/7650. Box 9, III E.10.
73. Sedgwick to Joseph Beete Jukes (December 15–17, 1855), CUL Add. mss. 7652/7650. Box 9, III E.11.
74. Sedgwick to Fanny Hicks (January 3, 1847), in Clark and Hughes, *Life and Letters* (1890), 2:108.
75. Clark and Hughes, *Life and Letters* (1890), 2:573.
76. Sedgwick, in ibid., 2:573–74. Such stories were favorites among the gentlemanly geological community.
77. *Life and Letters* (1890), 1:256, 2:145.
78. For examples relating to other geologists such as William Buckland and Edward Forbes, see Shortland, "Darkness Visible" (1994), 37 and n201.
79. See Sommer, "Romantic Cave" (2003).
80. Geikie, *Life of Roderick Murchison*, 1:195–96.
81. For an account of Sedgwick's lecturing to the British Association, see Brown, "Squibs and Snobs" (1992), 187.
82. Clark and Hughes, "Introduction," in *Life and Letters* (1890), 1:3.
83. Crosbie Smith summarizes, following Secord, the "differing geological styles of Sedgwick and Murchison": "Sedgwick's style was characterised above all by his

primary concern with geological structures and by correspondingly less emphasis on palaeontological techniques. Sedgwick's style, in short, was geometrical, and his geological language that of lines, dips, strikes and synclines. By contrast, Murchison, ex-soldier and foxhunter, aimed for a rapid grasp of the fundamental features of a district" and "placed a consequent emphasis on palaeontology." "Geologists and Mathematicians" (1985), 52.

84. Geikie, *Life of Roderick Murchison* (1875), 2:231.

85. See Ormond, *Monarch of the Glen* (2005), 19–39.

86. Reginald Henry Graham recalls visiting "the Grange, where at that time lived the second Lord and Lady Ashburton. I was often there and met many celebrities—Landseer, the Carlyles, Sir Roderick Murchison, Charles Kingsley, Venables (of the Saturday Review), Laurence Oliphant, the Brookfields, and others." Graham, *Fox-Hunting Recollections* (1908), 9.

87. Geikie, *Life of Roderick Murchison* (1875).

88. Ibid., 1:8n1.

89. See Oldroyd, "Sir Archibald Geikie" (1980).

90. See Secord, *Controversy in Victorian Geology* (1986), 306–10.

91. Geikie, *Story of a Boulder* (1858), 164–65.

92. Canto 1:16. In fact, the passage is doubly intertextual, since Geikie's *Boulder* was powerfully shaped by the writings of his mentor, Hugh Miller, whose own geological writings tended much further toward narrative than those of the Geological Society gentlemen I am concerned to outline here. I am indebted to Ralph O'Connor for pointing this out to me.

93. Canto 1:2. The sound of the huntsmen and the thunder echoing round the glens is also described as louder than the sound of "sylvan war," which has also rung through this place at Uam-var (canto 1:4).

94. Canto 1:3.

95. Grenier, *Tourism and Identity in Scotland* (2005), 83.

96. Ibid., 53.

97. Sedgwick, to Rev. John Sedgwick (September 26, 1829), in Clark and Hughes, *Life and Letters* (1890), 1:355.

98. Sedgwick to Mrs Alison (November 15, 1834), in ibid., 1:431.

99. Sedgwick, in ibid., 2:33.

100. Porter, "Gentlemen and Geology" (1978), 819.

101. Sedgwick to Thomas Ainger (February 19, 1825), in Clark and Hughes, *Life and Letters* (1890), 1:265.

102. Sedgwick to Kate Malcolm (October 25, 1855), in ibid., 2:302.

103. George IV dressed in Highland garb for the occasion. Scottish cultural distinctiveness did not mean the same thing as political independence, of course, and the procession orchestrated by Scott can be seen as a conciliatory measure designed to retain and identify something recognizably "Scottish" even as it validated and confirmed the Union.

104. Geikie, *Life of Roderick Murchison* (1875), 91.

105. Jameson, *Mineralogy of the Scottish Isles* (1800), 36–37, 62–65.
106. Young, *Mind over Magma*, chap. 5.
107. See Eyles, "John MacCulloch, F.R.S.," (1937); Cumming, "John MacCulloch, F.R.S. at Addiscombe" (1980); Cumming, "Description of the Western Islands" (1977); Flinn, "John MacCulloch, M.D., F.R.S." (1981).
108. See Cumming, "John MacCulloch, F.R.S. at Addiscombe" (1980) and "John MacCulloch" (1982).
109. MacCulloch, *Description of the Western Islands* (1819), 1:384.
110. Ibid., 2:345–46. MacCulloch, *Geological Classification* (1821), 64–65.
111. MacCulloch, *Description of the Western Islands* (1819), 2:385, 386.
112. Ibid., 2:366.
113. Ibid., 2:387.
114. Ibid., 2:312–13.
115. Ibid., 2:313–18.
116. Ibid., 3:59.
117. Ibid., 3:61, 59–60.
118. Ibid., 2:316. See Cumming, "John MacCulloch, F.R.S., at Addiscombe" (1980).
119. MacCulloch, *Description of the Western Islands* (1819), 1:450.
120. Ibid., 2:311.
121. Garside, "Scott, the Romantic Past" (1972), 260.
122. MacCulloch, *Highlands and Western Islands* (1824), 1:192.
123. Ibid., 1:193.
124. Ibid., 1:194.
125. Ibid., 1:195. For tourism to Fingal's Cave following, not Scott's writings, but those of Joseph Banks, see Shortland, "Darkness Visible" (1994), 5–8. This reminds us that Scott was tapping into preexisting popular enthusiasm for wild or cavernous landscapes in his writing, rather than creating that enthusiasm from scratch. Shortland also remarks the "banalization" of Fingal's Cave in contemporary literary and artistic representations, noting that Turner's "contrivance" of the scene for his Scott illustrations is due to its having "been prepared to fit Scott's verses" on the cave in *The Lord of the Isles* (10).
126. MacCulloch, *Highlands and Western Islands* (1824), 1:196.
127. Ibid., 1:196.
128. Ibid., 1:196–202.
129. Sedgwick and Murchison, "Geological Relations" (1829) and "Structure and Relations" (1829).
130. MacCulloch, *Highlands and Western Islands* (1824), 1:450.
131. Ibid., 1:8.
132. Sedgwick to Whewell (July 15, 1827), in Clark and Hughes, *Life and Letters* (1890), 1:302.
133. Sedgwick, in ibid., 1:305. The description of this passage even ends with a reference by Clark and Hughes to the full title of Scott's *Waverley; or, 'Tis Sixty Years*

Since: "Such were some of the difficulties a geologist had to face in the Highlands sixty years since" (305).

134. MacCulloch, *Highlands and Western Islands* (1824), 2:25.
135. Scott, *Rob Roy* (1998), 344.
136. MacCulloch, *Highlands and Western Islands* (1824), 2:130.
137. Ian Duncan discusses "romance" as "the aesthetic signifier of a specific historical process: the appropriation, and reinvention, of 'a common cultural heritage' as individual literary property," "a cultural genealogy for the educated imagination." Partly this romance was constructed as puerile narcissism to be grown out of and partly it signifies the lost but "ethically true" "cultural identity." Duncan, *Modern Romance* (1992), 58, 59. Gary Kelly talks about "romance" in a similar way to me and to Duncan in *English Fiction of the Romantic Period* (1989), 145.
138. MacCulloch, *Description of the Western Islands* (1819), 2:311.
139. Boud, "Early Geological Maps" (1975), 187.
140. Sedgwick, in Clark and Hughes, *Life and Letters* (1890), 1:305.
141. Cumming, "Geological Maps" (1981), 255–71.
142. Sedgwick and Murchison, "Geological Relations" (1829). Lyell's own "tone bordering on ridicule" here is slightly misleading—it was certainly not that MacCulloch was not interested in fossils—in fact, he wrote the essay on "Organic Remains" for Brewster's *Edinburgh Encyclopaedia* (1832), 15:1–74. But he was opposed to the use of fossils for stratigraphic purposes.
143. Lyell, "Anniversary Address" (1836), 359.
144. Sedgwick, "Appendix," in *Guide to the Lakes* (1842), 21.
145. Cooper, "Portrait in Miniature," Lapworth Museum, L2A.3, 6.
146. Lapworth, as quoted in ibid., 8.
147. Ibid.
148. As quoted in ibid., 2.
149. See introduction to the present volume. For more on Thorpe as a proxy for Sedgwick, see chapter 4 below.
150. See Cooper, "Portrait in Miniature," 5.
151. Elles, "Chapters in the Life of Charles Lapworth," Lapworth Museum, L2A.4, 6–7.
152. Ibid., 7.
153. Lapworth, "The Silurian Age in Scotland. Lecture I.," Lapworth Museum, L28.3C, 1.
154. Ibid.
155. Ibid.
156. Ibid.
157. Ibid
158. Scott, *Minstrelsy of the Scottish Border* (1802), 1.
159. Lapworth, "Silurian Age."
160. Scott, *Marmion* (1808), 73.
161. Lapworth, "Silurian Age."

162. Ibid.
163. Ibid.
164. Ibid.
165. Elles, "Chapters in the Life," 7, 15.
166. See Lapworth, "On the Tripartite Classification" (1879), for the paper in which he proposed these terms to the geological community. For a brief account of the reluctant acceptance of his nomenclature, see Bassett, "One Hundred Years" (1979). Most importantly, see Secord, *Controversy in Victorian Geology* (1986), 307–11.
167. Again, I am indebted to Ralph O'Connor for pointing out this connection to Hugh Miller.

Chapter Three

1. Ruskin won for "Salsette and Elephanta" (1839), Matthew Arnold for "Cromwell" (1843), and Oscar Wilde for "Ravenna" (1878).
2. Charles Lyell Sr. to Dawson Turner (April 24, 1816), in Leonard Wilson, *Charles Lyell* (1972), 38.
3. See Wilson, *Charles Lyell* (1972), 58.
4. "Lyell, to his father, Venice, August 26 1818," in K. M. Lyell, *Life, Letters and Journals* (1881), 1:105. See "Lyell to his father, October 21 1817," in ibid., 1:54, for the only other of his published poems. It is possible that there is other poetry in the Lyell archive at Kinnordy, though unfortunately I have not been able to gain access to this archive.
5. The "sensational" number of copies produced of *Vestiges of the Natural History of Creation* was over twenty-three thousand in a similar period. As Secord notes, however, statistics alone are not sufficient for measuring the impact of a work. For an almost comprehensive list of the reviews published of the first edition of *Principles* see Secord, *Principles of Geology* (1997), 459–61.
6. K. M. Lyell, *Life, Letters and Journals*, 1:56.
7. See Edmonds and Douglas, "William Buckland, F.R.S." (1976), 141–67.
8. "Lyell to Dawson Turner, 21 March 1816," cited in Wilson, *Charles Lyell* (1972), 41.
9. Wilson, *Charles Lyell* (1972), 50.
10. See ibid., 53–54 (54).
11. Thanks to Anne Secord for talking with me about this issue.
12. See Rupke, "Oxford's Scientific Awakening" (1997).
13. Buckland's inaugural lecture was published as *Vindiciæ Geologicæ* (1820).
14. See Pemberton and Frey, "William Buckland and His 'Coprolitic Vision'" (1991).
15. On Buckland's cave geology see Rupke, *Great Chain of History* (1983), 29–95, and particularly the following works by Sommer: "'Amusing Account'" (2004); "Romantic Cave?" (2003); and *Bones and Ochre* (2007).

16. See Buckland, "On the Discovery of Coprolites" (1835).

17. "Lyell to Gideon Mantell, 8 February 1822," in K. M. Lyell, *Life, Letters and Journals*, 1:115.

18. Mozley, *Reminiscences* (1882), 1:179. Mozley's text is well known for its inaccuracies, but this is at least a reflection of his own feelings about Buckland. See also Edmonds and Douglas, "William Buckland, F.R.S." (1976).

19. Mozley, *Reminiscences* (1882), 1:179.

20. See O'Connor, *Earth on Show* (2007), 71–117, and Sommer, "Romantic Cave?" (2003), for much fuller treatments of Buckland's use of the sublime, the magical, and the wondrous in his writings and performances. Nonetheless, Shortland "Darkness Visible" (1994) argues that "Buckland's enthusiasm for caves or their buried treasures had little in common with the manipulated delights offered to the Romantics, and his descriptions shared nothing with Romantic effusions to sublimity and sensation. Buckland's language is blunt and coarse and his idiom hewn from the rock. The particular observations and measurements are what attracted him, not the sensation or atmosphere," and that "mining and manual labour form his world" instead (36).

21. Much of my discussion here takes its cue from O'Connor, "Epic of Earth History," (2009), 207–23.

22. For Buckland's presentation of geology in consonance with classical learning see Rupke, *Great Chain of History* (1983), 51–63.

23. O'Connor, "Epic of Earth History" (2009); Lightman, *Victorian Popularizers of Science* (2007), 219–94.

24. O'Connor, "Epic of Earth History" (2009), 208.

25. Buckland, *Geology and Mineralogy* (1836), 1:113.

26. For a good, though dated, summary of different views on direction in the fossil record, see Bowler, *Fossils and Progress* (1976), 15–92.

27. See Cannon, "Charles Lyell" (1976), 112.

28. See *Earth on Show*, 249–50, where O'Connor writes that "although these passages were much expanded from their Regency counterparts, most of them still formed only a small (if rhetorically crucial) part of the text they inhabited" and notes that "any sense of narrative progression" in Buckland's *Geology and Mineralogy* "has to be inferred from the order in which the topics are discussed," "the large-scale historical aspect of [his] arrangement . . . left implicit."

29. For further discussion of this see A. Buckland, "Show and Tell" (2009), 114–17.

30. Buckland, *Geology and Mineralogy* (1836), 1:12.

31. Ibid., 1:114.

32. See Buckland, *Vindiciæ Geologicæ* (1820).

33. Buckland, *Geology and Mineralogy* (1836), 1:21.

34. Ibid., 1:596. O'Connor, in *Earth on Show* (2007), shows that much of this epic and sublime language was not published in print by Buckland until the 1830s, though it was a feature of his rhetoric in the more controlled spheres of his lectures.

35. For nineteenth- and early twentieth-century discussions of Milton's cosmology, see Orchard, *Astronomy of Milton's "Paradise Lost"* (1896); Paterson, "Astronomy of Milton" (1909); and Gilbert, "Milton's Textbook of Astronomy," *PMLA* (1923). Discussion continued throughout the century; more recent articles include Zivley, "Satan in Orbit" (1997), and Martin, "'What If the Sun Be Centre to the World?'" (2001).

36. Lyell, "Scrope's *Geology*" (1827), 473. Lyell also quotes Milton approvingly in "Transactions of the Geological Society" (1826), 538. Again, see O'Connor, "Epic of Earth History" (2009), esp. 214–19, and *Earth on Show*, chap. 4, for a discussion of Lyell's prediction and how it was fulfilled by subsequent writers.

37. Wilson, *Charles Lyell* (1972), suggests that poetry was important to Lyell in his youth (see esp. 27–28) but that his disappointment with having failed to win the Newdigate Prize at his second attempt "was perhaps softened by the fact that his enthusiasm had now been kindled by a new interest," geology (41). Wilson suggests that an embittered Lyell entered the prize for the third time in 1818 only "as a matter of form," and that "it must have been with a sense of pleasant relief, therefore, that Charles turned to his second course of Buckland's lectures" (58). Wilson documents Lyell's early interests in literature, but after Oxford his literary pursuits are rarely given attention. Bailey, *Charles Lyell* (1962), notes but does not comment on Lyell's literary interests (13, 31–38); they are not mentioned at all in Bonney, *Charles Lyell and Modern Geology* (1895).

38. Lyell, *Principles* (1830), 1:85.

39. Ibid., 1:86.

40. Ibid., 2:3.

41. On Lyell's *vera causa* methods, see Laudan, "Role of Methodology" (1982), 216. Also see Laudan, *From Mineralogy to Geology* (1987), chap. 9; Secord, "Introduction" (1997), xxi–xxiii; and Camardi, "Charles Lyell and the Uniformity Principle" (1999). Laudan's view, with which I concur, is that Lyell's *Principles* was not an argument about the historical pattern of earth history (i.e., that the earth *had actually* always operated as it did in the present) but an argument that the only method by which the past could be safely reasoned about was by analogy with observable processes. This is contrary to the more common view evinced by Rudwick; see, for instance, his "Uniformity and Progression" (1971) and *Worlds before Adam* (2007), 304–5.

42. For Rudwick's most forceful articulation of the different kinds of uniformitarianism, see "Uniformity and Progression" (1971), 209–27. See also Hooykaas, *Natural Law* (1963).

43. Whewell, *History of the Inductive Sciences* (1837), 3:596–97.

44. Laudan, "Ideas and Organisations" (1977), 527–38.

45. [Whewell], "*Principles of Geology* by Charles Lyell," *British Critic* (1831): 182.

46. See Rudwick, "Uniformity and Progression" (1971), 209–27.

47. See "Lyell, to George Poulett Scrope, June 14 1830," in K. M. Lyell, *Life, Letters and Journals* (1881), 1:268.

48. Lyell, *Principles* (1830), 1:169, 105.

49. Gould, *Time's Arrow* (1987), 98–179; Ospovat, "Lyell's Theory of Climate" (1977); Rudwick, "Uniformity and Progression" (1971). For the influential steady-state model, see also Rudwick, "Lyell and the *Principles of Geology*" (1998).

50. Cannon, "Charles Lyell" (1976), 110–11. Bartholomew "Non-Progress of Non-Progression" (1976).

51. See my introduction, notes 77 and 78. CHECK.

52. Smith summarizes, following Secord, that "mainstream geology was above all a practical activity not concerned with speculative cosmological theory but rather with geology as natural history and hence with *classification*." Smith, "Geologists and Mathematicians" (1985), 52.

53. Rudwick, *Great Devonian Controversy* (1985), 46.

54. Buckland, *Geology and Mineralogy* (1836), 1:27.

55. "Sedgwick to Charles Lyell, 20 September 1835," in Clark and Hughes, *Life and Letters* (1890), 1:447

56. Lyell, *Principles of Geology* (1830), 1:409.

57. Ibid., 1:409.

58. Ibid., 1:409.

59. Ibid., 1:196. Lyell's delight in catastrophe narrative, including the example of the Tivoli flood, is analyzed as part of his *vera causa* strategy by O'Connor, *Earth on Show*, chap. 4.

60. Lyell, *Principles of Geology* (1830), 1:197.

61. Ibid., 1:449–59.

62. Dolan, "Representing Novelty" (1998).

63. Secord, "Introduction" (1997), xviii.

64. Lyell, *Principles* (1832), 2:26–31. Lyell ignored some of the findings of this research that did reveal changes; see Corsi, "Importance of French Transformist Ideas" (1978), 233. Much of my discussion of the second volume is indebted to Corsi's essay (221–44).

65. See Bartholomew, "Non-Progress of Non-Progression" (1976).

66. Lyell, *Principles* (1832), 2:49–60.

67. On domestication see ibid., 2:39–48.

68. Ibid., 3:38.

69. Ibid., 1:147, 150.

70. For examples of this notion of the "alphabet" see [Whewell], "*Principles of Geology*," (1831), 181, and "A Geological Manual," (1832), 43.

71. Lyell, *Principles* (1833), 3:7.

72. For Lyell's explanation of his "subdivisions of the tertiary epoch," see ibid., 3:52–61. Pages 62–323 contain his detailed description of these strata. For a more detailed account of the structure of *Principles*, see Rudwick, "Charles Lyell's Dream" (1978).

73. Lyell, *Principles* (1833), 3:32.

74. Ibid., 3:33–34.

75. Porter, "Charles Lyell and the Principles" (1976), 96; Ospovat, "Distortion of Werner" (1976); Cannon, "Charles Lyell" (1976), 108. Also see Cannon, "Problem of Miracles" (1960), and "Uniformitarian-Catastrophist Debate," (1960); Hooykaas, *Natural Law* (1959); Page, "Diluvialism and Its Critics" (1969). Wilson is the exception to this revisionist rule: "Geology on the Eve" (1980).

76. John Thackray, "Charles Lyell and the Geological Society" (1998).

77. For references to Lyell as a member of the Geological Society, see the anonymous reviews in the *Edinburgh Literary Journal* (1830), 117, *Gentleman's Magazine* (1832), and *London Literary Gazette* (1830), the last of which claimed that "Mr. Lyell's aim appears to have been that of carrying into effect, so far as the laborious research of an individual can accomplish, the excellent principles pursued by the Geological Society, that of collecting and collating facts" (505). In a diplomatic piece of writing Lyell's friend Scrope wrote in the *Quarterly Review* (1835), "The foundation was in fact thus laid by Messrs. Greenough, MacCulloch, Buckland, Conybeare, and other active members of the Geological Society, for the building which Mr. Lyell, in a happy moment, undertook to raise" (407). This list is by no means exhaustive.

78. Lyell, "Journal, to Mary Horner, Rotterdam 4 August 1831," in K. M. Lyell, *Life, Letters and Journal* (1881), 1:326. This speaks of the British context. Lyell's work was less influential on the continent, though, in French translation, it appears to have been useful to Italian geologists as a guide to the tertiary strata of Italy. Vaccari, "Lyell's Reception" (1998).

79. The notion that British geology had its own distinctive agendas and a long-held and coherent research program, along with its own institutional bases, and that these were of importance to Lyell, is well established by historians. See, for instance, Morrell, "London Institutions" (1976), which argues that much of what was important about British geology in these decades had little to do with Lyell's kinds of arguments. For a less radical view of this research tradition in the 1820s, see Rudwick, "Strategy of Lyell's *Principles of Geology*" (1970), and, for his most recent discussions of Lyell in his British and European contexts, *Worlds before Adam* (2007), esp. 243–96. Again, the only historian to reject this view is Wilson. See particularly "Geology on the Eve" (1980).

80. Lyell, "Scientific Institutions" (1826a); "Transactions of the Geological Society" (1826b); "State of the Universities" (1827).

81. "Lyell to Scrope, June 14 1830," in K. M. Lyell, *Life, Letters and Journals* (1881), 1:268.

82. "Lyell to Scrope, September 11 1830," in ibid., 1:296.

83. See Burns, "John Fleming" (2007); Laudan, "Role of Methodology" (1982), 226–27.

84. As quoted in Burns, "John Fleming" (2007), 220, 218.

85. Ibid., 219.

86. "Charles Lyell to Gideon Mantell, April 1829," in K. M. Lyell, *Life, Letters and Journal* (1881), 1:252.

87. De la Beche, *Geological Manual* (1832), 131.
88. "Charles Lyell to Fleming, February 3 1830," in K. M. Lyell, *Life, Letters and Journal* (1881), 1:260.
89. "Charles Lyell to Fleming, February 3 1830," in ibid., 1:261.
90. O'Connor, "Epic of Earth History" (2009), 215.
91. Rossetti, *Comento Analitico* (1826–1827). For a brief description, see Caesar, *Dante* (1989), 501–3. For its supporters, Friederich, "Dante through the Centuries" (1949), 51–52. Rossetti and Lyell Sr. were good friends; Rossetti named his son Gabriel Charles Dante Rossetti, who became the poet Dante Gabriel Rossetti, partly after Lyell.
92. [Lyell], "Rossetti's Dante" (1830), 437. Lyell talks about the gergo at length but does not mention the Egyptian and Masonic history of this sect, which is described more fully in his father's work, Charles Lyell Sr., *Poems of the Vita nuova and Convito* (1842), xlviii.
93. Rossetti, *La Divina Commedia* (1826–1827).
94. Rossetti, *Disquisizioni Sullo Spirito Antipapale* (1832); translated by Caroline Ward as *Disquisitions on the Antipapal Spirit* (1834).
95. See Hallam, *Remarks* (1832), 58. More positive reviews can be found in anonymous reviews in the *Oriental Herald* across three issues in 1827 and the *Literary Gazette* (1827, 1828).
96. Hallam, *Remarks* (1832), 54. One reviewer compared Dante to Father Hardouin, an eighteenth-century scholar from Brittany who had declared that the greater part of the classical corpus, including the *Aeneid*, the *Odes* of Horace, and the histories of Livy and Tacitus, was the forgery of monks; see the comments of the reviewer in *Edinburgh Review* (1832), 531–51 (550). For a particularly hostile review, see "*La Divina Commedia*" in *Foreign Review* 3 (1828), 175–95.
97. [Lyell], "Rossetti's Dante" (1830), 433.
98. Ibid., 441.
99. Ibid., 444–45, 447.
100. "Charles Lyell, to his father, London, April 1828," in K. M. Lyell, *Life, Letters and Journal* 1:181.
101. Roy Porter discusses this Niebuhrian quotation in "Charles Lyell and the Principles" (1976), 96–97. In "Transposed Concepts" (1997), Martin Rudwick argues that the quotation summarizes "all that Lyell hoped to achieve in geology: to 'scatter the cloud' of poetic mystery that had been spread over the earth's past" and to bring the earth's past into light (79). On Niebuhr, see Ledger-Lomas, *Selective Affinities* (forthcoming 2013), chap. 1, and Vance, "Niebuhr in England" (2000). The discussion of the translation in the *Quarterly Review* prompted Hare to reply in *A Vindication of Niebuhr's "History of Rome"* (1829). For other historical models with which Lyell compared geology, see Rudwick, "Historical Analogies" (1977), 89–107, and *Worlds before Adam* (2009), 301–5.
102. Ledger-Lomas, *Selective Affinities* (forthcoming 2013).
103. Niebuhr, *History of Rome* (1828), 3.

104. Ibid., 4–5, 25.
105. Lyell, *Principles* (1830), 1:84.
106. Lyell, in K. M. Lyell, *Life, Letters and Journals* (1881), 1:24–25.
107. Lyell, in ibid., 1:25–26.
108. Lyell, in ibid., 1:26.
109. Lyell, *Principles* (1830), 1:37.
110. Ibid.
111. Ibid., 1:62.
112. Ibid.
113. Dante, *Divine Comedy* (1994), 10.
114. Lyell, *Principles* (1830), 1:82–83.
115. Ibid., 1:184.
116. Ibid., 1:431–32.
117. Ibid., 1:238.
118. Ibid., 1:331.
119 Ibid., 1:530.
120. Tucker, *Epic* (2008), 191. See Ralph O'Connor, *Earth on Show*, 174–76 and chap. 8, for an alternative discussion of Byron and Lyell, and of Byron as used by later geologists.
121. Tucker, *Epic* (2008), 192.
122. Ibid., 234.
123. Ibid., 236–51.
124. Lyell, *Principles* (1832), 2:79, 81.
125. For a contemporary example of this, in spectacular prose style, see Davy, *Consolations in Travel* (1830). There is not room here to discuss that work's progressionist narrative of earth history, to which Lyell's text is a diametric opposite.
126. Ibid.

Chapter Four

1. Ralph O'Connor argues that such plots were pushed into the foreground by a broad variety of journalists and other professional writers even as the gentlemen of the Geological Society shied away from them. O'Connor, *Earth on Show* (2007).
2. The shift resembles the concurrent turn in historical writing from universal histories of mankind, which had proposed "that all human activity is under one guiding principle and can be told as one story," to a new focus between 1780 and 1830 on source criticism, a narrowing of geographical range, and a commitment among many historical writers to empirical methods, which, like geology, built on the methods and traditions of antiquarianism. This is Mazlish's definition of what he calls "ecumenical" history. See Mazlish, "Ecumenical, World and Global History," 42. For the decline in global histories in this period see Stuchtey and Fuchs, *Writing World History* (2003), 4–5. Martin Rudwick has written most strongly on this subject in *Bursting the Limits of Time*, esp. 181–94, arguing that claims that geology drew from

philosophical and conjectural history are misconceived and that empirical methods in geology owed more to antiquarianism, chorography, and the compilation of accurate "annals" of world history by biblical chronologers. Nonetheless, a similar point may be made, as in the previous note: universal histories were eschewed by the elites but reprinted in their thousands by educational publishers well into the 1860s.

3. See Secord, "Geological Survey" (1986), esp. 232-34.

4. For a detailed discussion of this map, with its lack of "geological purpose" or "stratigraphic levels," see Rappaport, "Geological Atlas" (1969), quotation 275. Oldroyd, in *Thinking about the Earth* (1996), states that eighteenth-century maps "gave the outcrops or locations of rocks, minerals and ores, chiefly of economic significance, and had little to do with telling a history of the earth" (108). See also Eyles, "Mineralogical Maps" (1972); Rudwick, "Visual Language" (1976), 159-62. The first map to do this was Cuvier and Brongniart's *Carte geognostique* (1811).

5. First published in Webster, "On the Fresh-Water Formations" (1814), here reproduced as the frontispiece to Phillips, *Guide to Geology* (1835). See Rudwick, "Walk with the Founding Fathers" (2009); Heringman, "Picturesque Ruin" (2009); De la Beche, *Sections and Views* (1830), plates 1 and 5.

6. The leading authority on Smith is Hugh Torrens; for key examples see his "Timeless Order" (2001) and *Memoirs of William Smith* (2003). Relatively recent popular accounts of this (inevitably hagiographic) can be found in Winchester, *Map That Changed the World* (2002), and Morton, *Strata* (2001). For a more scholarly account see Eyles, "William Smith (1769-1839)"(1969).

7. See Laudan, "Ideas and Organisations" (1977), for Greenough's map. She suggests that it may not have been merely snobbishness that kept the Geological Society from collaborating with the practical man Smith on the map, but disagreement over the appropriate use of fossils in determining the age and deposition of rocks. Greenough, committed to the strictly empirical research program of the Geological Society in its early years, felt Smith's use of characteristic fossils to be "anathema" (535).

8. See Porter, "Gentlemen and Geology," 825-29. Murchison, for example, was keen to secure an economic significance for the sequence of strata he had "discovered," which he named the "Silurian" system, ensuring that his classifications of the strata would make the Silurian mark the endpoint of the coal-bearing beds. This secured the posterity of his system because it made the Silurian strata useful to surveyors and coal prospectors. See Rudwick, *Great Devonian Controversy* (1985), 57. Lucier argues against Porter's view in "Plea for Applied Geology" (1999), 6. See also Veneer, "Practical Geology" (2009), which suggests that economic and practical considerations were important in the early Geological Society but were later sacrificed (albeit not entirely) to stratigraphic research (which, of course, had its own economic implications, though they were not usually direct). All of this counters Porter, "Industrial Revolution" (1973).

9. Quotation from Secord, "Geological Survey" (1986), 243. For gentlemanly geology, see Porter, "Gentlemen and Geology," 825-29.

10. See Rudwick, "Visual Language"; Cook (1995), "From False Starts."

11. Anon., "Geological Map of Europe" (1856), 582. See, for instance, De Strzelecki, *Physical Description* (1845), which was reviewed by an anonymous reviewer in *Tait's Edinburgh Magazine* (1845) and by [Brougham] in *Quarterly Review* (1845).

12. Anon., "Stanford's *Geological Atlas*" (1904), 559, describes "Stanford's Atlas, an updated version of Reynolds." The two maps were the same size. See Freeman, "Tracks to a New World" (2001), esp. 52–56. Murchison's map included "the lines of the chief steam voyages" for the use of travelers across Europe, as well as railway lines (Anon., "Geological Map of Europe" (1856), 583. John Phillips produced maps with lines of railway track on them even when those lines had not yet been built: see Edmonds and Douglas, "John Phillips's Geological Maps" (1950), 362. See Bourne, *History and Description* (1846), combining timetables, lithographs of railway sections and buildings, and a geological map, for a further example.

13. [Johnston], "Physical Atlas" (1849), 331.

14. Ibid., 332.

15. Ibid., 332. The suggestion that the development of human history or the characteristics of human populations were shaped by geological features was widespread in the nineteenth century. See Oldroyd, "Sir Archibald Geikie" (1980), 448–49, for an example by a prominent later nineteenth-century geologist and historian of geology; see chapter 6 on George Eliot's "Natural History of German Life" and W. H. Riehl for another important example.

16. [Johnston], "Physical Atlas" (1849), 334–35.

17. Ibid., 335.

18. [Watts], "Yorkshire" (1859), 328.

19. Ibid., 332.

20. Ibid., 345.

21. Ibid., 349–50.

22. Ibid., 339.

23. Ibid., 351.

24. Bentley, *English Regional* (1941), 11. This geological complexity furnished a great part of "the material for English regional literature," she continued (11).

25. Ibid., 17.

26. Heywood, "Introduction" (2002), 19.

27. Homans, "Repression and Sublimation" (1978). A debate followed this article, in which the argument that nature was "repressed" or otherwise absent in the novel was disputed: see Homans and Conger, "Nature in Wuthering Heights" (1978). See Homans, *Bearing the Word* (1986), for further elaboration. While key readings of the novel have relied on the notion that Heathcliff and Cathy represent wild "nature," opposed to patriarchal, artificial, or coercive forms of culture, the lovers have equally often been allied with the supernatural, the Gothic, the excessively romantic, and the fairy tale. For Catherine falling from the rebellious and Edenic relationship with Heathcliff into the patriarchal structures of the Grange, see Gilbert and Gubar, *Madwoman in the Attic* (2000), 248–308, esp. 248–87. See Yaeger,

Honey-Mad Women (1988), for criticism and amplification Gilbert and Gubar, in which the text's laughter is powerfully reinstated (177–206). For "the raw, inhuman reality of anonymous natural energies" in the love between Catherine and Heathcliff, see Van Ghent, *English Novel* (1961), 157. Cecil in *Early Victorian Novelists* (1934) also describes the pair as "expression of the same spiritual principle" (156), and "Catherine's love" as "sexless" (147), though not quite supernatural, for the spiritual "is a natural feature of the world as[Brontë] sees it" (159). More recently, Anne Williams has attempted a reconciliation of the "realist" and "Romantic" readings of the text by associating it with Wordsworth rather than with Byron and the Gothic, in "Natural Supernaturalism" (1985).

28. Dean "'Through Science to Despair'" (1985), 121. See, for examples, Anon., "Wuthering Heights" (1848a), 141, in which it is stated that the "terrific story, associated with an equally fearful and repulsive spot . . . should have been called *Withering Heights*, for any thing from which the mind and body would more instinctively shrink, than the mansion and its tenants, cannot be easily imaged. 'Wuthering,' however, as expressive in provincial phraseology of 'the frequency of atmospheric tumults out of doors' must do, however much the said tumults may be surpassed in frequency and violence by the disturbances that occur in doors." See also Anon., "Wuthering Heights" (1848b), 21: "Whoever has traversed the bleak heights of Hartside or Cross Fell, on his road from Westmoreland to the dales of Yorkshire, and has been welcomed there by the winds and rain on a 'gusty day,' will know how to estimate the comforts of Wuthering Heights in wintry weather."

29. Brontë, *Wuthering Heights* (1965), 46.

30. Not only that, but earlier literary works had been named, if not for geological processes, then at least for geological events, such as Heinrich von Kleist's *The Earthquake in Chile* (1807).

31. This idea has been suggested by Christopher Heywood: see his "Yorkshire Slavery," 198; "Yorkshire Background" (1993); and, for further links between Sedgwick and the Brontës, "Introduction" (2002), 74–76.

32. See Heywood, "Yorkshire Slavery," 184–98, and "Yorkshire Background" (1993), 817–30.

33. Swan, *Valuable Library* (1873).

34. Ibid.

35. Heywood, "Yorkshire Slavery" (1987), 187. Another reading that restores the plantation context to Brontë's novel is von Sneidern, "*Wuthering Heights* and the Liverpool Slave Trade" (1995), 171–96, which argues that the novel imagines a plantation colony in the heart of Yorkshire.

36. Howitt, *Rural Life* (1838), 1:316.

37. Ibid., 1:318.

38. Ibid., 1:318.

39. Ibid., 1:319.

40. Whone, "Where the Brontës Borrowed Books" (1950); Dewhirst, "Rev. Patrick Brontë" (1965).

41. Brontë, *Wuthering Heights* (1965), 77.
42. Eagleton, *Myths of Power* (1988), 116, 114, 112.
43. Sedgwick, *Memorial* (1868), 14.
44. Ibid., viii-x.
45. Ibid., 77.
46. Gaskell, *Life of Charlotte Brontë* (2009), 27.
47. See Barker, "Haworth Context" (2002), for an account of Gaskell's fictionalization of the active and well-connected Haworth as an isolated moorland village.
48. [Lewes], "Wuthering Heights" (1850). This article in fact defends *Wuthering Heights* against such charges.
49. Anon., "Wuthering Heights and Agnes Grey" (1851), 448, 449.
50. Ibid., 449.
51. Anon., "Novels, Novel Readers" (1859), 239–51 (250).
52. Anon., "Wuthering Heights" (1848b), 483.
53. Ibid., 21.
54. Anon., "Tenant of Wildfell Hall" (1848) 21, 22.
55. This despite the fact that the novel's very use of the word "wuthering" drew, like many a regional novel, on antiquarian literature, topographical and picturesque travel literature and illustration, and the use of local dialect—all features that are usually taken as definitive of the "regional novel." See for instance Snell, *Regional Novel* (1998), 9–12.
56. Snell, *Regional Novel* (1998), 35.
57. Armstrong, "Emily's Ghost" (1992), 247.
58. Garofalo, "Impossible Love" (2008), 821. See also Duncan, "Provincial or Regional Novel" (2002), 324.
59. Homans, "Repression and Sublimation" (1978), 13.
60. Brontë, *Wuthering Heights* (1965), 110.
61. Ibid., 77, 222, 76.
62. Gose, *"Wuthering Heights"* (1966), 9, 15, 11.
63. Brontë, *Wuthering Heights* (1965), 122.
64. Phillips, *Rivers* (1853), 15.
65. Phillips, *Illustrations* (1835), 31.
66. Ibid., 31.
67. Ibid.
68. Phillips, *Rivers* (1853), 15. Phillips was widely quoted in travel guides to Yorkshire. This passage, for instance, was later quoted in Murray, *Handbook for Travellers* (1882), x.
69. Bevan, *Tourists' Guide* (1877), 3.
70. Sedgwick, "Description" (1836).
71. Ibid., 94.
72. Ibid.
73. See Porter, *Making of Geology* (1977), 209–10.
74. Sedgwick, *Memorial* (1868), 14.

75. Ibid.
76. Sedgwick names this source in a footnote; ibid., 85.
77. For Taylor's references to Lyell's *Principles* see *Words and Places*(1865), 50, 350–58.
78. Ibid., 5.
79. Ibid., 3.
80. Sedgwick, *Memorial* (1868), xv, 103, 104.
81. Ibid., 104.
82. Sedgwick, "Description" (1836), 101.
83. Ibid., 90.
84. Clark and Hughes, *Life and Letters* (1890), 1:6. A guidebook in 1910 also reproduced this imagery, describing Sedgwick as "a most *primitive* man, of the solid, ancient rock of humanity. He appears like a great boulder-stone of granite, such as he describes (in his letters addressed to Wordsworth in Wordsworth's *Guide to the Lakes* from 1842), transported from Shap Fell over the hills of Yorkshire, dropped here in our lowland country, and here fixed for life." He is "primitive in his name, Adam; primitive in his nature; in his noble, rugged simplicity; a dalesman of the north; primitive in his love of all ancient good things and ways; primitive in his love of nature, and of his native rock from which he was hewn; primitive in his loyalty to truth, and hatred of everything false and mean." Thompson, *Sedbergh, Garsdale and Dent* (1910), 283–84.
85. Sedgwick, "Three Letters," in Wordsworth, *Complete Guide to the Lakes* (6th ed., 1842), 23.
86. Ibid., 24.
87. Ibid.
88. Ibid., 25.
89. Ibid., 39. For the most detailed account of Sedgwick's friendship/acquaintance with Wordsworth, see Wyatt, *Wordsworth and the Geologists* (1995), 76–84. There are intermittent discussions of the *Guide to the Lakes* throughout.
90. Miall, "On a System of Anticlinals" (1859–1868), 578, 581.
91. Ibid., 579.
92. Ibid., 584.
93. Ibid.
94. See Morrell, *John Phillips* (2005).
95. Ibid., 157–75.
96. Ibid., 101. See also Edmonds and Douglas, "John Phillips's Geological Maps" (1950). Discussion of the *Encyclopaedia Metropolitana* map is on 363.
97. Morrell, *John Phillips* (2005), 228.
98. See, for examples, Brown, *Tourist Rambles* (1878), 55; Murray, *Handbook for Travellers in Yorkshire* (1882), lx; and Bevan, *Tourist's Guide* (1877), which begins by stating, "The western boundaries of the Riding are formed by grand masses of mountain, which have been made classic by the brush of Turner and by the geologi-

cal investigations and writings of the late Professor Phillips," and then discusses the geology of the region (1).

99. See Morrell, *John Phillips* (2005), 111.
100. Ibid.
101. Secord, "Geological Survey" (1986), 246. On chromolithography, see Cook, "From False Starts" (1995), 160–61 (quotation on 164). See also Allen, "Standardization of Mapping Practices" (1997), for how these shading principles had evolved (esp. 609–12).
102. Phillips, *Rivers* (1853), viii.
103. Ibid., 1.
104. Ibid.
105. Ibid., 171.
106. The use of geological terms to describe dialects was widespread in the nascent study of philology—and vice versa—oftentimes to suggest that certain dialects were not barbarous corruptions of English but that English was a corruption of older languages more perfectly preserved in dialect form. See Stitt, *Metaphors of Change* (1998), for the use of geological metaphors in linguistic writings of the nineteenth century; for the "analogy" of geology "with linguistics" in geological writing (specifically in *Principles of Geology*), see Rudwick, "Transposed Concepts" (1997), 82–84.
107. Sedgwick, *Memorial* (1868), 86.
108. Ibid., 90.
109. Ibid., 91, 92.
110. Ibid., 80, 81.
111. Ibid., 92.
112. Ibid., 105.
113. Phillips, *Rivers* (1853), 201.
114. Ibid., 196, 197.
115. Secord, "King of Siluria" (1982).
116. Taylor, *Words and Places* (1865), 9.
117. Here we might also note with Eagleton that, at the end of *Wuthering Heights*, "the gentry," represented by the Lintons, "have reached out to the stout yeomanry [of the Heights] and infused their own overbred civility with something of that racy vigour. Or, to put the matter differently, the British are once more busily appropriating the more admirable qualities of the Celt." That desire to appropriate "the more admirable qualities of the Celt" was clearly shared by Phillips. See Eagleton, *Heathcliff and the Great Hunger* (1995), 15.
118. Banks, *Wooers and Winners* (1880), 1:28.
119. See Heywood, "Introduction" (2000).
120. Banks, *Wooers and Winners* (1880), 3:219–20. These men in "real life" were "Mr. Denny, Mr. Farrer, and other gentlemen." See Dawkins, *Cave-Hunting* (1874), 84.
121. Banks, *Wooers and Winners* (1880), 3:222.
122. Ibid., 3:242.

123. For key reports to local and national societies, see C. R. Smith, "Caves,"; Smith and Jackson, "Roman Remains" (1842). Also see Dawkins, "Exploration of Caves" (1870), "Report on the Results" (1872), and "Report of the Committee" (1874). Tiddeman, "Discovery of Extinct Mammals" (1872), "Older Deposits" (1873), and "Works and Problems" (1875). More widely available accounts include Dawkins, *Macmillan's Magazine* (1871); Brown, *On Foot Round Settle* (1896), 62–63; Howson, *Illustrated Guide* (1850); and Dawkins, *Cave-Hunting* (1874).

124. Talus is defined as "a sloping mass of detritus lying at the base of a cliff or the like, and consisting of material which has fallen from its face" (OED). Dawkins, *Cave-Hunting* (1874), 87.

125. Ibid., 112–15.

126. Ibid., 116–17.

127. Ibid., 118–21.

128. Ibid., 124 (image on 117).

129. Ibid., 124–25.

130. Ibid., 124.

131. Michael Shortland argues in "Darkness Visible" (1994) that caves were, at least in Romantic culture, suggestive of "a certain time and temporality that were soothing. The cave not only pointed backwards but locked into a series of traditional British images of the ancient, the serene, and the sacred" (5).

132. Banks, *Wooers and Winners* (1880), 2:217; 109.

133. Ibid., 2:65.

134. Ibid., 2:296, 297.

135. Again, see Shortland, "Darkness Visible" (1994), for earlier, Romantic configurations of the cave as "an escape from the city and from civilization . . . a site in which the realm of private feeling, under erosion from the power of industrialization and city public business, could be nurtured" (14). Certainly this imagined function of the cave continues to operate here, though the cave's Romantic offering of a confrontation with a wild, inaccessible, and remote kind of nature has been transmuted into a domestic retreat from the machinations of the urban world. The image of the healing cave is also, as Shortland suggests, pitted against the hellish underground represented by coal mines (24–28). That opposition is also found in Banks, *Wooers and Winners*, with the disaster of the flooded Osmanthorpe colliery at its center.

136. Sedgwick, *Memorial* (1868), 78.

137. Banks, *Wooers and Winners* (1880), 1:1–2.

138. Ibid., 1:35.

138. Ibid., 1:37.

140. Ibid., 1:285; 2:110. For the visit of Agassiz and Buckland, see 2:290–97.

Chapter Five

1. Kingsley, "To His Wife, 4 August, 1851," in *Charles Kingsley* (1877), 1:294–96.

2. Kingsley, *Charles Kingsley* (1877), 2:153.

3. Kingsley, "Lecture at Wellington College," in ibid., 2:165.
4. Examples of Kingsley's letters to the geological elite are published in ibid., 2:168–70, 172–75. Fanny writes that "the thought of giving importance to the society by adding honorary members now occurred to the president [Kingsley], and he wrote to Sir Charles Lyell, Sir Philip Egerton, Dr. Hooker, Professors Huxley, Tyndall, Hughes, &c., whose distinguished names are all enrolled in the Chester Natural Science Society" (150).
5. Kingsley, *Town Geology* (1872), xl.
6. Kingsley, *Glaucus* (1855) and *Madam How and Lady Why* (1870).
7. Kingsley, *Town Geology* (1872), 239. While there have been several studies focusing on Kingsley's scientific interests there has been no sustained attention to his "favourite study." For work on *Glaucus* and on marine studies, botany, and natural history, see Merrill, *Romance of Victorian Natural History* (1989), esp. 215–35; Müller, "Spiritual Evolution" (1986); O'Gorman, "'More Interesting Than All the Books, Save One'" (2000). An early Kingsley biographer, Pope-Hennessy, also singles out marine biology for special attention in chapter 10 of *Canon Charles Kingsley* (1948), 129–49.
8. Beer, *Darwin's Plots* (1983), 133–39; Johnston, "*Water Babies*" (1959); Fasick, "Charles Kingsley's Scientific Treatment of Gender" (1994).
9. Kingsley, "To Rev. F. D. Maurice," in *Charles Kingsley* (1877), 2:171–72 (171). For a reading of Kingsley's work in the light of his involvement in Christian Socialism, see Hartley, *Novels of Charles Kingsley* (1977).
10. Gray, *Natural Selection* (1861).
11. Fanny Kingsley, in Kingsley, *Charles Kingsley* (1877), 2:156.
12. Anon., "Mr. Kingsley's *Water-Babies*" (1864), 6. For an account of contemporary reviews of *The Water-Babies*, see Johnston, "Kingsley's Debt to Darwin" (1959), 215–19.
13. Darwin, *Origin of Species*, 2nd ed. (1860), 481.
14. Cazamian, *Social Novel in England* (1973), 254.
15. Rauch, *Useful Knowledge* (2001), 176; Cunningham, "Soiled Fairy" (1985), 122.
16. Of all the criticism I have read on Kingsley, only Beer sees Kingsley as a self-conscious author interested in "jettisoning" literary "conventions," "deliberately and emphatically calling attention to the conventions of novel-writing in order to show how different from the realities of life are the expectations which books arouse." See Beer, "Charles Kingsley" (1965).
17. Kingsley, "To the Same [Ludlow], Eversley, July 1848," in *Charles Kingsley* (1877), 1:180.
18. "Kingsley to George Brimley, Esq., Eversley, 1857," in ibid., 2:43.
19. Ibid.
20. Ibid., 2:44.
21. "Charles to Thomas Cooper, Torquay 1854," in ibid., 1:380.
22. Kingsley, *Madam How* (1870), 145.

23. Charles Kingsley, *Yeast* (1851), 364. Further references are given after quotations in the text.

24. Beer, *Darwin's Plots* (1983); Levine, *Darwin and the Novelists* (1988).

25. Sussman, *Victorian Masculinities* (1995), 55. For the most significant discussions of Kingsley's ideology of masculinity, see Hall, *Fixing Patriarchy* (1996) and *Muscular Christianity* (1994); Girouard, *Return to Camelot* (1981); and Vance, *Sinews of the Spirit* (1985). Tosh, *A Man's Place* (1999), also touches on Kingsley and masculinity.

26. Vance, *Sinews of the Spirit* (1985), 110. Vance objects to the use of the term "muscular Christianity" on the grounds that Kingsley himself rejected it, preferring "Christian manliness" because it emphasizes the central importance of Christianity and Christian values in Kingsley's masculine ethos. Though this chapter sees the male body as central to Kingsley's articulation of Christian virtue, it also attempts to take into account Vance's warning not to emphasize Kingsley's interest in physicality at the expense of his desire to advance spiritual and moral causes.

27. Hall, *Fixing Patriarchy* (1996), 69.

28. Kingsley, *Glaucus* (1855), 37. Further references are given after quotations in the text.

29. Girouard, *Return to Camelot* (1981), 143.

30. Porter, "Gentlemen and Geology" (1978), 819.

31. Secord, "King of Siluria" (1982), 422.

32. Kingsley, *Town Geology* (1870), 223.

33. Clark and Hughes, *Life and Letters* (1890), 1:214–15.

34. Geikie, *Life of Sir Roderick I. Murchison* (1875), 1:89, 91.

35. Kingsley, "Lecture at Wellington College" (1863), in *Charles Kingsley* (1877), 2:163, 162. For Kingsley's involvement with Wellington College, see Newsome, *History of Wellington College* (1959).

36. Smith, *Fact and Feeling* (1994), 14 (see also 19).

37. [Sedgwick], "Vestiges" (1845), 4.

38. Geikie, *Life of Sir Roderick I. Murchison* (1875), 1:89.

39. Hall has also noted several images of "castration anxiety" in Kingsley's work, in *Fixing Patriarchy* (1996), 77–78.

40. The notion here that the descriptive element of the novel actually produces its narrative momentum is partly attributable to Bal, "Over-Writing as Un-Writing" (2006).

41. Vance, *Sinews of the Spirit* (1985), 81. Kingsley was forced to end the serial version of the novel early in 1849 when its readers complained about Lancelot's "blank materialism" (94). Though Kingsley expressed dissatisfaction at this imposition on his novel writing, he retained this ending in all subsequent editions, justifying it in his epilogue, and even provided additional dialogue between Barnakill and Lancelot in which they discuss the "condition of England" in more detail, without changing the shape of the ending itself.

42. Levine, *Dying to Know* (2002).

43. Kingsley, *Alton Locke* (1850), 2: 224.

44. Kingsley's biographer writes of the novel's "foreshadowing of the idea of evolution many years before the publication of the *Origin of Species*." Chitty, *Beast and the Monk* (1974), 135–36. See also Vance, *Sinews of the Spirit* (1985), 113. Though Klaver notes that "Kingsley was amongst those who deprecated the book" when it was first published, and that "to Kingsley the book would henceforth be damned along with all manifestations of materialism and atheism," when he comes to analyze the dream sequence of *Alton Locke* he regards it as an example of the "fascinating instances of how educated Victorians pondered the development theory which reached them from (Lyell's representation of) Lamarck through to Chambers's *Vestiges of Creation*." His struggle to explain the dreams in terms of this evolutionary context is evident in the following passage, in which three different evolutionary and geological models of the past are invoked (Darwinian evolution, Chambers' transmutation/Lamarckist successive creations): "More important than the idea that these early dreams anticipate Darwinian evolution (which strictly speaking they do not), is the representation of instinctive physical strength and a struggling consciousness of moral purpose in man," the representation of "successive creations on a perfect (divine) plan," and the fact that the "dreams show why Kingsley thought Darwinism so attractive when he, and the world, became acquainted with it ten years later." Klaver, *Apostle of the Flesh* (2006), 104, 225.

45. Gallagher, *Industrial Reformation* (1985), 99; Guy, *Victorian Social-Problem Novel* (1996), 118.

46. Cooper was an autodidact shoemaker, poet, and Chartist who taught himself Latin, Greek, and French and became a journalist, author, lecturer, and advocate of religious radicalism. He attracted Kingsley's admiration when he wrote the nine-hundred-stanza *Purgatory of Suicides* (1845) while serving a prison sentence on charges of sedition, the same sentence served by Alton, also a self-taught tailor-poet and Chartist.

47. Müller, "Spiritual Evolution" (1986), 24.

48. Kingsley, "To Thomas Cooper, 1856," in *Charles Kingsley* (1877), 1:385–86; "To Thomas Cooper, December 4, 1856," in ibid., 1:391.

49. Kingsley, *Two Years Ago* (1857), 1:265, 266. Further references to this volume are given after quotations in the text.

50. For an extended discussion of Elsley's effeminacy and its relation to science and poetry, see Blair, "Spasmodic Affections" (2004).

51. See the introduction to this book.

52. Kingsley, *Alton Locke* (1850), 1:7. Further references to this volume are given after quotations in the text.

53. Vance, *Sinews of the Spirit* (1985), 82.

54. Gallagher, *Industrial Reformation* (1985), 91.

55. For Kingsley as "negotiator" see Hall, *Fixing Patriarchy* (1996), 63.

56. For an alternative discussion of the idea of the volcano in Kingsley's work, see Rosen, "Volcano and the Cathedral" (1994). Here Rosen argues that the volcano represents the "unrepressed" "force of *thumos* . . . spiritual, primal, animal, potent, and potentially destructive," a "sacred primal fire" that has been "quenched" by an effeminate "civilization" and must be unleashed in order to save society, "a saving sacred primitivism" (31). I see the volcano not only as symbolic of this essentially masculine energy but as a space in which masculinity can be tested, strengthened, and purified.
57. Kingsley, *Alton Locke* (1850), 2:160–61. All further references to *Alton Locke* in this chapter are from volume 2, and are given after quotations in the text.
58. O'Connor, *Earth on Show* (2007), 391–432.
59. Ibid., 397.
60. Rudwick, *Scenes from Deep Time* (1992), 249.
61. O'Connor, *Earth on Show* (2007), 391–432.
62. Kingsley, *Town Geology* (1870), 4.
63. Murray, *Handbook for Travellers* (1882), xliii.
64. This debate is the subject of Rudwick, *Great Devonian Controversy* (1987).
65. Kingsley, *Two Years Ago* (1857), 1:63.
66. Kingsley, *Westward Ho!* (1855), 193–94.
67. Kingsley, *Two Years Ago* (1857), 3:2.
68. Ibid., 3:2.
69. Ibid., 3:22.
70. Ibid., 3:23.
71. Ibid., 3:23–24.
72. As quoted in Pointon, "Geology and Landscape Painting" (1997), 106.
73. Kingsley, *Two Years Ago* (1857), 3:46–47.
74. Ibid., 3:47–48.
75. Ibid., 3:48.
76. Ibid., 3:49.
77. Ibid., 3:55.
78. Ibid., 3:92.
79. Ibid.
80. Ibid., 3:92–93.
81. Ibid., 3:93.
82. Ibid., 3:132.
83. Ibid.
84. Ibid., 3:137.
85. Ibid., 3:135.
86. Ibid., 3:136.
87. Ibid., 3:137.
88. Ibid., 3:273.
89. Ibid., 3:278.
90. Ibid., 3:279–80.

Chapter Six

1. Shuttleworth, "Language of Science" (1985), 280.
2. Shuttleworth, *George Eliot* (1984), 81.
3. Ibid., 52. This ending is often seen as a deviation from realism, "topping off realism with romance," as Rosemary Ashton summarizes in *Mill on the Floss* (1990), 21.
4. See Klaver, "'I Will Ferry Thee Across'" (2004), for the view that the flood is catastrophic and unites religion and science. See Smith, *Fact and Feeling* (1994), for the view that it is uniformitarian and unites romance and realism (257, 121, 120). Arac similarly writes that "Uniformitarianism has not wholly removed the possibility of catastrophe" at the end of *The Mill on the Floss*, as exemplified by the flood. Arac, "Rhetoric and Realism" (1979), 681. In *The Great Tradition* (1983), Leavis writes, "The flooded river has no symbolic or metaphysical value. It is only the dreamed-of perfect accident that gives us the opportunity for the dreamed-of heroic act—the act that shall vindicate us against a harshly misjudging world, bring emotional fulfillment and (in others) changes of heart, and provide a gloriously tragic curtain" (60).
5. See Cosslett, *"Scientific Movement"* (1982), 4–5. One of many such comments in Beer's work is as follows: "In the uniformitarian ordering of *Middlemarch* events, however seemingly catastrophic, are prepared for by the slight incipient movements, crumblings, pressures, erosions, and siltages observable to an immeasurably patient eye." Beer, *Darwin's Plots* (1983), 181.
6. Quotation from "George Eliot to Charles Bray," in Haight, *George Eliot Letters* (1954), 2:203. Eliot here is referring to her reading of *all* Whewell's works, not simply the *History*. For Eliot's early reading of geological texts, see Haight, *George Eliot* (1968), 29. Also see Ashton, *George Eliot* (1997). See also "George Eliot to Martha Jackson, 4 March 1841," and "George Eliot to Maria Lewis, 4 September 1839," in Haight, *George Eliot Letters* (1978), 8:8, 1:34.
7. For Lyell's *Principles of Geology* as an "anti-narrative," see chapter 3 above and Secord, "Introduction" (1998), xviii.
8. Whewell, *History of the Inductive Sciences* (1837), 3:596–97.
9. See chapter 3 above.
10. Shuttleworth, *George Eliot* (1984), 53. Lyell, in any case, defended "the theory of the fixity of species" until the twelfth edition of *Principles of Geology* was published in 1867. See chapter 3 above for much more extended discussion of this issue.
11. Eliot, in Collins, "Questions of Method" (1980), 390. Collins estimates, based on handwriting evidence and on the size of the paper Eliot uses, that this note was written between 1870 and 1876.
12. "George Eliot to Mr and Mrs Charles Bray and Sara Sophia Hennell, 22 October 1851," in Haight, *George Eliot Letters*, 1:370.
13. "George Eliot to Sara Sophia Hennell, 9 June 1852," in ibid., 2:33–34.
14. [Forbes], "Future of Geology" (1852). Further references are given after quotations in the text.
15. See chapter 2 above.

16. Eric L. Mills, "Forbes, Edward (1815–1854)," *Oxford Dictionary of National Biography* (Oxford: Oxford University Press, 2004), http://www.oxforddnb.com/view/article/9824 (accessed August 16, 2007).

17. Spencer, "Illogical Geology" (1863) :104.

18. William Baker states in *The Libraries of George Eliot and George Henry Lewes* (1981) that Eliot and Lewes owned Spencer's *Essays*, though this source, based on an inventory compiled by Lewes's granddaughter of books her mother (Lewes's daughter-in-law) owned, is problematic. The library she cataloged consisted mainly of books owned by Lewes and Eliot, but also included books added after their deaths. Nonetheless, it is clear that Eliot and Lewes read Spencer's work and almost certainly owned these volumes. For more, see Paxton, *George Eliot and Herbert Spencer* (1991).

19. Spencer, "Illogical geology" (1863), 2:58.

20. Ibid, 59.

21. Eliot, "Historic Imagination" (1884), 296.

22. [Eliot], "Silly Novels" (1856), 442–61. Further references will be given after quotations in the text.

23. Nonetheless, of course, Eliot does theorize about novels here, giving herself license to theorize by repudiating theory, just as she gives herself license to write fictions by repudiating "silly" "lady novelists." Given this discrepancy, it is unsurprising that narratalogists such as D. A. Miller and Franco Moretti disagree over the status of storytelling in Eliot's fiction. For Miller the community of Middlemarch sees "all narratives" as "offences against its established order," attempting to efface its "story-worthiness," and *Middlemarch* dramatizes "the impossibility of transcendence" of this regime. Characters may desire to become the heroes or heroines of stories that transcend the narrow provincial boundaries in which they live, but "'reality,' or the realist text, unmoors." Miller, *Narrative and Its Discontents* (1981), 120, 113, 152, 141 Eliot's realism, for Miller, dramatizes the lack of story available to those leading a "provincial life." Practical realism eschews the exorbitance of narrative with its tendency to excess. But for Moretti, Eliot's "realism" causes her, in *Middlemarch*, to dramatize failure, and her commitment to ongoing change through progress makes her the sole Victorian novelist who "engage[s] in narrative proper." Moretti, *Way of the World* (1987), 201. Contradictory definitions of narrative abound here, but the point is that Eliot's fictions mark a reflective and conscious engagement with the powers, possibilities, and pitfalls of certain kinds of plotted thinking.

24. [Eliot], "Natural History of German Life" (1856). Further references are given after quotations in the text. Riehl, *Die Gebürgeliche Gesellschaft* (1856); Riehl, *Land und Leute* (1855). Several critics have linked Eliot's method with that of the "natural historian" or "sociologist" through their readings of this essay: Graver labels Eliot a "natural historian" in her early works; see *George Eliot and Community* (1984), 28–39. J. Hillis Miller states that even *Middlemarch* (1872) was "modeled on the sociologist's [Riehl's] respect for individual fact" in his essay "Optic and Semiotic in *Middlemarch*" (1975), 127.

25. This is argued by McDonagh in "Space, Mobility, and the Novel" (2007), 61.
26. Byatt, "Introduction" (2003), xvi.
27. Graver, *George Eliot and Community* (1984), 38.
28. This tendency to use geological formations as a tool for sociological, anthropological, or historical discussion, explaining the characters of different populations, is discussed in more detail in chapter 4 above.
29. The trope was central to an earlier tradition of geological writing: Georges Baron Cuvier in his "Discours Préliminaire" to *Ossemens fossiles* (1812) had written, "When a traveller crosses fertile plains, where the regular course of tranquil rivers sustains abundant vegetation, and where the land—crowded with numerous people and ornate with flourishing villages, rich cities, and superb monuments—is never disturbed unless by the ravages of war or by the oppression of powerful men, he is not tempted to believe that nature has also had its civil wars, and that the surface of the globe has been upset by successive revolutions and catastrophes. But these ideas change as soon as he seeks to excavate this ground that today is so peaceful, or to climb onto the hills that border onto the plain. His ideas enlarge, as it were, with his viewpoint." Translation from Rudwick, *Georges Cuvier* (1997), 186. Thank you to Ralph O'Connor for pointing out this link to me.
30. See Law, "Water Rights" (1992), 58–60.
31. For the critical reception of the novel in the periodical press in the 1860s, see Carroll, *George Eliot* (1995).
32. "George Eliot to Martha Jackson, 4 March 1841," in Haight, *George Eliot Letters* (1978), 8:8. The Treatise and Lyell's book seem to have followed a program of geological reading, in which Eliot also "swallowed without much mastication 'The Doctrine of the Deluge,' by the Revd. V. Harcourt," in which he "supports or rather shakes a weak position" on the connection of myths of the deluge to the Christian baptism "by weak arguments." "George Eliot to Maria Lewis, 22 November 1839," in ibid., 1:34. She also "read Dr. P. Smith's work on the connexion between Scripture and Geology," and remarked that "the main subject of the work," an "interpretation of the Mosaic records, is fully satisfactory to me." "George Eliot to Maria Lewis, September 1841," in ibid., 1:110. Smith had argued for a nonglobal local deluge, an ancient earth, and extinction, reinterpreting Scripture by the light of science rather than the other way round. By 1860, of course, Eliot no longer found such arguments satisfactory. Nonetheless, it is significant that in her own early life she was interested in accounts of the flood, as Maggie is. See Harcourt, *Doctrine of the Deluge* (1838), and J. P. Smith, *On the Relation* (1839).
33. Topham, "Beyond the 'Common Context'" (1998), 255, 256.
34. Secord, *Victorian Sensation* (2000), 59.
35. Eliot, *Daniel Deronda* (1984), 18. Further references are given after quotations in the text.
36. Irwin, *George Eliot's Daniel Deronda Notebooks* (1996), 315.
37. Eliot, in ibid., 317.
38. [Scrope], "Wiltshire" (1858), 108–38 (113).

39. Ibid., 114.

40. The comparative anatomist Richard Owen sent Eliot his *Memoir on the Megatherium* (1861) as a gift when he read *The Mill on the Floss*.

41. See Rosemary Sweet, *Antiquaries* (2004), esp. introduction, for examination (and rejection) of this stereotype, for which she lays the blame at Scott's door. Also see chapter 1 of *Novel Science*.

42. For discussion of antiquarianism see chapter 1 above.

Chapter Seven

1. Dickens, at a "Banquet in his honour: Hartford, 7 February 1842," in Fielding, *Speeches of Charles Dickens* (1988), 24. For an earlier published version of this chapter see A. Buckland, "'Poetry of Science'" (2007). The version here is able to take account of more recent work by Zimmerman and especially O'Connor in ways that the earlier account was not.

2. O'Gorman, *Victorian Novel* (2002), 252.

3. Hill, "Books That Dickens Read" (1949), 203. We might ask, of course, just what "would be expected of a man of the world" in the early to mid-Victorian period, or in 1949 when Hill was writing. See also Sanders, *Charles Dickens* (2003), for a typically sparse account of Dickens and science.

4. For the wonder and romance of science see Metz, "Science in *Household Words*" (1978) (quotation on 129); Ostry, "'Social Wonders'" (2001). Ostry is much more keen to point out anxieties expressed about science, as well as enthusiastic responses, than Metz. For "the difficulty" of matching the journals with Dickens and his novels, see Drew, *Dickens the Journalist* (2003), 106. Lai also demonstrates in some detail that "we can almost never be sure about deducing Dickens's own opinions from articles written by his contributors unless we have external evidence." Lai, "Fact or Fancy" (2001), 41.

5. Zimmerman, *Excavating Victorians* (2008), 143–75. Though I think Zimmerman's readings are interesting and persuasive, and agree with her argument that dust and ruin are central images in this archaeological and geological discourse, she, like many other critics, relies heavily on a notion of "uniformitarianism" that is different from that suggested by Lyell. She writes, for instance, that "Playfair's redaction of Hutton's *Theory of the Earth* popularly introduced uniformitarianism, and the paradigm of natural history and temporal progress irrevocably shifted" (145), which is anachronistic, since the term was not invented until 1837 and was never a Kuhnian "paradigm" in this sense. She also suggests that Dickens was "acutely aware of the paradox central to the uniformitarian view: while small actions are overwhelmed by the immense expanse of time, they evidently affect great change" (145). Small events in themselves do not affect great change in Lyell's *Principles*—he simply suggests that ordinary causes, given a long enough span of time in which to continuously act, are sufficient for explaining great change. The difference may seem pedantic, but it is key,

for Zimmerman sees geology as a spur to narrative, and my aim is to show the places in which geology was a problem for it.

6. Levine, *One Culture* (1987); Morris, "Taste for Change" (2000); Fulweiler, "'Dismal Swamp'" (1994); Morgentaler, *Dickens and Heredity* (2000). See also Fontana, "Darwinian Sexual Selection" (2005), which studies a novel published after the publication of *Origin of Species* but before Darwin fully articulated his ideas on sexual selection in *The Descent of Man* (1871).

7. O'Connor, *Earth on Show* (2007), 261.

8. Ibid. On geology as a "surrogate reality" and a substitute for travel, see also Wilcox, "Introduction" (1988), esp. 37–40.

9. See O'Connor, *Earth on Show* (2007), 265–83, for a detailed account of "the panoramic imagination" in the nineteenth-century imagination, esp. 274–80 on this as it relates specifically to geology. See also ibid., chaps. 9 and 10.

10. Lightman, *Victorian Popularizers of Science* (2007), 168; on the visual dimensions of science, see pages 167–218 more generally. See also the important collection edited by Lightman and Fyfe, *Science in the Marketplace* (2007).

11. Secord, *Victorian Sensation* (2000), 439–40. For the classic study of geological visual culture see Rudwick, *Scenes from Deep Time* (1992). These new approaches are summed up best in Fyfe and Lightman, "Introduction" (2007), in which they remind us of the importance of publishers, museums, conversaziones, and public lectures in science. "The sciences," they note, "could thus be encountered in an incredibly wide variety of venues and in a multiplicity of forms, often incorporating elements from print, display, and the spoken word at the same time. An emphasis on scientific elites and professionalization ignores all of these crucial dimensions of nineteenth-century science" (7).

12. For Dickens and shows, see Schlicke, *Dickens and Popular Entertainment* (1985); Eigner, *Dickens Pantomime* (1989); Vlock, *Dickens, Novel Reading* (1998); John, *Dickens's Villains* (2001); and G. Smith, *Dickens and the Dream of Cinema* (2003). For work on science and shows specifically relating to geology and evolutionary biology, see O'Connor, *Earth on Show* (2007); Secord, "Monsters at the Crystal Palace" (2004) and *Victorian Sensation* (2000), esp. 437–70; Morus, Schaffer, and Secord, "Scientific London" (1992); Altick, *Shows of London* (1978); J. Smith, *Charles Darwin and Victorian Visual Culture* (2006); and Goodall, *Performance and Evolution* (2002).

13. See, in Dickens, *Letters* (1965–2002), his letters to Sir Roderick Murchison, dated August 113, 1852 (6:739); June 4, 1855, (7:638); June 21, 1855 (7:654); April 3, 1870 (12:505); and March 4, 1857 (12:675–76). These letters reveal that Dickens read Murchison's *Address to the Royal Geographical Society of London* (1852).

14. Lightman, *Victorian Popularizers* (2007), 167.

15. Schor, "Dickens and Plot" (2006), 90. Here is an example: "The artistic fault of *Little Dorrit* is that it is not a tale. It neither begins nor ends—it has no central interest, no legitimate catastrophe, and no modelling of the plot into a whole" (as

quoted in Zimmerman, *Excavating Victorians* [2008], 154). We should not be misled into thinking that this "catastrophe" is a geological reference.

16. Butt and Tillotson, *Dickens at Work* (1957); Stone, *Dickens's Working Notes* (1987).

17. Miller, *Charles Dickens* (1958), 292.

18. Miller, *Novels behind Glass* (1995), 124.

19. Schor, "Dickens and Plot" (2006), 95.

20. Miller, *Charles Dickens* (1958), 291, 295. See also Zimmerman, *Excavating Victorians* (2008), 161–75, for a reading of excavation in this novel, in which Zimmerman argues that "John Harmon waits until the ruins have been removed or reconstructed to begin his new life" (175).

21. Morgentaler, *Dickens and Heredity* (2000), 161; Fielding, "Dickens and Science?" (1996), 206–7; Levine, *Darwin and the Novelists* (1998), 122.

22. Secord, "Monsters at the Crystal Palace" (2004), 155.

23. Ibid., 153.

24. Desmond, *Hot-Blooded Dinosaurs* (1976), 24–27; Altick, *Shows of London* (1978), 484.

25. "Charles Dickens to Professor Richard Owen, 19 October 1852" in Dickens, *Letters*, 6:779–80.

26. "Charles Dickens to Professor Richard Owen, 4 November 1865," in ibid., 11:105.

27. [Hart], "Nature's Greatness in Small Things" (1857); [Linton], "First Idea of Everything" (1858); [Mann], "Everything after Its Kind" (1858). See Desmond, *Archetypes and Ancestors* (1982), for a discussion of Owen's "archetype" and its social, political, and scientific functions. Owen, "Leaf from the Oldest of Books" (1856).

28. Dawson, *Darwin, Literature and Victorian Respectability* (2008), 7.

29. Dickens, "Poetry of Science" (1937), 179.

30. For additional accounts of panoramas and dioramas in Victorian culture, see Hyde (ed.), *Panoramania!* (1998); Stafford and Terpak, *Devices of Wonder* (2001); and Mannoni, Nekes, and Warner, *Eyes, Lies and Illusion* (2004).

31. Altick, *Shows of London* (1978), 324.

32. Ibid., 325.

33. Ibid., 326. See also Colligan, *Canvas Documentaries* (2003), 123–40, on Vesuvius at the Cremorne gardens.

34. Dickens, *Speeches Literary and Social* (1870), 123.

35. [Wreford], "Earthquake Experiences" (1858), 553–58. See Altick, *Shows of London* (1978), 553–58.

36. Altick, *Shows of London* (1978), 157–58.

37. Ibid., 158–61.

38. Ibid., 170.

39. Dickens, *Pictures from Italy* (1846), 233–55.

40. Bulwer-Lytton, *Last Days of Pompeii* (1834); [Dulton], "Up Vesuvius" (1852); [Wreford], "Vesuvius in Eruption" (1855).

41. Dickens, *Pictures from Italy* (1846), 234–35.
42. [Dickens], "Some Account" (1850), 77.
43. Wilcox, "Introduction" (1998), 38.
44. Cruchley, *Cruchley's London* (1865), 310.
45. Yapp, *Official Catalogue* (1851).
46. Babbage, in Altick, *Shows of London* (1978), 457.
47. Quoted in Schlicke, *Dickens and Popular Entertainment* (1985), 201. From Dickens, "Amusements of the People" (1850), 13–15, 57–60.
48. [Wills and Sala], "Fairyland in Fifty Four" (1853).
49. Secord, "Monsters at the Crystal Palace" (2004), 147, 143.
50. Chesterton, *Appreciations and Criticisms* (1911), 118.
51. Dickens, *Dombey and Son* (2001), 67. Further references are given after quotations in the text.
52. Dickens, *Pictures from Italy* (1846), 247.
53. Ibid., 250.
54. Ibid., 245.
55. Schivelbush, *Railway Journey* (1980), 172.
56. Ibid., 63. For his account of railway perception as panorama, see 55–72.
57. Flint also notes the relationship between Dickens's presentation of Italy in *Pictures from Italy* and the forms of panorama and diorama. Flint, *Victorians and the Visual Imagination* (2000), 146–47.
58. Dickens, *Our Mutual Friend* (1998), 69. Further references will be given after quotations in the text.
59. Freeman, *Victorians and the Prehistoric* (2004), 48.
60. Dickens, *Bleak House* (1996), 11. Further references will be given after quotations in the text.
61. Sutherland, "Visualising Dickens" (2006), 124.
62. Ibid.
63. Zimmerman opens her study of Dickens's use of "geology and archaeology to reveal evanescence and to assert coherence" with the quotation from the opening chapter of *Bleak House*, but she does not discuss the novel in detail. Zimmerman, *Excavating Victorians* (2008), 144.
64. Levine, *Darwin and the Novelists* (1988), 122.
65. Frank, *Victorian Detective Fiction* (2003), 71–99 (78).
66. Meckier, *Hidden Rivalries* (1987), 249. This criticism also reiterates a problem set up by Eliot's partner G. H. Lewes in his complaints that Dickens was not enough of a man of science to be a genuine "realist." Lewes, of course, was involved in defining Victorian high realism in the image of his own work, and of Eliot's. See Winyard and Furneaux, "Introduction" (2010).
67. Turner, *Mechanism and the Novel* (1993), 97.
68. See, for example, [Hunt], "Hunterian Museum" (1850).
69. [Morley], "Our Phantom Ship" (1851), 492.
70. [Buckland], "Old Bones" (1854), 83–84.

71. See introduction, above for a discussion of Professor Dingo.

72. See Wilkinson, "*Bleak House*" (1967), 238, for an apocalyptic reading of this imagery.

73. Altick, *Shows of London* (1978), 505, 173–83.

74. In ibid., 505.

75. Darwin, *Origin of Species* (1859); Temple et al., *Essays and Reviews* (1860); Lyell, *Antiquity of Man* (1863).

76. Anon., "England Long and Long Ago" (1860).

77. Fulweiler, "'Dismal Swamp'" (1994), 54.

78. Anon., "England Long and Long Ago" (1860), 562.

79. J. H. Miller, *Charles Dickens* (1958), 289, 292.

80. Ginsburg, *Economies of Change* (1996), 142. Ginsburg uses *Our Mutual Friend* as a caveat to Moretti's castigation of the English novel, and particularly of Dickens, for its inability to imagine transformation; see Moretti, *Way of the World* (1987), esp. 181–228. See also Poovey, "Reading History in Literature" (1993) on speculation and *Our Mutual Friend*.

81. Fulweiler, "'Dismal Swamp'" (1994), 51. Also see Morris, "Taste for Change" (2000).

82. Cotsell, *Companion* (1986), 27, 146.

83. [Hunt], "Hunterian Museum" (1850), 125.

84. See, in Dickens, *Letters* (1965–2002), his letters to Professor Richard Owen, dated April 9, 1862 (10:68); August 7, 1862 (10:117); December 15, 1863 (10:327); and July 12, 1865 (11:69–70). In the last, Dickens thanks Owen for the copy of his *Memoir of the Gorilla*, telling him it is an "admirable treatise" and noting "how much interest it has awakened in me, and how often it has set me a-thinking." Also see his November 4, 1865, letter to Owen (11:105).

85. For work other than Fulweiler's on Owen and *Our Mutual Friend*, see Bown, "What the Alligator Didn't Know" (2010), which cites my earlier version of this essay.

86. See Torrens, "Politics and Palaeontology" (2012); Desmond, *Hot-Blooded Dinosaurs* (1990), 19. It should be noted, of course, that Owen only created the new name; the creatures themselves had already been constructed as spectacles by such writers, lecturers, and collectors as Gideon Mantell. The word "dinosaur" was not widely used until the end of the nineteenth century.

87. I am indebted to Gowan Dawson's work on this topic, as published in "Literary Megatheriums" (2011) and as articulated at the 2009 Dickens Day conference at Birkbeck College, University of London.

88. Eagleton, *English Novel* (2005), 149, 145.

Appendix

1. Lyell, in K. M. Lyell, *Life, Letters and Journals* (1881), 1:55–56.

BIBLIOGRAPHY

Ackerberg-Hastings, A. 2002. Analysis and synthesis in John Playfair's *Elements of geometry*. *British journal for the history of science* 35:43–72.
Alberti, S. J. M. M. 2002. Placing nature: Natural history collections and their owners in nineteenth-century provincial England. *British journal for the history of science* 35:291–311.
———. 2005. Objects and the museum. *Isis* 96:559–71.
Allen, D. E. 1994. *The naturalist in Britain: A social history*. Princeton, NJ: Princeton University Press.
———. 1997. The lost limb: Geology and natural history. In *Images of the earth: Essays in the history of the environmental sciences*, edited by L. Jordanova and R. Porter, 203–14. Oxford: Alden Press.
Allen, P. M. 1997. Standardization of mapping practices in the British Geological Survey. *Computers and geosciences* 23:609–12.
Altick, R. 1957. *English common reader: A social history of the mass reading public, 1800–1900*. Chicago: University of Chicago Press.
———. 1972. *Victorian studies in scarlet*. London: Dent.
———. 1978. *The shows of London*. Cambridge, MA: Belknap.
Anger, S., ed. 2001. *Knowing the past: Victorian literature and culture*. Ithaca, NY: Cornell University Press.
Anon. 1826. *La divina commedia* de Dante Alighieri; con *Comento analitico* do Gabriele Rossetti. *Literary gazette: A weekly journal of literature, science, and the fine arts* 468:8.
Anon. 1827. *La divina comedia* de Dante Alighieri, &c. *Literary gazette: A weekly journal of literature, science, and the fine arts* 532:201.
———, 1827. Rossetti's elucidation of the mysteries of Dante. *Oriental herald and journal of general literature* 13:517–22; 14:112–21, 187–91.
———. 1828. *La divina comedia* de Dante Alighieri; con *Comento analitico* di Gabriele Rossetti. *Literary gazette: A weekly journal of literature, science, and the fine arts*, no. 578, 104.
———. 1828. *La divina commedia. Foreign Review* 3:175–95.
———. 1830. Literary criticism: *Principles of geology. Edinburgh literary journal* 93:115–17.
———. 1830. Principles of geology. *Athenaeum*. 152:595–97.

———. 1830. Principles of geology. *London literary gazette* 707:505–6.
———. 1832. A geological manual. *Gentleman's magazine* 102:43–47.
———. 1845. New South Wales and Van Diemen's Land. *Quarterly review* 76:488–521.
———. 1845. Physical description of New South Wales and Van Diemen's Land, accompanied by a geological map, sections, and diagrams, and figures of the organic remains. *Tait's Edinburgh magazine* 12:471.
———. 1846. Australia. *North British review* 4:281–312.
———. 1847. Our library table. *Athenaeum* 1052:1324–25.
———. 1848. The tenant of Wildfell Hall. *Examiner* 2113:483–84.
———. 1848a. Wuthering Heights. *New monthly magazine and humorist* 82: 140–41.
———. 1848b. *Wuthering Heights*. A novel. By Ellis Bell. *Examiner* 2084:20–21.
———. 1851. *Wuthering Heights* and *Agnes Grey*. *Bentley's miscellany* 29:448–49.
———. 1853. Our honeymoon. *Punch's almanack for 1853*, 51. London: Punch.
———. 1856. Geological map of Europe. *Athenaeum* 1489:582–83.
———. 1859. Novels, novel readers, and novel writers. *Scottish review* 7:239–51.
———. 1860. England long and long ago. *All the year round* 50:562.
———. 1860. How long will our coal last? *All the year round* 47:488–90.
———. 1861. The story of the Geological Survey. *Chambers's journal of popular literature, science and arts* 414:356–59.
———. 1864. Mr. Kingsley's *Water-Babies*. *Times*, January 26, 6.
———. 1865. Mary Anning, the fossil finder. *All the year round* 13:60–63.
———. 1904. Stanford's *Geological atlas*. *Geological magazine* 1:559–61.
Arac, J. 1979. Rhetoric and realism in nineteenth-century fiction: Hyperbole in *The mill on the floss*. *English literary history* 46:673–92.
Armstrong, N. Emily's ghost: The cultural politics of Victorian fiction, folklore, and photography. *Novel: A forum on fiction* 25:245-67.
———. 1999. *Fiction in the age of photography: The legacy of British realism*. Cambridge, MA: Harvard University Press.
Arnold, M. 1863. *Cromwell: A prize poem, recited in the theatre, Oxford, June 28, 1843*. Oxford: T. and G. Shrimpton.
Ashton, R. 1990. *The mill on the floss: A natural history*. Boston: Twayne.
———. 1997. *George Eliot: A life*. London: Penguin.
Bacon, F. 1808. *The two books of Francis Bacon: Of the proficience and advancement of learning, divine and human*. London: Payne.
——— 1858. *Works of Francis Bacon*. London: Longman.
Bailey, E. 1962. *Charles Lyell*. London: Nelson.
Baker, W. 1977. *The George Eliot G. H. Lewes library: An annotated catalogue of their books*. New York: Garland.
———. 1981. *The libraries of George Eliot and George Henry Lewes*. Victoria, BC: University of Victoria.

Bakhtin, M. 1981. *The dialogic imagination: Four essays*. Austin: University of Texas Press.

Bal, M. 2006. Over-writing as un-writing: Descriptions, world-making, and novelistic time. In *The novel*, edited by F. Moretti, 1:571–610. Princeton, NJ: Princeton University Press.

Balee, S. 1993. English critics, American crisis, and the sensation novel. *Nineteenth-century contexts* 17:213–41.

Banks, I. 1880. *Wooers and winners; or, Under the scars; a Yorkshire story*. London: Hurst and Blackett.

Barber, L. 1980. *The hey-day of natural history, 1820–1870*. London: Cape.

Barker, J. 1994. *The Brontes*. London: Weidenfeld and Nicolson.

———. 2002. The Haworth context. In *The Cambridge companion to the Brontes*, edited by H. Glen, 13–33. Cambridge: Cambridge University Press.

Barnes, J. 1992. The mineral collection of John Ruskin in the Ruskin Gallery. *Journal of mines and minerals* 11:26–29.

Barrell, J. 1998. Geographies of Hardy's Wessex. In *The regional novel in Britain and Ireland 1800–1990*, edited by K. D. M. Snell, 99–118. Cambridge: Cambridge University Press.

Barthes, R. 1967. The reality effect. *The rustle of language*. Oxford: Blackwell. Repr. 1986.

Bartholomew, M. 1976. The non-progress of non-progression: Two responses to Lyell's doctrine. *British journal for the history of science* 9:166–74.

Bassett, M. G. 1979. One hundred years of Ordovician geology. *Episodes: Journal of international geoscience* 2:18–21.

Baumgarten, M. 2001. Fictions of the city. In *The Cambridge companion to Charles Dickens*, edited by J. O'Jordan, 106–19. Cambridge: Cambridge University Press.

Beaumont, M. ed. 2007. *Adventures in realism*. Malden, MA: Blackwell.

Becher, H. W. 1980. William Whewell and Cambridge mathematics. *Historical studies in the physical sciences* 11:1–48.

Beer, G. 1965. Charles Kingsley and the literary image of the countryside. *Victorian studies* 8:243–54.

———. 1970. *The romance*. London: Methuen.

———. 1983. *Darwin's plots: Evolutionary narrative in Darwin, George Eliot and nineteenth-century fiction*. London: Routledge & Kegan Paul.

———, ed. 1989. *Arguing with the past: Essays in narrative from Woolf to Sidney*. London: Routledge.

Bentley, P. E. 1941. *The English regional novel*. London: Allen & Unwin.

Bersani, L. 1976. *A future for Astyanax: Character and desire in literature*. Boston: Little, Brown.

Bevan, G. P. 1877. *Tourists' guide to the West Riding of Yorkshire*. London: Stanford.

Blair, K. 2004. Spasmodic affections: Poetry, pathology and the spasmodic hero. *Victorian poetry* 42:473–90.

Blundell, D. J., and A. C. Scott, eds. 1998. *Lyell: The past is the key to the present*. London: Geological Society of London.
Bonney, T. G. 1895. *Charles Lyell and modern geology*. London: Cassell.
Bono, J. J. 2010. Making knowledge: History, literature, and the poetics of science. *Isis* 101:555–59.
Boud, R. C. 1975. The early geological maps of the isle of Arran, 1807–1858. *Cartographica: The international journal for geographic information and geovisualization* 12:179–93.
Boumelha, P. 1988. Realism and the ends of feminism. In *Grafts: Feminist cultural criticism*, edited by S. Sheridan. London: Verso.
Bourne, J. C. 1846. *The history and description of the Great Western Railway, including its geology, and the antiquities of hte district through which it passes: Accompanied by a plan and section of the railway, a geological map, and by numerous views, of its principal viaducts, bridges, tunnels, stations, and of the scenery, and antiquities in its vicinity*. London: Bogue.
Bowden, A. J. 2009. Geology at the crossroads: Aspects of the geological career of Dr John MacCulloch. In *The making of the Geological Society of London* (GSL Special Publications no. 317), edited by C. L. E. Lewis and S. J. Knell, 255–78. London: Geological Society of London.
Bowen, J., and R. L. Patten, eds. 2006. *Charles Dickens studies*. Basingstoke: Palgrave.
Bowler, P. J. 1976. *Fossils and progress: Palaeontology and the idea of progressive evolution in the nineteenth century*. New York: Science History Publications.
———. 1988. *The non-Darwinian revolution: Reinterpreting a historical myth*. Baltimore: Johns Hopkins University Press.
———. 2005. Revisiting the eclipse of Darwinism. *Journal of the History of Biology* 38:19–32.
Bown, N. 2010. What the alligator didn't know: Natural selection and love in *Our mutual friend*. *19: Interdisciplinary studies in the long nineteenth century* 10. www 19.bbk.ac.uk.
Boylan, P. B. 2009. The Geological Society and its official recognition, 1824–1828. In *The making of the Geological Society of London* (GSL Special Publications no. 317), edited by C. L. E. Lewis and S. J. Knell, 319–30. London: Geological Society of London.
Brantlinger, P. 1982. What is "sensational" about the "sensation novel"? *Nineteenth-century fiction* 37:1–28.
Brantlinger, P., and W. B. Thesing, eds. 2002. *A companion to the Victorian novel*. Oxford: Blackwell.
Brett-Surman, M. K., T. R. Holtz Jr., and J. O. Farlow, eds. 2012. *The complete dinosaur*. Bloomington: Indiana University Press.
Brewster, D., ed. 1830. *The Edinburgh encyclopaedia*. Edinburgh: Blackwood.
———. 1850. Astronomy and geology. *London journal, and weekly record of literature, science and art* 11:415.

Bricmont, J., and A. D. Sokal. 1998. *Intellectual impostures: Postmodern philosophers' abuse of science.* London: Profile.

Briggs, A. 1996. *Victorian things.* London: Folio Society.

Brock, M. G., and M. C. Curthoys, eds. 1997. *Nineteenth-century Oxford, part one.* Oxford: Clarendon Press.

Bronte, E. 1965. *Wuthering Heights.* London: Penguin.

Brooke, J. H. 1997. The natural theology of the geologists: Some theological strata. In *Images of the earth: Essays in the history of the environmental sciences,* edited by L. Jordanova and R. Porter, 53–74. Oxford: Alden Press.

Brooke, J. H., and G. Cantor. 1998. *Reconstructing nature: The engagement of science and religion.* Edinburgh: Edinburgh University Press.

Brooks, P. 2005. *Realist vision.* New Haven, CT: Yale University Press.

[Brougham, H.] 1845. Strzelecki on New South Wales and Van Diemen's Land. *Quarterly review* 76:488–521.

Brown, G. H. 1896. *On foot round Settle; Also a chapter on the plants of the district by R.F.T. and F.P.L.* Settle: Lambert.

Brown, J. 1878. *Tourist rambles in Yorkshire, Lincolnshire, Durham, Northumberland, and Derbyshire.* London: Simkin.

Browne, J. 1992. Squibs and snobs: Science in humorous British undergraduate magazines around 1830. *History of science* 30:165–97.

Buckland, A. 2007. "The poetry of science": Charles Dickens, geology, and visual and material culture in Victorian London. *Victorian literature and culture* 35:679–94.

———. 2007. "The truth lies hidden deep in mines": Victorian geology and the realist novel. DPhil diss., English literature, University of Oxford.

———. 2008. "A product of Dorsetshire": Geology and the material imagination of Thomas Hardy. *19: Interdisciplinary studies in the long nineteenth century* 6. www.19.bbk.ac.uk.

———. 2009. "Pictures in the fire": The Dickensian hearth and the concept of history. *Romanticism and Victorianism on the net* 53. www.ron.umontreal.ca.

———. 2009. Show and tell: The dramatic story of nineteenth-century geological science. *Studies in history and philosophy of science* 40:114–17.

———. 2011. "Hardy and the physical sciences." In *Hardy in context,* edited by P. Mallett. Cambridge: Cambridge University Press.

Buckland, A., and A. Vaninskaya. 2009. Introduction: Epic's historic form. *Journal of Victorian Culture* 14:163–72.

[Buckland, F. T.] 1854. Old bones. *Household words* 8:83–84.

Buckland, W. 1820. *Vindiciae geologicae: The connexion of geology with religion, explained in an inaugural lecture delivered before the University of Oxford, May 15, 1819, on the endowment of the readership in geology.* Oxford: Oxford University Press.

———. 1835. On the discovery of coprolites, or fossil faeces, in the lias at Lyme Regis, and in other formations. *Transactions of the Geological Society of London* 3:223–36.

———. 1836. *Geology and mineralogy: Considered with reference to natural theology.* London: Pickering.

Buckley, J. H. 1975. *The worlds of Victorian fiction.* Cambridge, MA: Harvard University Press.

Bullen, J. B. 1986. *The expressive eye: Fiction and perception in the work of Thomas Hardy.* Oxford: Clarendon Press.

Bulwer-Lytton, E. 1834. *Last days of Pompeii.* London: Bentley.

Burd, V. A. 2008. Ruskin and his "good master," William Buckland. *Victorian literature and culture* 36:299–315.

Burns, J. 2007. John Fleming and the geological deluge. *British journal for the history of science* 40:205–25.

Butt, J., and K. Tillotson. 1957. *Dickens at work.* London: Methuen.

Byatt, A. S. 2003. Introduction. In George Eliot, *The mill on the floss.* London: Penguin.

Cadbury, D. 2001. *The dinosaur hunters: A study of scientific rivalry and the discovery of the prehistoric world.* London: Fourth Estate.

Caesar, M. 1989. *Dante: The critical heritage, 1314(?)–1870.* London: Routledge.

Cahan, D., ed. 2003. *From natural philosophy to the sciences: Writing the history of nineteenth-century science.* Chicago: University of Chicago Press.

Camardi, G. 1999. Charles Lyell and the uniformity principle. *Biology and philosophy* 14:537–60.

Campbell, I., and D. Hutchison. 1978. A question of priorities: Forbes, Agassiz, and their disputes on glacier observations. *Isis* 69:388–99.

Cannon, W. F. 1960. The problem of miracles in the 1830s. *Victorian studies* 4:4–32.

———. 1960. The uniformitarian-catastrophist debate. *Isis* 51:38–55.

———. 1976. Charles Lyell, radical actualism, and theory. *British journal for the history of science* 9:104–20.

Cantor, G. 1997. Revelation and the cyclical cosmos of John Hutchinson. In *Images of the earth: Essays in the history of the environmental sciences,* edited by L. Jordanova and R. Porter, 17–35. Oxford: Alden Press.

Carignan, M. 2003. Analogical reasoning in Victorian historical epistemology. *Journal of the history of ideas* 64:445–64

Carlyle, T. 1845. *Oliver Cromwell's letters and speeches: With elucidations by Thomas Carlyle.* London: Chapman and Hall.

Carroll, D. 1995. *George Eliot: The critical heritage.* London: Routledge.

Carroll, J. 2004. *Literary Darwinism: Evolution, human nature, and literature.* New York: Routledge.

Cazamian, L. 1973. *The social novel in England, 1830–1850: Dickens, Disraeli, Mrs. Gaskell, Kingsley.* London: Routledge & Kegan Paul.

Cecil, D. 1934. *Early Victorian novelists: Essays in revaluation.* London; Constable.

Chadarevian, S. de, and N. Hopwood, eds. 2004. *Models: The third dimension of science.* Stanford, CA: Stanford University Press.

[Chambers, R.] 1844. *Vestiges of the natural history of creation.* London: Churchill.

Chapple, J. A. V. 1986. *Science and literature in the nineteenth century*. London: Churchill.

Chatman, S. 1978. *Story and discourse: Narrative structure in fiction and film*. Ithaca, NY: Cornell University Press.

Chesterton, G. K. 1906. *Charles Dickens*. London: Methuen.

Childers, J. W. 2001. Industrial culture and the Victorian novel. In *The Cambridge companion to the Victorian novel*, edited by D. David, 77–96. Cambridge: Cambridge University Press.

Chitty, S. 1974. *The beast and the monk: A life of Charles Kingsley*. London: Hodder & Stoughton.

Clark, J. W., and T. M. Hughes, eds. 1890. *The life and letters of the Rev. Adam Sedgwick*. Cambridge: Cambridge University Press.

Cockburn, H. A. 1910. An account of the Friday Club, written by Lord Cockburn, together with notes on certain other social clubs in Edinburgh. *The book of the Old Edinburgh Club* 3.

Cohen, I. B. 1985. *Revolution in science*. Cambridge, MA: Belknap.

Cohen, J. R. 1980. *Charles Dickens and his original illustrators*. Columbus: Ohio State University Press.

Collins, K. K. 1980. Questions of method: Some late unpublished essays. *Nineteenth-century fiction* 35:385–405.

Conybeare, W. D., and W. Phillips. 1822. *Outlines of the geology of England and Wales. Part 1: With an introductory compendium of the general principles of that science, and comparative views of the structure of foreign countries*. London: Phillips.

Cook, E. T., and A. Wedderburn, eds. 1903. *The works of John Ruskin*. London: George Allen.

Cook, K. S. 1995. From false starts to firm beginnings: Early colour printing of geological maps. *Imago mundi* 47:155–72.

Cooper, T. 1845. *The purgatory of suicides: A prison-rhyme in ten books*. London: How.

Cooter, R., and R. Pumfrey. 1994. Separate spheres and public places: Reflections on the history of science popularization and science in popular culture. *History of science* 32:237–67.

Cordle, D. 1999. *Postmodern postures: Literature, science and the two cultures debate*. Aldershot: Ashgate.

Corsi, P. 1978. The importance of French transformist ideas for the second volume of Lyell's *Principles of geology*. *British journal for the history of science* 11:221–44.

———. 2005. Before Darwin: Transformist concepts in European natural history. *Journal of the History of Biology* 38:67–83.

Cosslett, T. 1982. *The "scientific movement" and Victorian literature*. New York: St. Martin's Press.

Cotsell, M. 1986. *The companion to Our mutual friend*. London: Allen & Unwin.

Cottom, D. 1987. *Social figures: George Eliot, social history, and literary representation*. Minneapolis: University of Minnesota Press.

Craig, C. 1996. *Out of history: Narrative paradigms in Scottish and English culture.* Edinburgh: Polygon.

Crary, J. 1990. *Techniques of the observer: On vision and modernity in the nineteenth century.* Cambridge, MA: MIT Press.

Crawford, R. 1992. *Devolving English literature.* Oxford: Clarendon Press.

Cruchley, G. F. 1865. *Cruchley's London in 1865: A handbook for strangers.* London: Cruchley.

Culler, D. A. 1977. *The poetry of Tennyson.* New Haven, CT: Yale University Press.

Cumming, D. A. 1977. A description of the western islands of Scotland: John MacCulloch's successful failure. *Journal of the Society for the Bibliography of Natural History* 8:270–85.

———. 1980. John MacCulloch, F.R.S., at Addiscombe: The lectureships in chemistry and geology. *Notes and records of the Royal Society of London* 34:155–83.

———. 1981. Geological maps in preparation: John MacCulloch on the western islands. *Archives of natural history* 10:255–71.

———. 1982. John MacCulloch: Pioneer of Precambrian geology. PhD diss., University of Glasgow.

Cunningham, V. 1985. Soiled fairy: *The water babies* in its time. *Essays in criticism* 35:121–48.

Curran, S. 2002. Women and the *Edinburgh review*. In *British Romanticism and the Edinburgh review: Bicentenary essays*, edited by D. Massimiliano and D. Wu, 195–208. Basingstoke: Palgrave.

Cuvier, G. 1812. Discourse préliminaire. In *Recherches sur les ossemens fossiles de quadrupeds, où l'on rétablit les caractères de plusieurs espèces d'animaux que les revolutions du globe paroissent avoir détruites.* Paris: Chez Deterville.

Daiches, D. 1956. *Literary essays.* Edinburgh: Oliver and Boyd.

———. 1971. Sir Walter Scott and history. *Études Anglaises* 24:458–77.

———. 1971. *Sir Walter Scott and his world.* London: Thames and Hudson.

Dale, P. A. 1977. *The Victorian critic and the idea of history: Carlyle, Arnold, Pater.* Cambridge, MA: Harvard University Press.

Dames, N. 2007. *The physiology of the novel.* Oxford: Oxford University Press.

Dante Alighieri, 1994. *Divine comedy: The vision of Dante.* Translated by Henry Cary. London: Dent.

Darby, H. C. 1948. The regional geography of Thomas Hardy's Wessex. *Geographical review* 38:426–63.

Darwin, C. 1859. *On the origin of species by means of natural selection; or, The preservation of favoured races in the struggle for life.* London: Murray.

———. 1860. *On the origin of species by means of natural selection; or, The preservation of favoured races in the struggle for life.* 2nd ed, London: Murray.

———. 1871. *The descent of man; and, Selection in relation to sex.* London: Murray.

Darwin, F., ed. 1887. *Life and letters of Charles Darwin: Including an autobiographical chapter.* London: Murray.

Daubeny, C. G. B., ed. 1869. *Fugitive poems connected with natural history and physical science*. Oxford: Parker.

David, D., ed. 2000. *The Cambridge companion to the Victorian novel*. Cambridge: Cambridge University Press.

Davis, L., I. Duncan, et al., eds. 2004. *Scotland and the borders of Romanticism*. Cambridge: Cambridge University Press.

Davy, H. 1830. *Consolations in travel; or, The last days of a philosopher*. London: Murray.

Dawkins, W. B. 1874. *Cave-hunting: Researches on the evidence of caves respecting the early inhabitants of Europe*. London: Macmillan.

———. 1870. Exploration of caves at Settle, Yorkshire. *Nature* 1 (April 21): 628–29.

———. 1871. Cave hunting. *Macmillan's magazine*. 24:357–66.

———. 1872. Report on the results obtained by the Settle Cave Exploration Committee out of Victoria Cave 1870. *Journal of the Anthropological Institute* 1:60–70.

———. 1874. Report of the committee appointed for the purpose of exploring the Settle Caves. In *Report of the forty-third meeting of the British Association for the Advancement of Science*, 250–51. London: Murray.

Dawson, G. 2006. Literature and science under the microscope. *Journal of Victorian culture* 11:301–15.

———. 2008. *Darwin, literature and Victorian respectability*. Cambridge: Cambridge University Press.

———. 2011. Literary megatheriums and loose baggy monsters. *Victorian Studies* 53:203–30.

De la Beche, H. T. 1830. *Sections and views: Illustrative of geological phenomena*. London: Treuttel & Würtz.

———. 1831. *A geological manual*. London: n.p.

De Strzelecki, P. E. 1845. *Physical description of New South Wales and Van Diemen's Land, accompanied by a geological map, sections, and diagrams, and figures of the organic remains*. London: Longman.

Dean, D. R. 1979. The word geology. *Annals of science* 36:35–43.

———. 1985. *Tennyson and geology*. Lincoln: Tennyson Society.

———. 1985. "Through science to despair": Geology and the Victorians. In *Victorian science and Victorian values: Literary perspectives*, edited by J. Paradis and T. Postlewait, 111–36. New Brunswick, NJ: Rutgers University Press.

———. 1992. *James Hutton and the history of geology*. Ithaca, NY: Cornell University Press.

———. 1997. *James Hutton in the field and in the study*. Delmar, NY: Scholars' Facsimiles and Reprints.

———. 2007. *Romantic landscapes: Geology and its cultural influence in Britain, 1765–1835*. Ann Arbor, MI: Scholars' Facsimiles and Reprints.

Dean, D. R., ed. 2004. *Coleridge and geology*. Ann Arbor, MI: Scholars' Facsimiles and Reprints.

Dennis, I. 1997. *Nationalism and desire in early historical fiction*. London: Macmillan.

Dentith, S. 1997. Generic diversity in Elizabeth Gaskell's *Mary Barton*. *Gaskell Society journal* 11:43–53.

———. 2007. Realist synthesis in the nineteenth-century novel: "That unity which lies in the selection of our keenest consciousness." In *Adventures in realism*, edited by M. Beaumont, 33–49. Malden, MA: Blackwell.

Desmond, A. 1977. *The hot-blooded dinosaurs*. London: Hutchinson. Repr. 1990.

———. 1982. *Archetypes and ancestors: Palaeontology in Victorian London, 1850–1875*. London: Blond & Briggs.

———. 1989. *The politics of evolution: Morphology, medicine, and reform in radical London*. Chicago: University of Chicago Press.

———. 2009. *Darwin's sacred cause: Race, slavery and the quest for human origins*. London: Allen Lane.

Dever, C. 1998. *Death and the mother from Dickens to Freud: Victorian fiction and the anxiety of origins*. Cambridge: Cambridge University Press.

Dewhirst, I. 1965. The Rev. Patrick Bronte and the Keighley Mechanics' Institute. *Bronte studies* 14:35.

Dickens, C. 1846. *Pictures from Italy*. London: Bradbury & Evans.

———. 1870. *Speeches literary and social*. London: John Camden Hotten.

———. 1937. The poetry of science. *The nonesuch Dickens: Collected papers*, 1:178–80. London: Bloomsbury.

———. 1965–2002. *The letters of Charles Dickens, 1820–1870*. Oxford: Clarendon Press.

———. 1998. *Bleak House*. Oxford: Oxford University Press.

———. 1998. *Our mutual friend*. Oxford: Oxford University Press.

———. 2001. *Dombey and son*. Oxford: Oxford University Press.

[Dickens, C.] 1850. A preliminary word. *Household words* 1:1–3.

———. 1850. Some account of an extraordinary traveller. *Household words* 1:73–77.

Dolan, B. P. 1998. Representing novelty: Charles Babbage, Charles Lyell, and experiments in early Victorian geology, *History of science* 36:299–327.

Dolin, T. 2005. *George Eliot*. Oxford: Oxford University Press

———. 2005. Introduction. In *A pair of blue eyes*, edited by A. Manford. Oxford: Oxford University Press.

Draper, R. P., ed. 1989. *The literature of region and nation*. Basingstoke: Macmillan.

Drew, J. M. L. 2003. *Dickens the journalist*. Basingstoke: Palgrave.

[Dulton,] 1852. Up Vesuvius. *Household words* 5:235–36.

Duncan, I. 1992. *Modern romance and transformations of the novel: The Gothic, Scott, Dickens*. Cambridge: Cambridge University Press.

———. 2002. The provincial or regional novel. In *A companion to the British novel*, edited by P. Brantlinger and W. B. Thesing, 318–35. Oxford: Blackwell.

———. 2007. *Scott's shadow: The novel in Romantic Edinburgh*. Princeton, NJ: Princeton University Press.

Eagleton, T. 1988. *Myths of power: A Marxist study of the Brontes*. London: Macmillan.
———. 1995. *Heathcliff and the great hunger: Studies in Irish culture*. London: Verso.
———. 2005. *The English novel*. Oxford: Blackwell.
Edmonds, J. M., and J. A. Douglas. 1950. John Phillips's geological maps of the British Isles. *Annals of science* 6:361–75.
———. 1976. William Buckland, F.R.S. (1784) and an Oxford geological lecture, 1823. *Notes and records of the Royal Society* 30:141–67.
Eigner, E. M. 1979. *The Dickens pantomime*. Berkeley: University of California Press.
Eliot, G. 1856. The natural history of German life. *Westminster review* 66:51–79.
———. 1856. Silly novels by lady novelists. *Westminster review* 66:442–61.
———. 1884. *Essays and leaves from a note-book*. Edinburgh: Blackwood.
———. 1884. The historic imagination. In *Essays and leaves from a notebook*, 296–98. Edinburgh: Blackwood.
———. 1980. *The mill on the floss*. Oxford: Clarendon Press.
———. 1984. *Daniel Deronda*. Oxford: Clarendon Press.
———. 1986. *Middlemarch: A study of provincial life*. Oxford: Clarendon Press.
———. 2001. *Adam Bede*. Oxford: Clarendon Press.
———. 2003. *The mill on the floss*. London: Penguin.
Enstince, A. 1979. *Thomas Hardy: Landscapes of the mind*. London: Macmillan.
Ermarth, E. D. 1998. *Realism and consensus in the English novel: Time, space and narrative*. Edinburgh: Edinburgh University Press.
Eyles, J. M. 1969. William Smith (1769–1839): A bibliography of his published writings, maps and geological sections, printed and lithographed. *Journal for the society of the bibliography of natural history* 5:87–109.
Eyles, V. A. 1937. John MacCulloch, F.R.S., and his geological map: An account of the first geological survey of Scotland. *Annals of science* 2:114–29.
———. 1972. Mineralogical maps as forerunners of modern geological maps. *Cartographic journal* 9:133–35.
Fasick, L. 1994. Charles Kingsley's scientific treatment of gender. In *Muscular Christianity: Embodying the Victorian age*, edited by D. Hall, 91–113. Cambridge: Cambridge University Press.
Ferngren, G. B., ed. 2000. *The history of science and religion in the Western tradition: An encyclopedia*. New York: Garland.
Ferris, I. 1991. *The achievement of literary authority: Gender, history, and the Waverley novels*. Ithaca, NY: Cornell University Press.
———. 2004. Melancholy, memory, and the "narrative situation" of history in post-Enlightenment Scotland. In *Scotland and the Borders of Romanticism*, edited by I. Duncan, L. Davis, and J. Sorensen. 77–93. Cambridge: Cambridge University Press.
Fielding, K. J., ed. 1988. *The speeches of Charles Dickens: A complete edition*. Hemel Hempstead: Harvester Wheatsheaf.

———. 1996. Dickens and science? *Dickens quarterly* 13:200–216.
Figuier, L. 1865. *The world before the deluge*. London: Chapman and Hall.
[Fitton, W.] 1817a. Transactions of the Geological Society. *Edinburgh review* 28:174–92.
———. 1817b. Transactions of the Geological Society. *Edinburgh review* 29: 70–94.
Flinn, D. 1981. John MacCulloch, M.D., F.R.S., and his geological map of Scotland: His years in the Ordnance, 1795–1826. *Notes and records of the Royal Society of London* 36:83–101.
Flint, K. 2000. *The Victorians and the visual imagination*. Cambridge: Cambridge University Press.
Fludernik, Monika. 1993. *The fictions of language and the languages of fiction: The linguistic representation of speech and consciousness*. London: Routledge.
Fontana, E. S. 2005. Darwinian sexual selection and Dickens's *Our mutual friend*. *Dickens quarterly* 12:36–42.
[Forbes, E.] 1852. The future of geology. *Westminster review* 58:67–92.
———. 1853. Geology, popular and artistic. *Dublin University magazine* 42: 338–49.
Forster, E. M. 2002. *Aspects of the novel*. New York: Rosetta.
Foster, S. 2006. Note on the text. In E. Gaskell, *Mary Barton: A tale of Manchester life*. Oxford: Oxford University Press.
Frank, L. 2003. *Victorian detective fiction and the nature of evidence: The scientific investigations of Poe, Dickens, and Doyle*. Basingstoke: Palgrave.
Franklin, J. J. 1999. *Serious play: The cultural form of the nineteenth-century realist novel*. Philadelphia: University of Pennsylvania Press.
Freedgood, E. 2006. *The ideas in things: Fugitive meaning in the Victorian novel*. Chicago: University of Chicago Press.
Freeman, M. 2001. Tracks to a new world: Railway excavation and the extension of geological knowledge in mid-nineteenth-century Britain. *British journal for the history of science* 34:51–65.
———. 2004. *Victorians and the prehistoric: Tracks to a lost world*. New Haven, CT: Yale University Press.
Frend, W. 1796–1799. *The principles of algebra*. London: Davis.
Friederich, W. P. 1949. Dante through the centuries. *Comparative literature* 1:44–54.
Fulweiler, H. W. 1994. "A dismal swamp": Darwin, design and *Our mutual friend*. *Nineteenth-century literature* 49:50–74.
Fyfe, A. 2004. *Science and salvation: Evangelical popular science publishing in Victorian Britain*. Chicago: University of Chicago Press.
———. 2007. Reading natural history at the British Museum and the Pictorial Museum. In *Science in the marketplace: Nineteenth-century sites and experiences*, edited by A. Fyfe and B. Lightman, 196–230. Chicago: University of Chicago Press.

Fyfe, A., and B. Lightman. 2007. *Science in the marketplace: Nineteenth-century sites and experiences.* Chicago: University of Chicago Press.
Gallagher, C. 1985. *The industrial reformation of English fiction: Social discourse and narrative form, 1832–1867.* Chicago: University of Chicago Press.
Gange, D. 2009. Odysseus in Eden: Gladstone's Homer and the idea of universal epic, 1850–1880. *Journal of Victorian culture* 14:190–206.
Gange, D., and M. Ledger-Lomas. 2011. *Cities of God.* Cambridge: University of Cambridge Press.
Garofalo, D. 2008. Impossible love and commodity culture in Emily Brontë's "Wuthering Heights." *English literary history* 75:819–40.
Garside, P. 1972. Scott, the romantic past, and the nineteenth century. *Review of English studies* 23:147–60.
———. 1991. Popular fiction and national tale: Hidden origins of Scott's *Waverley*. *Nineteenth-century literature* 46:30–53.
———. 1999. Walter Scott and the "common" novel, 1808–1819. *Cardiff corvey: Reading the romantic text*, no. 3. www.cardiff.ac.uk/encap/journals/corvey/.
Gaskell, E. 2006. *Mary Barton: A tale of Manchester life.* Oxford: Oxford University Press.
———. 2006. *Mary Barton: A tale of Manchester life.* London: Pickering & Chatto.
———. 2009. *The life of Charlotte Bronte.* Stroud: Amberley.
Geikie, A. 1858. *The story of a boulder; or, Gleanings from the note-book of a field geologist.* Edinburgh: Constable.
———. 1875. *Life of Sir Roderick I. Murchison.* London: Murray.
———. 1897. *The founders of geology.* New York: Dover. Repr. 1962.
———. 1924. *A long life's work: An autobiography.* London: Macmillan.
Genette, G. 1988. *Narrative discourse revisited.* Ithaca, NY: Cornell University Press.
Geological Society of London. 1808. *Geological inquiries.* London: Phillips.
Ghiselin, M. The Darwinian revolution as viewed by a philosophical biologist. *Journal of the history of biology* 38:123–26.
Ghose, E. B. 1963. Psychic evolution: Darwinism and initiation in *Tess*. *Nineteenth-century fiction* 18:261–72.
Gilbert, A. H. 1923. Milton's textbook of astronomy. *Proceedings of the Modern Languages Associaton* 38:297–307.
Gilbert, S. M., and S. Gubar. 2000. *The madwoman in the attic: The woman writer and the nineteenth-century literary imagination.* New Haven, CT: Yale University Press.
Gill, S. 1998. *Wordsworth and the Victorians.* Oxford: Clarendon Press.
Gilmartin, S. 2000. Geology, genealogy and church restoration in Hardy's writing. In *The achievement of Thomas Hardy*, edited by P. Mallett, 22–40. Basingstoke: Macmillan.
Gilmour, R. 1989. Regional and provincial in Victorian literature. In *The literature of region and nation*, edited by R. P. Draper, 51–60. Basingstoke: Macmillan.

Ginsburg, M. P. 1996. *Economies of change: Form and transformation in the nineteenth-century novel.* Stanford, CA: Stanford University Press.
Girouard, M. 1981. *The return to Camelot: Chivalry and the English gentleman.* London: Yale University Press.
Glen, H., ed. 2002. *The Cambridge companion to the Brontes.* Cambridge: Cambridge University Press.
Goldsmith, O. 2006. *The vicar of Wakefield.* Oxford: Oxford University Press.
Goodall, J. 2002. *Performance and evolution in the age of Darwin.* London: Routledge.
Goode, J. 1988. *Thomas Hardy: The offensive truth.* Oxford: Blackwell.
Gordon, E. O., ed. 1894. *The life and correspondence of William Buckland, D.D., F.R.S.: Sometime dean of Westminster, twice president of the Geological Society, and first president of the British Association.* London: Murray.
Gose, E. B. J. 1966. *Wuthering Heights*: The heath and the hearth. *Nineteenth-century fiction* 21:1–19.
Gosse, P. H. 1853. *A naturalist's rambles on the Devonshire coast.* London: Van Voorst.
Gossin, P. 2007. *Thomas Hardy's novel universe: Astronomy, cosmology and gender in the post-Darwinian world.* Burlington: Ashgate.
Gould, P. J. 1987. *Time's arrow, time's cycle: Myth and metaphor in the discovery of geological time.* Cambridge, MA: Harvard University Press.
Graham, R. 1908. *Fox-hunting recollections.* London: n.p.
Grant, R. 1997. Hutton's theory of the earth. In *Images of the earth: Essays in the history of the environmental sciences,* edited by L. Jordanova and R. Porter, 37–51. Oxford: Alden Press.
Graver, S. 1984. *George Eliot and community: A study in social theory and fictional form.* Berkeley: University of California Press.
Gray, A. 1861. *Natural selection not inconsistent with natural theology: A free examination of Darwin's treatise on the origin of species and of its American reviewers.* London: Trubner.
Grenier, K. H. 2005. *Tourism and identity in Scotland, 1770–1914: Creating Caledonia.* Aldershot: Ashgate.
Guettard, J.-E., A.-L. Lavoisier, et al. 1780. *Atlas et description mineralogique de la France... premiere parti.* Paris: Didot l'aine.
Gully, A. L. 1993. Sermons in stone: Ruskin and geology. In *John Ruskin and the Victorian eye,* edited by H. Whelchel, 158–83. New York: Abrams.
Gunrau, M. 1978. The emergence of geology as a scientific discipline. *History of science* 16:280–90.
———. 2009. The rise of geology as a science in Germany around 1800. In *The making of the Geological Society of London* (GSL Special Publications no. 317), edited by C. L. E. Lewis and S. J. Knell, 163–77. London: Geological Society of London.
Guy, J. M. 1996. *The Victorian social-problem novel.* Basingstoke: Macmillan.

Haight, G. S., ed. 1954–1978. *The George Eliot letters*. London: Oxford University Press.
———. 1968. *George Eliot: A biography*. London: Clarendon Press.
Hall, D., ed. 1994. *Muscular Christianity: Embodying the Victorian age*. Cambridge: Cambridge University Press.
Hall, D. 1994. Muscular Christianity: Reading and writing the male social body. In *Muscular Christianity: Embodying the Victorian age*, ed. D. Hall, 3–17. Cambridge: Cambridge University Press.
———. 1996. *Fixing patriarchy: Feminism and mid-Victorian male novelists*. London: Macmillan.
Hallam, A. H. 1832. *Remarks on Professor Rossetti's Disquizioni sullo spirito antipapale*. London: n.p.
Halperin, J. 1980. Leslie Stephen, Thomas Hardy, and *A pair of blue eyes*. *Modern language review* 75:738–45.
Hamblyn, R. 1994. Landscape and the contours of knowledge: The literature of travel and the sciences of the earth in eighteenth-century Britain. PhD diss., English literature, University of Cambridge.
Harcourt, V. L. 1838. *The doctrine of the deluge: Vindicating the scriptural account from the doubts which have recently been cast upon it by geological speculations*. London: Longman, Orme, Green, Brown and Longmans.
Hardy, T. 1998. *A pair of blue eyes*. London: Penguin.
———. 1999. *Return of the native*. London: Penguin.
———. 2005. *A pair of blue eyes*. Oxford: Oxford University Press.
Hare, Julius Charles. 1829. *A vindication of Niebuhr's "History of Rome": From the charges of the Quarterly Review*. Cambridge: Taylor.
Harman, P. M. 1985. *Wranglers and physicists: Studies on Cambridge physics in the nineteenth century*. Manchester: Manchester University Press.
Harrison, J. F. C. 1954. *A history of the Working Men's College, 1854–1954*. London: Routledge.
[Hart, E. A.] 1857. Nature's greatness in small things. *Household words* 16:511–13.
Hart, F. R. 1966. *Scott's novels: The plotting of historic survival*. Charlottesville: University Press of Virginia.
———. 1978. *The Scottish novel: A critical survey*. London: Murray.
Hartley, A. J. 1977. *The novels of Charles Kingsley: A Christian Social interpretation*. Folkestone: Hour-Glass Press.
Harvey, J. R. 1970. *Victorian novelists and their illustrators*. London: Sidgwick & Jackson.
Hawkins, T. 1840. *The book of the great sea-dragons, icthyosauri and plesiosauri, godeolim taninim, of Moses: Extinct monsters of the ancient earth*. London: Pickering.
Herbert, S. The Darwinian revolution revisited. *Journal of the history of biology* 38:51–66.
Heringman, N., ed. 2003. *Romantic science: The literary forms of natural history*. Albany: State University of New York Press.

———. 2004. *Romantic rocks, aesthetic geology*. Ithaca, NY: Cornell University Press.

———. 2009. "Very vain is science' proudest boast": The resistance to geological theory in early nineteenth-century England. In *Emergence of modern geology and evolutionary thought from the scientific revolution to the Enlightenment*, edited by G. D. Rosenborg, 235–45. Boulder, CO: Geological Society of America.

———. 2009. Picturesque ruin and geological antiquity: Thomas Webster and Sir Henry Englefield on the Isle of Wight. In *The making of the Geological Society of London* (GSL Special Publications no. 317), edited by C. L. E. Lewis and S. J. Knell, 299–318. London: Geological Society of London.

Hevly, B. 1996. The heroic science of glacier motion. *Osiris* 11:66–86.

Hewison, R. 1996. "Paradise lost": Ruskin and science. In *Time and tide: Ruskin and science*, edited by M. Wheeler, 29–44. London: Pilkington.

Heywood, C. 1987. Yorkshire slavery in *Wuthering Heights*. *Review of English studies* 150:184–98.

———. 1993. A Yorkshire background for *Wuthering Heights*. *Modern languages review* 88:817–30.

———. 2002. Introduction. In *Wuthering Heights*, 18–90. Peterborough, ON: Broadview.

Higonnet, M. R. 1993. *The sense of sex: Feminist perspectives on Hardy*. Urbana: University of Illinois Press.

Hill, T. W. 1949. Books that Dickens read. *Dickensian* 45:201–7.

Hilton, T. 2002. *John Ruskin*. New Haven, CT: Yale University Press.

Hodge, J. Against "revolution" and "evolution." *Journal of the history of biology* 38:101–21.

Hollingsworth, K. 1963. *The Newgate novel, 1830–1847: Bulwer, Ainsworth, Dickens and Thackeray*. Detroit: Wayne State University Press.

Homans, M. 1978. Repression and sublimation of nature in *Wuthering Heights*. *PMLA* 93:9–19.

———. 1986. *Bearing the word: Language and female experience in nineteenth-century women's writing*. Chicago: University of Chicago Press.

Homans, M., and S. M. Conger. 1978. Nature in *Wuthering Heights*. *PMLA* 93:1003–4.

Hooykaas, R. 1963. *Natural law and divine miracle: A historical-critical study of the principle of uniformity in geology, biology and theology*. Leiden: Brill.

———. 1963. *The principle of uniformity in geology*.

———. 1970. *Catastrophism in geology: Its scientific character in relation to actualism and uniformitarianism*. Amsterdam: North-Holland Publishing Company.

[Horne, R. H.] 1850. The true story of a coal fire. *Household words* 1:26–31, 68–72, 90–96.

Howitt, W. 1838. *The rural life of England*. London: Longman.

Howson, W. 1850. *An illustrated guide to the curiosities of Craven, with a geological introduction; notices of the dialect, a list of the fossils; and a local flora*. London: Whitaker.

Hull, D. Deconstructing Darwin: evolutionary theory in context. *Journal of the history of biology* 38:137–52.

[Hunt, F. K.] 1850. The Hunterian museum. *Household words* 2:277–82.

Hunter, P. 1990. *Before novels: The cultural contexts of eighteenth-century English fiction*. New York: Norton.

Hutton, J. 1785. *Abstract of a dissertation . . . concerning the system of the earth, its duration, and stability*. Edinburgh. Repr. in G. W. White and V. A. Eyles, *James Hutton* (Darien: Hafner, 1970).

———. 1788. *Theory of the earth; or, An investigation of the laws observable in the composition, dissolution and restoration of the land upon the globe*. Repr. in G. W. White and V. A. Eyles, *James Hutton* (Darien: Hafner, 1970).

———. 1795. *Theory of the earth, with proofs and illustrations*. 2 vols. Edinburgh: William Creech.

Huxley, T. H. 1881. Science and culture. *Science and culture, and other essays*. London: Macmillan.

Hyde, R. 1988. *Panoramania! The art and entertainment of the "all-embracing" view*. London: Trefoil.

Ingham, P. 1980. Hardy and the wonders of geology. *Review of English studies* 31: 59–64.

Inkster, I., and J. Morrell, eds. 1983. *Metropolis and province: Science in British culture, 1780–1850*. London: Hutchinson.

Irwin, J., ed. 1996. *George Eliot's Daniel Deronda notebooks*. Cambridge: Cambridge University Press.

Irwin, M. 2000. *Reading Hardy's landscapes*. Basingstoke: Macmillan.

Jahn, Manfred. 2005. Narratology: A guide to the theory of narrative. http://www.uni-koeln.de/~ame02/pppn.htm.

Jameson, R. 1800. *Mineralogy of the Scottish Isles: With mineralogical observations made in a tour through different parts of the mainland of Scotland, and dissertations upon peat and kelp*. Edinburgh: White.

Jardine, N., J. A. Secord, et al. 1996. *Cultures of natural history*. Cambridge: Cambridge University Press.

[Jeffrey, F., and W. Scott.] 1817. Tales of my landlord. *Edinburgh review* 28:193–259.

Jenkins, A. 2007. *Space and the "march of mind": Literature and the physical sciences in Britain 1815–1850*. Oxford: Oxford University Press.

John, J. 2001. *Dickens's villains: Melodrama, character, popular culture*. Oxford: Oxford University Press.

John, J., and A. Jenkins, eds. 2000. *Rethinking Victorian culture*. Basingstoke: Macmillan.

Johnson, B. 1977. The perfection of species and Hardy's Tess. In *Nature and the Victorian imagination*, edited by U. C. Knoepflmacher and G. B. Tennyson, 259–77. Berkeley: University of California.

Johnson, E. 1970. *Sir Walter Scott: The great unknown*. London: Hamilton.
Johnston, A. 1959. The water babies: Kingsley's debt to Darwin. *English: The journal of the English Association* 12:215–19.
[Johnston, J. F. W.] 1849. The physical atlas. *Edinburgh review* 89:327–52.
Jones, L. 1975. Thomas Hardy's "Idiosyncratic mode of regard." *English literary history* 3:433–59.
Jordan, J. O., ed. 2001. *The Cambridge companion to Charles Dickens*. Cambridge: Cambridge University Press.
Jordanova, L., and R. Porter, eds. 1997. *Images of the earth: Essays in the history of the environmental sciences*. Oxford: Alden Press.
Kalikoff, B. 1986. *Murder and moral decay in Victorian popular literature*. Ann Arbor, MI: UMI Research Press.
Kargon, R. H. 1977. *Science in Victorian Manchester: Enterprise and expertise*. Manchester: Manchester University Press.
Kay-Robinson, D. 1972. *Hardy's Wessex reappraised*. Newton Abbot: David & Charles.
Kelly, G. 1989. *English fiction of the Romantic period, 1789–1830*. London: Longman.
Kettle, A., ed. 1972. *The nineteenth-century novel: Critical essays and documents*. London: Heinemann Educational.
King, A. M. 2003. Taxonomical cures: The politics of natural history and herbalist medicine in Elizabeth Gaskell's *Mary Barton*. In *Romantic science: The literary forms of natural history*, edited by N. Heringman, 255–70. Albany: State University of New York Press.
Kingsley, C. 1850. *Alton Locke: Tailor and poet*. London: Chapman and Hall.
———. 1851. *Yeast: A problem*. London: Parker.
———. 1855. *Glaucus; or, Wonders of the shore*. Cambridge: Macmillan.
———. 1855. *Westward ho!; or, The voyages and adventures of Sir Amyas Leigh, Knight, of Burrough, in the county of Devon, in the reign of her most glorious majesty Queen Elizabeth*. Cambridge: Macmillan.
———. 1857. *Two years ago*. Cambridge: Macmillan.
———. 1870. *Madam How and Lady Why; or, First lessons in earth lore for children*. London: Bell and Daldy.
———. 1872. *Town geology*. London: Strahan & Co.
Kingsley, F., ed. 1877. *Charles Kingsley: His letters and memories of his life*. London: King.
Klancher, J. 1987. *The making of English reading audiences, 1790–1832*. Madison: University of Wisconsin Press.
Klaver, J. M. I. 1997. *Geology and religious sentiment: The effect of geological discoveries on English society and literature between 1829 and 1859*. Leiden: Brill.
———. 2004. "I will ferry thee across". The meaning of fluvialism in George Eliot's *The mill on the floss*. In *The mill on the floss*, edited by N. Henry, 629–32. Boston: Houghton Mifflin.
———. 2006. *The apostle of the flesh: A critical life of Charles Kingsley*. Leiden: Brill.

Klonk, C. 1996. *Science and the perception of nature: British landscape art in the late eighteenth and early nineteenth centuries*. New Haven, CT: Yale University Press.

Knell, S. 2000. *The culture of English geology, 1815–51: A science revealed through its collecting*. Aldershot: Ashgate.

Knoepflmacher, U. C., and G. B. Tennyson, eds. 1977. *Nature and the Victorian imagination*. Berkeley: University of California Press.

Kolbl-Ebert, M. 2004. Charlotte Murchison. In *The dictionary of nineteenth-century British scientists*, edited by B. Lightman, 3:1440–42. Bristol: Thoemmes Continuum.

———. 2009. George Bellas Greenough's "Theory of the earth" and its impact on the early Geological Society. In *The making of the Geological Society of London* (GSL Special Publications no. 317), edited by C. L. E. Lewis and S. J. Knell, 115–28. London: Geological Society of London.

Kramer, D., ed. 1999. *Cambridge companion to Thomas Hardy*. Cambridge: Cambridge University Press.

Kucich, J. 2002. Scientific ascendancy. In *A companion to the Victorian novel*, edited by P. Brantlinger and W. B. Thesing, 119–36. Oxford: Blackwell.

Kuhn, T. 1996. *The structure of scientific revolutions*. Chicago: University of Chicago Press.

Lai, S.-F. 2001. Fact or fancy: What can we learn about Dickens from his periodicals *Household words* and *All the year round*? *Victorian periodicals review* 34:41–53.

Lapworth, C. 1879. On the tripartite classification of the lower Palaeozoic rocks. *Geological magazine* 6:1–15.

Laudan, R. 1977. Ideas and organisations in British geology: A case study in institutional history. *Isis* 68:527–38.

———. 1982. The role of methodology in Lyell's science. *Studies in history and philosophy of science* 13:215–50.

———. 1987. *From mineralogy to geology: The foundations of a science, 1650–1830*. Chicago: University of Chicago Press.

Law, J. 1992. Water rights and the "crossing o' breeds": Chiastic exchange in *The mill on the floss*. In *Rewriting the Victorians: Theory, history and the politics of gender*, edited by L. M. Shires, 52–69. New York: Routledge.

Leavis, F. R. 1962. *Two cultures? The significance of C. P. Snow*. London: Chatto & Windus.

———. 1983. *The great tradition*. London: Penguin.

Leavis, Q. D. 1969. A fresh approach to *Wuthering Heights*. In *Lectures in America*. New York: Pantheon.

Ledger-Lomas, M. Forthcoming 2013. *Selective affinities: England and Protestant Germany, 1815–1870*.

Lennox, J. Darwin's methodological evolution. *Journal of the history of biology* 38:85–99.

Levine, C. 2003. *The serious pleasures of suspense: Victorian realism and narrative doubt*. Charlottesville: University of Virginia Press.

Levine, G. 1981. *The realistic imagination: English fiction from Frankenstein to Lady Chatterley*. Chicago: University of Chicago Press.
———. 1988. *Darwin and the novelists: Patterns of science in Victorian fiction*. Cambridge, MA: Harvard University Press.
———. 1988. The novel as scientific discourse: The example of Conrad. *Novel: A forum on fiction* 21:220–27.
———. 1993. By knowledge possessed: Darwin, nature, and Victorian narrative. *New literary history* 24:363–91.
———, ed. 1993. *Realism and representation: Essays on the problem of realism in relation to science, literature, and culture*. Madison: University of Wisconsin Press.
———, ed. 2001. *The Cambridge companion to George Eliot*. Cambridge: Cambridge University Press.
———. 2001. *Daniel Deronda*: A new epistemology. In *Knowing the past: Victorian literature and culture*, edited by S. Anger, 52–73. Ithaca, NY: Cornell University Press.
———. 2007. *Dying to know: Scientific epistemology and narrative in Victorian England*. Chicago: University of Chicago Press.
———. 2008. *Realism, ethics and secularism: Essays on Victorian literature and science*. Cambridge: Cambridge University Press.
———. 2008. Why science isn't literature: The importance of differences. In *Realism, ethics and secularism: Essays on Victorian literature and science*, 165–81. Cambridge: Cambridge University Press.
[Lewes, G. H.] 1850. Wuthering Heights. *Leader* 1:953.
Lewis, C., and S. Knell, eds. 2009. *The making of the Geological Society of London*. GSL Special Publications no. 317. London: Geological Society of London.
Lightman, B. 2007. *Victorian popularizers of science: Designing nature for new audiences*. Chicago: University of Chicago Press.
[Linton, E. L.] 1858. The first idea of everything. *Household words* 17:481–84.
Livingstone, D., et al., ed. 1999. *Evangelicals and science in historical perspective*. Oxford: Oxford University Press.
Livingstone, D. N. 2003. *Putting science in its place: Geographies of scientific knowledge*. Chicago: University of Chicago Press.
Lockhart, J. G. 1819. *Peter's letters to his kinsfolk*. Edinburgh: Blackwood.
Lohrli, A. 1973. *Household words: A weekly journal, 1850–1859*. Toronto: University of Toronto Press.
Lucas, J. 1966. Mrs. Gaskell and brotherhood. In *Tradition and tolerance in nineteenth-century fiction: Critical essays on some English and American novels*, edited by D. Howard, J. Lucas, and J. Goode, 141–205. London: Routledge & Kegan Paul.
———. 1980. *The literature of change: Studies in the nineteenth-century provincial novel*. Sussex: Harvester.
Lucier, P. 1999. A plea for applied geology. *History of science* 37:1–36.

Lukacs, G. 1950. *Studies in European realism: A sociological survey of the writings of Balzac, Stendhal, Zola, Tolstoy, Gorki and others.* London: Hillway Publishing.

———. 1971. *Theory of the novel: A historico-philosophical essay on the forms of great epic literature.* Cambridge, MA: MIT Press.

———. 1976. *The historical novel.* Harmondsworth: Penguin.

———. 1983. *The historical novel.* Preface by Fredric Jameson. Boston: Beacon.

Lyell, C. 1830–1833. *Principles of geology: Being an attempt to explain the former changes of the earth's surface, by reference to causes now in operation.* London: Murray.

———. 1836. Anniversary address of the President, February 19 1834. *Proceedings of the Geological Society of London* 2:357–90.

———. 1838. *Elements of geology.* London: Murray.

———. 1851. *A manual of elementary geology; or, The ancient changes of the earth and its inhabitants, as illustrated by geological monuments.* London: Murray.

———. 1863. *The geological evidences of the antiquity of man: With remarks on the theories of the origin of species by variation.* London: Murray.

———. 1867–1868. *Principles of geology; or, The modern changes of the earth and its inhabitants, considered as illustrative of geology.* London: Murray.

———. 1997. *Principles of geology.* London: Penguin.

[Lyell, C.] 1826a. Scientific institutions. *Quarterly review* 34:153–79.

———. 1826b. Transactions of the Geological Society. *Quarterly review* 34: 507–40.

———. 1827. Scrope's geology of central France. *Quarterly review* 36:437–83.

———. 1827. State of the universities. *Quarterly review* 36:216–68.

———. 1830. Rossetti's Dante. *Foreign and quarterly review* 5:419–49.

Lyell, C., Sr. 1842. *The poems of the Vita nuova and Convito of Dante Alighieri.* London: Molini.

Lyell, K. M., ed. 1881. *Life, letters and journals of Sir Charles Lyell.* London: Murray.

MacCulloch, J. 1819. *A description of the western islands of Scotland, including the Isle of Man. Comprising an account of their geological structure; with remarks on their agriculture, scenery, and antiquities.* London: Hurst, Robinson, and Co.

———. 1821. *A geological classification of the rocks: With descriptive synopses of the species and varieties, comprising the elements of practical geology.* London: Longman.

———. 1824. *The Highlands and Western Isles of Scotland, containing descriptions of their scenery and antiquities with an account of the political history and ancient manners: Founded on a series of annual journeys between the years 1811 and 1821 ... in letters to Sir Walter Scott.* London: Longman, Hurst, Rees. Orme, Brown and Green.

Makdisi, S. 1998. *Romantic imperialism: Universal empire and the culture of modernity.* Cambridge: Cambridge University Press.

Mallett, P. 2004. Introduction. In *Palgrave advances in Thomas Hardy studies*, edited by P. Mallett, 1–13. Basingstoke: Palgrave.

———, ed. 2004. *Palgrave advances in Thomas Hardy studies.* Basingstoke: Palgrave.

[Mann, C. W.] 1858. Everything after its kind. *Household words* 17:451–53.
Manning, S. 2004. Antiquarianism, the Scottish science of man, and the emergence of modern disciplinarity. In *Scotland and the borders of Romanticism*, edited by L. Davis, I. Duncan, and J. Sorensen, 57–76. Cambridge: Cambridge University Press.
Mannoni, L., W. Nekes, et al. 2004. *Eyes, lies and illusions: The art of deception.* Aldershot: Lund Humphries.
Mantell, G. A. 1827. *Illustrations of the geology of Sussex: Containing a general view of the geological relations of the south-eastern part of England.* London: Lupton Relfe.
———. 1838. *The wonders of geology.* London: Relfe and Fletcher.
———. 1848. *The wonders of geology; or, A familiar exposition of geological phenomena.* London: Bohn.
Maresca, T. 1974. *Epic to novel.* Columbus: Ohio State University Press.
Martin, C. 2001. "What if the sun be centre to the world?" Milton's epistemology, cosmology, and paradise of fools reconsidered. *Modern philology* 99:231–65.
Massimiliano, D., and D. Wu, eds. 2002. *British Romanticism and the Edinburgh review: Bicentenary essays.* Basingstoke: Palgrave.
Mazlish, B. 1998. Ecumenical, world and global history. In *World history: Ideologies, structures, and identities*, edited by P. Pomper, R. H. Elphick, and R. T. Vann, 41–52. Malden, MA: Blackwell.
McCaw, N. 2000. *George Eliot and Victorian historiography: Imagining the national past.* Basingstoke: Palgrave.
McCracken-Flesher, C. 2007. *Possible Scotlands: Walter Scott and the story of tomorrow.* Oxford: Oxford University Press.
McDonagh, J. 2007. Space, mobility and the novel: "The spirit of place is a great reality." In *Adventures in realism*, edited by M. Beaumont, 50–67. Malden, MA: Blackwell.
McGowan, J. P. 1986. *Representation and revelation: Victorian realism from Carlyle to Yeats.* Columbia: University of Missouri Press.
McKeon, M. 1987. *Origins of the English novel.* Baltimore: Johns Hopkins University Press.
———. 2000. Generic transformation and social change: Rethinking the rise of the novel. In *Theory of the novel: A historical approach*, edited by M. McKeon, 382–99. Baltimore: Johns Hopkins University Press.
———, ed. 2000. *Theory of the novel: A historical approach.* Baltimore: Johns Hopkins University Press.
———. 2005. Prose fiction: Great Britain. In *The Cambridge history of literary criticism: The eighteenth century*, edited by H. D. Nisbet, G. A. Kennedy, and C. Rawson. Cambridge: Cambridge University Press.
McNeil, K. 2007. *Scotland, Britain, empire: Writing the Highlands, 1760–1860.* Columbus: Ohio State University Press.

Meckier, J. 1987. *Hidden rivalries in Victorian fiction: Dickens, realism, and revaluation*. Lexington: University Press of Kentucky.

Meisel, M. 1983. *Realizations: Narrative, pictorial, and theatrical arts in nineteenth-century England*. Princeton, NJ: Princeton University Press.

Meisel, P. 1972. *Thomas Hardy: The return of the repressed: A study of the major fiction*. New Haven, CT: Yale University Press.

Menke, R. 2000. Fiction as vivisection: G. H. Lewes and George Eliot. *English literary history* 67:617–53.

Merrill, L. 1989. *The romance of Victorian natural history*. Oxford: Oxford University Press.

Metz, N. A. 1978. Science in *Household words*: "The poetic . . . passed into our common life." *Victorian periodicals newsletter* 11:121–33.

Miall, L. C. 1859–1868. On a system of anticlinals in South Craven. *Proceedings of the Geological and Polytechnic Society of the West Riding of Yorkshire* 4:577–88.

Miller, A. J. 1995. *Novels behind glass: Commodity culture and Victorian narrative*. Cambridge: Cambridge University Press.

Miller, D. A. 1981. *Narrative and its discontents: Problems of closure in the traditional novel*. Princeton, NJ: Princeton University Press.

Miller, H. 1841. *The Old Red Sandstone; or, New walks in an old field*. Edinburgh: Johnstone.

———. 1857. *The testimony of the rocks; or, Geology in its bearings on the two theologies, natural and revealed*. Edinburgh: Constable.

Miller, J. H. 1958. *Charles Dickens: The world of his novels*. Cambridge, MA: Harvard University Press.

———. 1971. Introduction to Bleak House. In *Bleak House*, edited by N. Page, 11–34. Harmondsworth: Penguin.

———. 1974. History and narrative. *English literary history* 41:455–73.

———. 1975. Optic and semiotic in *Middlemarch*. In *The worlds of Victorian fiction*, edited by J. H. Buckley, 125–45. Cambridge, MA: Harvard University Press.

Millgate, M. 1982. *Thomas Hardy: A biography*. Oxford: Oxford University Press.

———. 2001. *Thomas Hardy's public voice*. Oxford: Clarendon Press.

Mills, E. L. 2004. Forbes, Edward (1815–1854). *Oxford dictionary of national biography*. Oxford: Oxford University Press.

Moretti, F. 1987. *The way of the world: The Bildungsroman in European culture*. London: Verso.

———. 1998. *Atlas of the European novel: 1800–1900*. London: Verso.

———. 2005. *Graphs, maps, trees*. London: Verson.

———, ed. 2006. *The novel*. Princeton, NJ: Princeton University Press.

Morgan, R. 1988. *Women and sexuality in the novels of Thomas Hardy*. London: Routledge.

Morgentaler, G. 2000. *Dickens and heredity: When like begets like*. Basingstoke: Macmillan.

[Morley, H.] 1851. Our phantom ship on an antediluvian cruise. *Household words* 3:492–96.
———. 1857. Hammering it in. *Household words* 15:59–60.
Morrell, J. 2005. *John Phillips and the business of Victorian science.* Aldershot: Ashgate.
Morrell, J., and A. Thackray. 1981. *Gentlemen of science: Early years of the British Association for the Advancement of Science.* Oxford: Clarendon Press.
Morrell, J. 1976. London institutions and Lyell's career: 1820–1841. *British journal for the history of science* 9:132–46.
———. 1988. The early Geological and Polytechnic Society: A reconsideration. *Annals of science* 45:153–67.
Morris, P. 2000. A taste for change in *Our mutual friend*: Cultivation or education? In *Rethinking Victorian culture*, edited by J. John and A. Jenkins, 180–95. Basingstoke: Macmillan.
Morton, J. L. 2001. *Strata: How William Smith drew the first map of the earth in 1801 and inspired the science of geology.* Stroud: Tempus.
Morton, P. 1974. Tess of the D'Urbervilles: A neo-Darwinian reading. *Southern review* 7:38–50.
Morus, I. 2006. Replacing Victoria's scientific culture. *19: Interdisciplinary studies in the long nineteenth century* 2. www.19.bbk.ac.uk.
Morus, I., S. Schaffer, et al. 1992. Scientific London. In *London: World city, 1800–1840*, edited by C. Fox, 129–42. New Haven, CT: Yale University Press.
Mozley, T. 1882. *Reminiscences chiefly of Oriel College and the Oxford Movement.* London: Longmans, Green & Co.
Müller, C. H. 1986. Spiritual evolution and muscular theology: Lessons from Kingsley's natural theology. *University of Cape Town studies in English: A journal of studies in English and the humanities* 15:24–34.
Murchison, R. 1839. *The Silurian system: Founded on geological researches in the counties of Salop, Hereford, Radnor, Montgomery, Caernarthen, Brecon, Pembroke, Monmouth, Gloucester, Worceser and Stafford; with descriptions of the coalfields and overlying formations.* London: Murray.
———. 1852. *Address to the Royal Geographical Society of London.* London: Clowes.
———. 1854. *Siluria: The history of the oldest known rocks containing organic remains; with a brief sketch of the distribution of gold over the earth.* London: Murray.
[Murchison, R.] 1841. Tours in the Russian provinces. *Quarterly review* 67:334–75.
———. 1850. Siberia and California. *Quarterly review* 87:395–434.
Murphy, P. T. 1994. *Toward a working-class canon: Literary criticism in British working-class periodicals.* Columbus: Ohio State University Press.
Murray, J. 1878. *A comparative view of the Huttonian and Neptunian systems of geology, in answer to the illustrations of the Huttonian theory of the earth, by Professor Playfair.* New York: Arno.
———. 1882. *Handbook for travellers in Yorkshire, and for residents of the county.* London: Murray.

Murray, V. 1998. *High society: A social history of the Regency period.* London: Viking.
Nairn, T. 1977. *Break-up of Britain: Crisis and neo-nationalism.* London: NLB.
Napier, M., ed. 1879. *Selection from the correspondence of the late Macvey Napier.* London: Macmillan.
Newsome, D. 1959. *A history of Wellington College.* London: John Murray.
Niebuhr, B. G. 1828. *History of Rome.* Cambridge: Taylor.
Nixon, C. L. 2009. *Novel definitions: An anthology of commentary on the novel, 1688–1815.* Claremont: Broadview.
O'Connor, J. G., and A. J. Meadows. 1976. Specialization and professionalization in British geology. *Social studies of science* 5:77–89.
O'Connor, R. 2007. *The earth on show: Fossils and the poetics of popular science, 1802–1856.* Chicago: University of Chicago Press.
———. 2007. Young earth creationists in early nineteenth-century Britain? Towards a reassessment of "scriptural geology." *History of science* 45:357–403.
———. 2009. Facts and fancies: The Geological Society of London and the wider public, 1807–1837. In *The making of the Geological Society of London* (GSL Special Publications no. 317), edited by C. Lewis and S. Knell, 331–40. London: Geological Society of London.
———. 2009. From the epic of earth history to the evolutionary epic in nineteenth-century Britain. *Journal of Victorian culture* 14:207–23.
O'Gorman, F. 1996. "The eagle and the whale?" Ruskin's arguments with John Tyndall. In *Time and tide: Ruskin and science*, edited by M. Wheeler, 45–64. London: Pilkington.
———. 2000. "More interesting than all the books, save one": Charles Kingsley's construction of natural history. In *Rethinking Victorian culture*, edited by J. John and A. Jenkins, 146–62. Basingstoke: Macmillan.
———, ed. 2002. *The Victorian novel.* Oxford: Blackwell.
Oldroyd, D. R. 1979. Historicism and the rise of historical geology. *History of science* 17:191–213.
———. 1980. Sir Archibald Geikie (1835–1924), geologist, romantic aesthete, and historian of geology. *Annals of science* 37:441–62.
———. 1990. *The highlands controversy: Constructing geological knowledge through fieldwork in nineteenth-century Britain.* Chicago: University of Chicago Press.
———. 1996. *Thinking about the earth: A history of ideas.* London: Athlone.
———. 1997. Some youthful beliefs of Sir Archibald Geikie, PRS, and the first publication of his "On the study of the sciences." *Annals of science* 54:69–86.
———. 2003. The earth sciences. In *From natural philosophy to the sciences: Writing the history of nineteenth-century science*, edited by D. Cahan, 88–128. Chicago: University of Chicago Press.
Oppenlander, E. A. 1984. *Dickens' All the year round: Descriptive index and contributor list.* Troy, NY: Whitston.
Orchard, T. N. 1896. *The astronomy of Milton's Paradise lost.* London: Longmans, Green, and Co.

Ormond, R. 2005. *The monarch of the glen: Landseer in the Highlands*. Edinburgh: National Galleries of Scotland.
Ortega y Gasset, J. 2000. *Meditations on Quixote*. Urbana: University of Illinois Press.
Ospovat, A. M. 1976. The distortion of Werner in Lyell's *Principles of geology*. *British journal for the history of science* 9:190–98.
———. 1977. Lyell's theory of climate. *Journal of the history of biology* 10:317–39.
———. 1980. The importance of regional geology in the geological theories of Abraham Gottlob Werner. *Annals of science* 37:433–40.
Ostry, E. 2001. "Social Wonders": Fancy, science, and technology in Dickens's periodicals. *Victorian periodicals review* 34:54–74.
Otis, L. 2010. Science surveys and histories of literature: Reflections on an uneasy kinship. *Isis* 101:570–77.
Oulton, C. 2003. *Literature and religion in mid-Victorian England from Dickens to Eliot*. Basingstoke: Palgrave.
Owen, R. 1854. *Geology and inhabitants of the ancient world*. London: Crystal Palace Library and Bradbury & Evans.
———. 1856. A leaf from the oldest of books. *Household words*. 13:500–502.
———. 1861. *Memoir on the megatherium; or, Giant ground-sloth of America*. London: Williams and Norgate.
———. 1894. *The life of Richard Owen*. London: Murray.
Page, D. 1856. *Advanced text-book of geology: Descriptive and industrial*. Edinburgh: Blackwood.
Page, L. E. 1969. Diluvialism and its critics in Great Britain in the early nineteenth century. In *Toward a history of geology*, edited by C. J. Schneer, 257–71. Cambridge, MA: MIT Press.
———. 1976. The rivalry between Charles Lyell and Roderick Murchison. *British journal for the history of science* 9:156–65.
Paradis, J., and T. Postlewait, eds. 1985. *Victorian science and Victorian values: Literary perspectives*. New Brunswick, NJ: Rutgers University Press.
Parerson, J. A. 1909. The astronomy of Milton. *Journal of the Royal Astronomical Society of Canada* 3:56–76.
Patten, R. L. 1977. "A surprising transformation": Dickens and the hearth. In *Nature and the Victorian imagination*, edited by U. C. Knoepflmacher and G. B. Tennyson, 153–70. Berkeley: University of California Press.
———. 1996. *George Cruickshank's life, times, and art*. Cambridge: Lutterworth.
Paxton, N. L. 1991. *George Eliot and Herbert Spencer: Feminism, evolutionism, and the reconstruction of gender*. Princeton, NJ: Princeton University Press.
Pemberton, S. G., and R. W. Frey. 1991. William Buckland and his "coprolitic vision." *Ichnos* 1 (4): 317–25.
Phillips, J. 1835. *Illustrations of the geology of Yorkshire*. 2nd ed. Tulu Wilson.
———. 1834. *A guide to geology*. London: Longman.
———. 1853. *The rivers, mountains, and sea-coast of Yorkshire; with essays on the climate, scenery and ancient inhabitants of the county*. London: Murray.

Pite, R. 2002. *Hardy's geography: Wessex and the regional novel.* Basingstoke: Palgrave.
Pittock, M. 1995. *The myth of the Jacobite clans: The Jacobite army in 1745.* Edinburgh: Edinburgh University Press.
Playfair, J. 1802. *Illustrations of the Huttonian theory of the earth.* Edinburgh: Cadell and Davies.
———. 1822. Biographical account of the late James Hutton, M.D. In *Works of John Playfair*, edited by J. G. Playfair, 4:33–120. Edinburgh: Constable.
[Playfair, J.] 1804. Donna Agnesi's *Analytical institutions. Edinburgh review* 3:401–10.
———. 1806. Mascheroni, *Geometrie du compas. Edinburgh review* 9:161–68.
———. 1807. Mesure d'un arc du meridien. *Edinburgh review* 9:373–91.
———. 1808. Buee, Sur les quantites imaginaires. *Edinburgh review* 12:306–18.
———. 1810. Laplace's *System of the world. Edinburgh review* 15:396–417.
———. 1810. Woodhouse's *Trigonometry. Edinburgh review* 17:122–35.
———. 1811. Cuvier on fossil bones. *Edinburgh review* 18:214–30.
———. 1811. Werner on the formation of veins. *Edinburgh review* 18:80–97.
———. 1812. Geographie mineralogique des environs de Paris. *Edinburgh review* 20:369–86.
———. 1812. Leslie's *Elements of geometry. Edinburgh review* 20:79–100.
———. 1814. Essay philosophique sur les probabilités. *Edinburgh review* 23:320–40.
———. 1816. Dealtry's *Principles of fluxions. Edinburgh review* 27:87–98.
Playfair, J., and F. Jeffrey. 1822. Biographical memoir. In *Works of John Playfair*, edited by J. G. Playfair, 1:ix–cv. Edinburgh: Constable.
Pointon, M. 1997. Geology and landscape painting in nineteenth-century England. In *Images of the earth: Essays in the history of the environmental sciences*, edited by L. Jordanova and R. Porter, 93–123. Oxford: Alden Press.
Pomper, P., R. H. Elphick, et al., eds. 1998. *World history: Ideologies, structures, and identities.* Malden, MA: Blackwell.
Poovey, M. 1993. Reading history in literature: Speculation and virtue in *Our mutual friend*. In *Historical criticism and the challenge of theory*, edited by J. L. Smarr, 42–80. Urbana: University of Illinois Press.
Pope, A. 1714. *The rape of the lock: An heroi-comical poem in five cantos.* London: Lintott.
Pope-Hennessy, U. 1948. *Canon Charles Kingsley.* London: Chatto & Windus.
Porter, R. 1973. The industrial revolution and the rise of the science of geology. In *Changing perspectives in the history of science: Essays in honour of Joseph Needham*, edited by M. Teich and R. Young, 320–43. London: Heinemann.
———. 1976. Charles Lyell and the principles of the history of geology. *British journal for the history of science* 9:92–103.
———. 1977. *The making of geology: Earth science in Britain 1660–1815.* Cambridge: Cambridge University Press.

———. 1978. Gentlemen and geology: The making of a scientific career 1660–1920. *Historical journal* 21:809–36.

Postlethwaite, D. 2000. George Eliot and science. In *The Cambridge companion to the Victorian novel*, edited by D. David. Cambridge: Cambridge University Press.

Pugh, S. 1990. *Reading landscape: Country, city, capital*. Manchester: Manchester University Press.

Pycior, H. 1984. Internalism, externalism, and beyond: Nineteenth-century British algebra. *Historia mathematica* 11:424–41.

Radford, A. 2002. The Victorian dilettante and *A pair of blue eyes*. In *Thomas Hardy year book*, 5–19. Guernsey: Toucan Press.

———. 2003. *Thomas Hardy and the survivals of time*. Aldershot: Ashgate.

Rappaport, R. 1969. The geological atlas of Guettard, Lavoisier, and Monnet: Conflicting views of the nature of geology. In *Toward a history of geology*, edited by C. J. Schneer, 272–87. Cambridge, MA: MIT Press.

Rauch, A. 2001. *Useful knowledge: The Victorians, morality, and the march of intellect*. Durham, NC: Duke University Press.

Richardson, A. 2004. Hardy and science: A chapter of accidents. In *Palgrave advances in Thomas Hardy studies*, edited by P. Mallett, 156–80. Basingstoke: Palgrave.

Riehl, W. H. 1855. *Land und leute*. Stuttgart: Cotta.

———. 1856. *Die geburgeliche gesellschaft*. Stuttgart: Cotta.

Rimmer, M. 1993. Club laws: Chess and the construction of gender in *A pair of blue eyes*. In *The sense of sex: Feminist perspectives on Hardy*, edited by M. R. Higonnet. Urbana: University of Illinois Press.

———. 2004. Hardy, Victorian culture and provinciality. In *Palgrave advances in Thomas Hardy studies*, edited by P. Mallett. Basingstoke: Palgrave.

Roller, D. H. D., ed. 1971. *Perspectives in the history of science and technology*. Norman: University of Oklahoma Press.

Romano, J. 1978. *Dickens and reality*. New York: Columbia University Press.

Rose, E. P. F. 2009. Military men: Napoleonic warfare and early members of the Geological Society. In *The making of the Geological Society of London* (GSL Special Publications no. 317), edited by C. Lewis and S. Knell, 219–41. London: Geological Society of London.

Rosen, D. 1994. The volcano and the cathedral: Muscular Christianity and the origins of primal manliness. In *Muscular Christianity: Embodying the Victorian age*, edited by D. Hall, 17–45. Cambridge: Cambridge University Press.

Rosenborg, G. D., ed. 2008. *Emergence of modern geology and evolutionary thought from the scientific revolution to the Enlightenment*. Boulder, CO: Geological Society of America.

Rossetti, G. 1826–1827. *Comento analitico sulla Divina commedia*. London: Murray.

———. 1832. *Disquizioni sullo spirit antipapale*. London: Murray.

———. 1834. *Disquisitions on the antipapal spirit which produced the reformation; its secret influence on the literature of Europe in general, and of Italy in particular*. Translated by Caroline Ward. London: Smith, Elder & Co.

Rudwick, M. 1962. Hutton and Werner compared: George Greenough's geological tour of Scotland in 1805. *British journal for the history of science* 1:117–35.

———. 1963. The foundation of the Geological Society of London. *British journal for the history of science* 1:325–55.

———. 1970. The strategy of Lyell's *Principles of geology*. *Isis* 61:5–33.

———. 1971. Uniformity and progression: Reflections on the structure of geological theory in the age of Lyell. In *Perspectives in the history of science and technology*, edited by D. H. D. Roller, 209–27. Norman: University of Oklahoma Press. Repr. in Rudwick, *Lyell and Darwin* (2005).

———. Poulett Scrope on the volcanoes of Auvergne: Lyellian time and political economy. *British journal for the history of science* 7:205–42. Repr. in Rudwick, *Lyell and Darwin* (2005).

———. 1975. Caricature as a source for the history of science: De la Beche's anti-Lyellian sketches of 1831. *Isis* 66:534–60. Repr. in Rudwick, *Lyell and Darwin* (2005).

———. Charles Lyell F.R.S. (1797–1875) and his London lectures on geology, 1832–33. *Notes and records of the Royal Society of London* 29:231–63.

———. 1976. Charles Lyell speaks in the lecture theatre. *British journal for the history of science* 9:147–55.

———. 1976. The emergence of a visual language for geological science, 1760–1840. *History of science* 14:149–95. Repr. in Rudwick, *New science of geology* (2004).

———. 1976. *Meaning of fossils: Episodes in the history of palaeontology*. New York: Science History Publications.

———. 1977. Historical analogies in the early geological work of Charles Lyell. *Janus* 64:89–107. Repr. in Rudwick, *Lyell and Darwin* (2005).

———. 1978. Charles Lyell's dream of a statistical palaeontology. *Palaeontology* 21:225–44. Repr. in Rudwick, *Lyell and Darwin* (2005).

———. 1985. *The great Devonian controversy: The shaping of scientific knowledge among gentlemanly specialists*. Chicago: University of Chicago Press.

———. 1985. *The meaning of fossils: Episodes in the history of palaeontology*. Chicago: University of Chicago Press.

———. 1988. A year in the life of Adam Sedgwick and company, geologists. *Archives of natural history* 15:243–68. Repr. in Rudwick, *New science of geology* (2004).

———. 1992. *Scenes from deep time: Early pictorial representations of the prehistoric world*. Chicago: University of Chicago Press.

———. 1997. *Georges Cuvier, fossil bones, and geological catastrophes*. Chicago: University of Chicago Press.

———. 1997. Transposed concepts from the human sciences in the early work of Charles Lyell. In *Images of the earth: Essays in the history of the environmental sciences*, edited by L. Jordanova and R. Porter, 77–91. Oxford: Alden Press.

———. 1998. Lyell and the *Principles of geology*. London: Geological Society of London.

———. 2004. Encounters with Adam, or at least the hyaenas: Nineteenth-century visual representations of the deep past. In *The new science of geology: Studies in the earth sciences in the age of revolution*, 231–51. Aldershot: Ashgate.

———. 2004. *The new science of geology: Studies in the earth sciences in the age of revolution*. Aldershot: Ashgate.

———. 2005. *Bursting the limits of time: The reconstruction of geohistory in the age of revolution*. Chicago: University of Chicago Press.

———. 2005. *Lyell and Darwin: Geologists' studies in the earth sciences in the age of reform*. Aldershot: Ashgate.

———. 2007. *Worlds before Adam: The reconstruction of geohistory in the age of reform*. Chicago: University of Chicago Press.

———. 2009. The early Geological Society in its international context. In *The making of the Geological Society of London* (GSL Special Publications no. 317), edited by C. L. E. Lewis and S. J. Knell, 145–53. London: Geological Society of London.

———. 2009. Walk with the founding fathers. In *The making of the Geological Society of London* (GSL Special Publications no. 317), edited by C. L. E. Lewis and S. J. Knell, 417–38. London: Geological Society of London.

Rupke, N. A. 1983. *The great chain of history: William Buckland and the English school of geology, 1814–1849*. Oxford: Clarendon Press.

———. 1997. Oxford's scientific awakening and the role of geology. In *Nineteenth-century Oxford, part one*, edited by M. G. Brock and M. C. Curthoys, 543–62. Oxford: Clarendon Press.

———. 1998. "The end of history" in the early picturing of geological time. *History of science* 36:61–90.

Ruse, Michael. 2005. The Darwinian revolution, as seen in 1979 and as seen twenty-five years later in 2004. *Journal of the history of biology* 38:3–17.

Rutland, W. R. 1962. *Thomas Hardy: A study of his writings and their background*. New York: Russell & Russell.

Sanders, A. 2003. *Charles Dickens*. Oxford: Oxford University Press.

Sawyer, P. L. 1981. Ruskin and Tyndall: The poetry of matter and the poetry of spirit. *Annals of the New York Academy of Sciences* 360:217–46.

Schaffer, S. 1991. The history and geography of the intellectual world. In *William Whewell: A composite portrait*, edited by S. Schaffer and M. Fisch, 201–31. Oxford: Oxford University Press.

Schivelbush, W. 1980. *The railway journey: Trains and travel in the nineteenth century*. Oxford: Blackwell.

Schlicke, P. 1985. *Dickens and popular entertainment*. London: Allen & Unwin.

Schneer, C. J., ed. 1969. *Toward a history of geology*. Cambridge, MA: MIT Press.

Scholes, R., J. Phelan, et al. 2006. *The nature of narrative*. Oxford: Oxford University Press.

Schor, H. M. 2006. Dickens and plot. In *Charles Dickens studies*, edited by J. Bowen and R. L. Patten, 90–110. Basingstoke: Palgrave.

Schweik, R. 1999. The influence of religion, science, and philosophy on Hardy's writings. In *The Cambridge companion to Thomas Hardy*, edited by D. Kramer, 54–72. Cambridge: Cambridge University Press.

Scott, W. 1802. *Minstrelsy of the Scottish border, consisting of historical and romantic ballads, collected in the southern counties of Scotland; with a few of modern date, founded upon local tradition.* Kelso: Ballantyne.

———. 1808. *Marmion: A tale of Flodden Field.* Edinburgh: Constable.

———. 1824. *St. Ronan's well.* Edinburgh: Constable.

———. 1932–1937. *The letters of Sir Walter Scott.* London: Constable.

———. 1986. *Waverley; or, 'Tis sixty years since.* Oxford: Oxford University Press.

———. 1998. *Rob Roy.* Oxford: Oxford University Press.

———. 1999. *Guy Mannering.* London: Penguin.

———. 2002. *The antiquary.* Oxford: Oxford University Press.

[Scrope, G. P.] 1835. Lyell's *Principles of geology. Quarterly review* 53:406–48.

———. 1858. Wiltshire. *Quarterly review* 103:108–38.

Secord, A. 1996. Artisan botany. In *Cultures of natural history*, edited by N. Jardine, J. A. Secord, and E. C. Spary, 378–93. Cambridge: Cambridge University Press.

———. 2005. Elizabeth Gaskell and the artisan naturalists of Manchester. *Gaskell Society journal* 19:34–51.

Secord, J. A. 1982. King of Siluria: Roderick Murchison and the imperial theme in geology. *Victorian studies* 25:413–42.

———. 1986. *Controversy in Victorian geology: The Cambrian-Silurian dispute.* Princeton, NJ: Princeton University Press.

———. 1986. The Geological Survey of Great Britain as a research school, 1839–1855. *History of science* 24:223–75.

———. 1997. Introduction. In *Principles of geology*, edited by J. A. Secord, ix–xliii. London: Penguin.

———. 2000. *Victorian sensation: The extraordinary publication, reception, and secret authorship of "Vestiges of the natural history of creation."* Chicago: University of Chicago Press.

———. 2004. Monsters at the Crystal Palace. In *Models: The third dimension of science*, edited by S. de Chadarevian and N. Hopwood, 138–69. Stanford, CA: Stanford University Press.

———. 2007. How scientific conversation became shop talk. In *Science in the marketplace: Nineteenth-century sites and experiences*, edited by A. Fyfe and B. Lightman, 23–59. Chicago: University of Chicago Press.

Seddon, G. 1973. Abraham Gottlob Werner: History and folk-history. *Australian journal of earth sciences* 20:381–95.

Sedgwick, A. 1829. On the structure and relations of the deposits contained between the primary rocks and the oolitic series in the north of Scotland. *Transactions of the Geological Society of London* 3:125–60.

———. 1834a. Anniversary address of the President, the Geological Society of London, 19 February 1830. *Proceedings of the Geological Society of London* 1:187–212.

———. 1834b. Anniversary address of the President, the Geological Society of London, 18 February 1831. *Proceedings of the Geological Society of London* 1:281–316.

———. 1836. Description of a series of longitudinal and transverse sections through a portion of the Carboniferous chain between Penigent and Kirkby Stephen. *Proceedings of the Geological Society* 1:318–20.

———. 1842. Three letters upon the geology of the Lake District. In W. Wordsworth, *A complete guide to the Lakes, comprising minute directions for the tourist, with Mr. Wordsworth's descriptions of the scenery of the country, &c. and three letters upon the geology of the Lake District by the Rev. Professor Sedgwick.* Kendal and London: Hudson and Nicholson; Longman; Whitaker.

———. 1850. *Discourse on the studies of the University of Cambridge.* London: Parker.

———. 1868. *Memorial by the trustees of Cowgill Chapel; with a preface and appendix, on the climate, history and dialects of Dent.* Cambridge: Cambridge University Press.

[Sedgwick, A.] 1845. Vestiges of the natural history of Creation. *Edinburgh review* 82:1–85.

Sedgwick, A., and R. Murchison. 1829. On the geological relations of the secondary strata in the Isle of Arran. *Transactions of the Geological Society of London* 3:21–36.

Shakespeare, W. 2006. *Hamlet.* London: Arden Shakespeare.

Shapin, S., and B. Barnes. 1977. Science, nature and control: Interpreting Mechanics' Institutes. *Social studies of science* 7:31–74.

Sharp, W. 1904. *Literary geography.* London: Offices of the Pall Mall Publications.

Shatto, S. 1988. *The companion to Bleak House.* London: Unwin Hyman.

Shattock, J. 1989. *Politics and reviewers: The Edinburgh and the Quarterly in the early Victorian age.* Leicester: Leicester University Press.

Sheridan, S., ed. 1988. *Grafts: Feminist cultural criticism.* London: Verso.

Shires, L. M., ed. 1992. *Rewriting the Victorians.* New York: Routledge.

Shortland, M. 1994. Darkness visible: Underground culture in the golden age of geology. *History of science* 32:1–61.

———. 1996. Bonneted mechanic and narrative hero: The self-modelling of Hugh Miller. In *Hugh Miller and the controversies of Victorian science*, edited by M. Shortland, 14–75. Oxford: Clarendon Press.

———, ed. 1996. *Hugh Miller and the controversies of Victorian science.* Oxford: Clarendon Press.

Shuttleworth, S. 1984. *George Eliot and nineteenth-century science: The make-believe of a beginning.* Cambridge: Cambridge University Press.

———. 1985. "The language of science and psychology in George Eliot's *Daniel Deronda*." Victorian science and Victorian values: Literary perspectives. In *Victorian science and Victorian values: Literary perspectives*, edited by J. Paradis and T. Postlewait, 269–98. New Brunswick, NJ: Rutgers University Press.

Simmons, J. R. 2002. Industrial and "condition of England" novels. In *A companion to the Victorian novel*, edited by P. Brantlinger and W. B. Thesing, 336–52. Oxford: Blackwell.

Small, H. 1996. *Love's madness: Medicine, the novel, and female insanity, 1800–1865*. Oxford: Clarendon Press.

Smith, C. 1985. Geologists and mathematicians: The rise of physical geology. In *Wranglers and physicists: Studies on Cambridge physics in the nineteenth century*, edited by P. M. Harman, 49–83. Manchester: Manchester University Press.

Smith, C. R. 1844. Caves in which Romano-British remains have been discovered, near Settle in Yorkshire. *Collecteana antiqua* 2:69–72.

Smith, C. R., and J. Jackson. 1842. Roman remains discovered in the caves near Settle in Yorkshire. *Archaeologia* 29:384–85.

Smith, G. 2003. *Dickens and the dream of cinema*. Manchester: Manchester University Press.

Smith, J. 1994. *Fact and feeling: Baconian science and the nineteenth-century literary imagination*. Madison: University of Wisconsin Press.

———. 2006. *Charles Darwin and Victorian visual culture*. Cambridge: Cambridge University Press.

Smith, J. P. 1839. *On the relation between the holy Scriptures and some parts of geological science*. London: Jackson & Walford.

Smocovitis, V. 2005. "It ain't over 'til it's over": Rethinking the Darwinian revolution. *Journal of the history of biology* 38:33–49.

Snell, K. D. M. 1998. *The regional novel in Britain and Ireland 1800–1990*. Cambridge: Cambridge University Press.

Snow, C. P. 1959. *The two cultures and the scientific revolution*. Cambridge: Cambridge University Press.

Soloman, E. 1969. The incest theme in *Wuthering Heights*. *Nineteenth-century fiction* 14:80–83.

Sommer, M. 2003. The romantic cave? The scientific and poetic quests for subterranean spaces in Britain. *Earth sciences history* 22:172–208.

———. 2004. "An amusing account of a cave in Wales": William Buckland (1784–1856) and the Red Lady of Paviland. *British journal for the history of science* 37:53–74.

———. 2007. *Bones and ochre: The curious afterlife of the Red Lady of Paviland*. Cambridge, MA: Harvard University Press.

Spencer, H. 1863. *Essays: Scientific, political, speculative*. London: Williams and Norgate.

———. 1863. Illogical geology. In *Essays: Scientific, political, speculative*, 2:57–104. London: Williams and Norgate.

Stafford, B. M., and F. Terpak. 2001. *Devices of wonder: From the world in a box to images on the screen*. Los Angeles: Getty.

Stafford, R. A. 1989. *Scientist of empire: Sir Roderick Murchison, scientific exploration and Victorian imperialism*. Cambridge: Cambridge University Press.

Steig, M. 1978. *Dickens and Phiz*. Bloomington: Indiana University Press.
Stephen, L. 1872. A bad five minutes in the Alps. *Fraser's magazine* 6:545–61.
Stevenson, J. A. 1988. "Heathcliff is me!" *Wuthering Heights* and the question of likeness. *Nineteenth-century literature* 43:60–81.
Stitt, M. P. 1998. *Metaphors of change in the language of nineteenth-century fiction: Gaskell, Scott, Kingsley*. Oxford: Clarendon Press.
Stone, H., ed. 1987. *Dickens's working notes for his novels*. Chicago: University of Chicago Press.
Stonehouse, J. H., ed. 1935. *Reprints of the catalogues of the libraries of Charles Dickens and W. M. Thackeray, etc.* London: Piccadilly Fountain Press.
Stuchtey, B., and E. Fuchs, eds. 2003. *Writing world history 1800–2000*. Oxford: Oxford University Press.
Sussman, H. L. 1995. *Victorian masculinities: Manhood and masculine poetics in early Victorian literature and art*. Cambridge: Cambridge University Press.
Sutherland, J. 2006. Visualizing Dickens. In *Charles Dickens studies*, edited by J. Bowen and R. L. Patten, 111–30. Basingstoke: Palgrave.
Swan, J. 1873. *The valuable library of scientific & miscellaneous books of the late Rev. Adam Sedgwick, L.L.D., F.R.S. . . . to be sold by auction*. Cambridge: John Swan.
Sweet, R. 2004. *Antiquaries: The discovery of the past in eighteenth-century Britain*. London: Hambledon and London.
Taylor, I. 1865. *Words and places; or, Etymological illustrations of history, ethnology, and geography*. London: Macmillan.
Taylor, R. H. 1982. *The neglected Hardy: Thomas Hardy's lesser novels*. London: Macmillan.
Teich, M., and R. Young, eds. 1973. *Changing perspectives in the history of science: Essays in honour of Joseph Needham*. London: Heinemann.
Temple, F., et al. 1860. *Essays and reviews*. London: Parker.
Tennyson, A. L. 1850. *In memoriam*. London: Moxon.
Thackeray, W. M. 1999. *Catherine: A story*. Michigan: University of Michigan Press.
[Thackeray, W. M.] 1839. Horae catnachianae. *Fraser's magazine* 19:408–9.
———. 1840. Going to see a man hanged. *Fraser's magazine* 22:154–45.
———. 1840. Review of *Jack Sheppard*. *Fraser's magazine* 21:227–45.
Thackray, J. 1998. Charles Lyell and the Geological Society. In *Lyell: The past is the key to the present*, edited by D. J. Blundell and A. C. Scott, 17–20. London: Geological Society of London.
Thiele, D. 2007. "That there Brutus": Elite culture and knowledge diffusion in the industrial novels of Elizabeth Gaskell. *Victorian literature and culture* 35:263–85.
Thompson, R. W. 1910. *Sedbergh, Garsdale and Dent: Peeps at the past history and present condition of some picturesque Yorkshire dales*. Leeds: Richard Jackson.
Thomson, P. 1977. *George Sand and the Victorians: Her influence and reputation in nineteenth-century England*. London: Macmillan.
Tiddeman, R. H. 1872. Discovery of extinct mammals in the Victoria Cave, Settle. *Nature* 7:127–28.

———. 1873. The older deposits in the Victoria Cave, Settle, Yorkshire. *Geological magazine* 10:11–16.

———. 1875. The works and problems of the Victoria Cave exploration. *Proceedings of the Geological and Polytechnic Society of the West Riding of Yorkshire* 6:77–92.

Tillotson, K. 1956. *Novels of the eighteen-forties*. Oxford: Clarendon Press.

Todhunter, I., ed. 1876. *William Whewell, D.D., Master of Trinity College, Cambridge: An account of his writings with selections from his literary and scientific correspondence*. London: Macmillan.

Tomalin, C. 2006. *Thomas Hardy: The time-torn man*. London: Penguin.

Topham, J. R. 1998. Beyond the "common context": The production and reading of the Bridgewater Treatises. *Isis* 89:233–62.

Torrens, H. 1997. Geological communication in the Bath area in the last half of the eighteenth century. In *Images of the earth: Essays in the history of the environmental sciences*, edited by L. Jordanova and R. Porter, 217–46. Oxford: Alden Press.

———. 2001. Timeless order: William Smith (1769–1839) and the search for raw materials 1800–1820. London: Geological Society of London.

———. 2002. *The practice of British geology, 1750–1850*. Aldershot: Ashgate.

———. 2003. *Memoirs of William Smith, LL.D., author of the "Map of the strata of England and Wales" by his nephew and pupil, John Phillips, F.R.S., F.G.S. first published in 1844 with an Introduction to the Life and Times of William Smith and the William Smith Lecture 2000 by Hugh Torrens*. Bath: Bath Royal Literary and Scientific Institution.

———. 2012. Politics and paleontology: Richard Owen and the invention of dinosaurs. In *The Complete Dinosaur*, 2nd ed., edited by M. K. Brett-Surman, T. R. Holtz Jr., ad J. O. Farlow, 24–53. Bloomington: Indiana University Press.

Tosh, J. 1999. *A man's place: Masculinity and the middle-class home in Victorian England*. New Haven, CT: Yale University Press.

Toulmin, S., and J. Goodfield. 1965. *The discovery of time*. London: Hutchinson.

Trumpener, K. 1993. National character, nationalist plots: National tale and historical novel in the age of Waverley, 1806–1830. *English literary history* 60:685–731.

Tucker, H. 2008. *Epic: Britain's heroic muse 1790–1910*. Oxford: Oxford University Press.

Turner, H. S. 2006. *The English Renaissance stage: Geometry, poetics, and the practical spatial arts, 1580–1630*. Oxford: Oxford University Press.

———. 2010. Lessons from literature for the historian of science (and vice versa). *Isis* 101:578–89.

Turner, M. A. 1993. *Mechanism and the novel: Science in the narrative process*. Cambridge: Cambridge University Press.

Tweedale, G. 1991. Geology and industrial consultancy: Sir William Boyd Dawkins, 1837–1929, and the Kent coalfield. *British journal for the history of science* 24:435–51.

Vaccari, E. 1998. Lyell's reception on the continent of Europe: A contribution to an open historiographical problem. In *Lyell: The past is the key to the present*, edited by D. J. Blundell and A. C. Scott, 3–15. London: Geological Society of London.

Van Ghent, D. 1961. *The English novel: Form and function.* New York: Harper & Row.

Vance, N. 1985. *The sinews of the spirit: The ideal of Christian manliness in Victorian literature and religious thought.* Cambridge: Cambridge University Press.

———. 2000. Niebuhr in England: History, faith and order. In *British and German historiography 1750–1950: Traditions, perceptions and transfers,* edited by B. Stuchtey and P. Wende, 83–98. Oxford: Oxford University Press.

Veneer, L. 2009. Practical geology in the Geological Society in its early years. In *The making of the Geological Society of London* (GSL Special Publications no. 317), edited by C. L. E. Lewis and S. J. Knell, 243–53. London: Geological Society of London.

Vine, S. 1994. The wuther of the other in Wuthering Heights. *Nineteenth-century literature* 49:339–59.

Vlock, D. 1998. *Dickens, novel reading, and the Victorian popular theatre.* Cambridge: Cambridge University Press.

Von Sneidern, M.-L. 1995. Wuthering Heights and the Liverpool slave trade. *English literary history* 62:171–96.

Watson, N. J. 2002. Introduction. In *The antiquary,* vii–xxvii. Oxford: Oxford University Press.

Watt, I. 1957. *Rise of the novel: Studies in Defoe, Richardson, and Fielding.* London: Chatto & Windus.

[Watts, H. E.] 1859. Yorkshire. *Westminster and foreign quarterly review* 71 : 327–57.

Wawn, A. 1982. Gunnlaugs saga ormstungu and the Theatre Royal Edinburgh 1812. *Scandinavica* 21:139–51.

———. 2000. *The Vikings and the Victorians: Inventing the old North in nineteenth-century Britain.* Cambridge: Brewer.

Webster, T. 1814. On the fresh-water formations in the Isle of Wight, with some observations on the strata over the chalk in the south-east part of England. *Transactions of the Geological Society of London* 2:161–254.

Weindling, P. J. 1997. Geological controversy and its historiography: The prehistory of the Geological Society of London. In *Images of the earth: Essays in the history of the environmental sciences,* edited by L. Jordanova and R. Porter, 248–71. Oxford: Alden Press.

Weinstein, M. A., ed. 1978. *The prefaces to the Waverley novels.* Lincoln: University of Nebraska Press.

Wheeler, M., ed. 1996. *Time and tide: Ruskin and science.* London: Pilkington.

Whewell, W. 1837. *History of the inductive sciences, from the earliest to the present times.* London: Parker.

———. 1850. *Of a liberal education in general: and with particular reference to the leading studies of the University.* London: Parker

[Whewell, W.] 1831. Principles of geology by Charles Lyell. *British critic* 9:180–206.

[White, J.] 1856. Two college friends. *Household words* 8:78–84, 105–16.

Whone, C. 1950. Where the Brontes borrowed books. *Bronte Society transactions* 11:344–58.

Wilcox, S. B. 1988. Introduction: Unlimiting the bounds of painting. In *Panoramania! The art and entertainment of the "all-embracing" view*, edited by R. Hyde, 13–44. London: Trefoil.

Wilde, O. 1878. *Ravenna, recited in the theatre, Oxford, June 26, 1878*. Oxford: Shrimpton & Son.

Wilkinson, A. Y. 1967. *Bleak House*: From Faraday to judgement day. *English literary history* 34:225–47.

Williams, A. 1985. Natural supernaturalism in "Wuthering heights." *Studies in philology* 82:104–27.

Williams, I. 1974. *The realist novel in England: A study in development*. London: Macmillan.

Williams, M. 1972. *Thomas Hardy and rural England*. London: Macmillan.

Williams, R. 1959. *Culture and society 1780–1950*. London: Chatto & Windus.

———. 1973. *The country and the city*. London: Chatto & Windus.

———. 1990. *Notes on the underground: An essay on technology, science, and the imagination*. Cambridge, MA: MIT Press.

[Wills, W. H., and G. A. Sala.] 1853. Fairyland in fifty four. *Household words* 7:313–17.

Wilson, L. G. 1972. *Charles Lyell, the years to 1841: The revolution in geology*. New Haven, CT: Yale University Press.

———. 1980. Geology on the eve of Charles Lyell's first visit to America, 1841. *Proceedings of the American Philosophical Society* 124:168–202.

———. 1998. *Lyell in America: Transatlantic geology, 1841–1853*. Baltimore: Johns Hopkins University Press.

———. 1998. Lyell: The man and his times. In *Lyell: The past is the key to the present*, edited by D. J. Blundell and A. C. Scott, 21–37. London: Geological Society of London.

Wilson, P. 1996. "Over yonder are the Andes": Reading Ruskin reading Humboldt. In *Time and tide: Ruskin and science*, edited by M. Wheeler, 65–84. London: Pilkington.

Winyard, B., and H. Furneaux. 2010. Introduction: Dickens, science and the Victorian literary imagination. *19: Interdisciplinary studies in the long nineteenth century* 10. www.19.bbk.ac.uk.

Wood, J. 2001. *Passion and pathology in Victorian fiction*. Oxford: Oxford University Press.

Wordsworth, W. 1842. *A complete guide to the Lakes, comprising minute directions for the tourist, with Mr. Wordsworth's descriptions of the scenery of the country, &c. and three letters upon the geology of the Lake District by the Rev. Professor Sedgwick*. 6th ed. Kendal and London: Hudson and Nicholson; Longman; Whitaker.

[Wreford, H. G.] 1855. Vesuvius in eruption. *Household words* 11:435–49.

———. 1858. Earthquake experiences. *Household words* 12:553–58.
Wyatt, J. 1995. *Wordsworth and the geologists*. Cambridge: Cambridge University Press.
Yaeger, P. 1988. *Honey-mad women: Emancipatory strategies in women's writing*. New York: Columbia University Press.
Yapp, G. W. 1851. *Official catalogue of the great exhibition of the works of industry of all nations*. London: Spicer Brothers.
Yeo, R. 2003. *Defining science: William Whewell, natural knowledge, and public debate in early Victorian Britain*. Cambridge: Cambridge University Press.
Young, D. A. 2003. *Mind over magma: The story of igneous petrology*. Princeton, NJ: Princeton University Press.
Zimmerman, V. 2008. *Excavating Victorians*. Albany: State University of New York Press.
Zivley, S. L. 1997. Satan in orbit: *Paradise lost*: IX:48–86. *Milton quarterly* 31:130–36.

INDEX

Adam Bede. See under Eliot, George
Agassiz, Louis, 108, 173, 174
algebra: continental, 36, 41, 43–44; imaginary symbols, 43; negative numbers, 43–44; pure language as, 63. *See also* method, scientific
All the Year Round. See under Dickens, Charles
Alton Locke. See under Kingsley, Charles
ammonites, 14. *See also* fossils
analysis. *See under* method, scientific
antiquarianism: *Daniel Deronda* and, 244; Dryasdust, 51, 244; Charles Lapworth and, 90; John MacCulloch and, 79; plot and, 50–54; *Principles of Geology* on, 112–13; Adam Sedgwick and, 93; *Wooers and Winners* on, 169; *Wuthering Heights* and, 142
Antiquary, The. See Scott, Walter
Arabian Nights, 49–50, 95–96, 105, 225
archaeology, 139, 167, 169–72
Arran, Isle of, 36, 45, 77–88, 366

BAAS. *See* British Association for the Advancement of Science (BAAS)
Babbage, Charles, 110, 258
Bacon, Francis: empirical methods of, 182–83, 185, 187, 190, 191–92, 196, 199; literary forms of science and, 43, 46, 48; thought and scientific reasoning, 115
Bakewell, Robert, 97

Banks, Isabella, *Wooers and Winners*, 13, 89, 141, 167–69, 172–76; discovery of Victoria Cave, 169, 172–73; geological language of, 172–73, 174; geological mapping in, 173–74, 175; metonym in, 173; realism of, 140–41, 173–76; Sedgwick as character in, 168–69; stratigraphic setting of, 168; *Wuthering Heights* and, 168–69, 173
basalt, 1, 10, 35, 97
Beaumont, Léonce Élie de, 157
Beche, Henry de la, 116, 153, 157, 208–10, 274, 284n23, 365, 368
Blackwood's Edinburgh Magazine, 40, 41
Bridgewater Treatises, 100–102, 107, 145
British Association for the Advancement of Science (BAAS), 59, 88, 108, 135, 160, 169
Brontë, Charlotte, 139–40, 158
Brontë, Emily, 139–40; metonymic writing of, 148–50, 167; Sills of Dent and, 143–46; *Wuthering Heights*, 19, 135, 140–52, 167–68, 275; *Wuthering Heights*, decline of yeomen in, 146–48; *Wuthering Heights*, erosion in, 140–41; *Wuthering Heights*, regional and topographical ambiguity of, 141–42, 146, 148–49, 150–51, 152, 167, 172, 174; Yorkshire and, 139–40, 145, 146, 148–49, 151, 158, 160, 163
Buch, Leopold von, 97
Buckland, Francis Trevelyan, 266

Buckland, William: Bridgewater Treatise, 2, 100–102, 107, 235–37; cave geology of, 99–100, 115, 169, 170; coprolitic geology of, 99–100; diluvialism of, 115–16; gentleman geologist, 14–15, 99; cosmology and, 100–102, 105, 107, 115–16; epic writing of, 100–102, 125, 114, 117, 125, 274; geological retrospects of, 204; Geological Society and, 105, 113, 115, 116, 117; *The Mill on the Floss* and, 235–37; professor of mineralogy and professor geology at Oxford, 97–102; synchronic approach to writing earth history of, 101, 118–19; *Wooers and Winners*, depiction of, 173, 174

Bulwer-Lytton, Edward, 256

Burnet, Thomas, 122–23. *See also* cosmogony; cosmology

Butler, Samuel, 123

Byron, Lord George Gordon, 19, 95, 96, 127–30; *Childe Harold's Pilgrimage*, 95, 127–28

cabinets, 4, 13, 92, 93, 200

catastrophism: Charles Dickens and, 263–64; George Eliot and, 221–24, 232–33, 245–46; misunderstood term, 25–26, 107–10, 113–14, 115–16, 222–24, 245; "romantic" and, 24–25, 221–22; scientific method and, 112–13; William Whewell, coiner of, 222

Childe Harold's Pilgrimage. *See* Byron, Lord George Gordon

classification. *See under* practice, geological

clubs, 14, 26, 49, 57, 66, 67, 69, 275; Athenaeum, 64; drinking at, 31–32, 40–41, 47, 54; Friday Club, 40–41; Geological Society, 31–34, 60, 65, 73, 88; quarterly reviews and, 67, 72; sociable narration and, 53–54; Wernerian Natural History Society, 39, 41, 97; Yorkshire Antiquarian Club, 160; Yorkshire Naturalists' Club, 160

collecting. *See under* practice, geological

columns, geological, 2, 3, 4–5, 7, 131–32; as form for disjointed or fragmented narrative, 26–27, 137–38, 264, 275; layered vision of history related to, 137–38, 163, 166, 167, 170–72, 175, 211, 219; mapping and, 133–34, 138, 161–62, 175; nomenclature and, 209–10; stories of the earth and, 101, 106, 124, 226–28; theoretical nature of, 105, 161–62, 226–28; *Two Years Ago* and, 208–20

Conybeare, William Daniel, 99, 105, 115–16, 159

Cooper, Thomas, 195–97

correlation. *See under* method, scientific

cosmogony: mocked in fiction, 47–48, 54; rejected in geology, 4, 7, 19, 43, 102, 106, 223, 225, 227; storytelling and, 15, 19, 43, 68, 106, 223; *Two Years Ago* and, 196–97, 215–16; Wernerian and Huttonian, 41; *Vestiges of the Natural History of Creation* as, 61, 68. *See also* cosmology

cosmology: William Buckland and tentative, 101–2; Thomas Burnet and, 122–23; evolutionary, 64–69; excluded from geology, 4, 5, 7, 18, 19, 43, 125, 224–27; Huttonian, 34–39, 42, 43, 55, 56, 57; plot and, 19–21, 43, 51, 61–69, 88, 89, 128–30; *Principles of Geology* on, 103–6, 122–24, 125–30; Walter Scott and, 54, 55; *Vestiges of the Natural History of Creation* as, 64–69. *See also* cosmogony

Creation, 68, 88, 128, 131, 183, 225;
biblical account of, 25, 68, 102, 235;
of new species, 100, 223, 226; of the
world, 13, 47–48, 68
Crystal Palace, 255; Great Exhibition
at, 257–58, 261, 262, 265, 266;
at Sydenham, 251, 253, 254, 262,
264–65; pictures of, 256
Cuvier, George, 4
cyclorama, 256, 261

Daniel Deronda. See under Eliot,
George
Dante, 15, 96, 100, 101, 103, 117–20,
123–24, 125–26. *See also* epic
Darwin, Charles: admiration of Scott,
56; Charles Dickens and, 270;
Charles Kingsley and, 180; Charles
Lyell and, 24; narrative and, 68;
Origin of Species, 24, 68, 180–81,
195, 242, 248; *Vestiges of the Natural
History of Creation* and, 68. *See also*
evolution
Dawkins, William Boyd, 170–72
deconstruction, 22, 23, 267–68
deduction. *See under* method, scientific
De Foe, Daniel, 70, 139
Devon, 12, 27, 166, 167, 372; Devonian
controversy, 208–10; *Two Years Ago*
and, 182, 208–20
Dickens, Charles, 14, 22, 24, 65, 69,
247–73; *All the Year Round*, 247,
268; *Bleak House*, 12, 27, 89, 250,
262–68, 269, 274; commercial
science of, 248–49; cycloramas,
dioramas, and panoramas, 248–49,
254–57, 259–60, 261, 263–65,
265–66; *Dombey and Son*, 27, 250,
259–62, 267–68, 269; evolutionary
plots of, 24, 247; Great Exhibition
and, 258; *Hard Times*, 258;
Household Words, 13, 247, 256–57,
265; *Little Dorrit*, 248; Hugh Miller
and, 257 and; Newgate novels and,
65; *Our Mutual Friend*, 27, 248,
250, 260, 268–72; Richard Owen
and, 250–54, 266, 271–73; *Pickwick
Papers*, 250; *Pictures from Italy*, 256,
257, 259, 261; plot in late novels of,
250–51; review of Robert Hunt,
Poetry of Science, 254, 269, 271; "uniformitarianism" and "catastrophism"
of, 24–25, 263–65, 322
diluvialists, 115–16
dinosaurs. *See* primeval lizards
diorama, 256–57, 260, 261, 265–66, 268;
distinct from plot of earth history,
264; double effect, 248–49; Great
Exhibition and, 258
*Discourse on the Studies of the University
of Cambridge. See under* Sedgwick,
Adam
Divine Comedy. See Dante; Rossetti,
Gabriele
Dombey and Son. See under Dickens,
Charles

Edinburgh Review: competition of, 65;
criticism of novels in, 48–49, 69;
William Fitton and, 57–58; gentle-
manly geology and, 56–57; J. F. W.
Johnston and, 135–38, 140; John
Playfair and, 36, 40, 41, 42, 43–45,
46, 62, 66; Adam Sedgwick and, 60,
62, 65, 67–69
Egyptian Hall, 254
Eifel region, 179, 209, 216–20
Elements of Geology. See under Lyell,
Charles
Eliot, George, 22; *Adam Bede*, 221,
228; anti-cosmogony and, 224–28;
William Buckland and, 235–36;
"catastrophism" of, 221–24, 232–34,
238; *Daniel Deronda*, 27, 229–30,

Eliot, George (*cont.*)
238–46; evolutionary plots of, 24–25; gaps, blanks, breaks, strata, in works of, 224–28, 231–32, 234–35, 239–40, 240–46; *The Mill on the Floss*, 27, 221, 228–38; "Natural history of German life," 229–30, 231–32; on observation vs. theory, 229–30; plot, seen as coercive by, 228–30, 233–34, 235–37, 238–39, 246; realism of, 221–22, 229, 231, 237, 239, 244–45, 246; *Scenes of Clerical Life*, 221; scientific narration in works of, 221–22, 229, 230, 232; *Silas Marner*, 221, 228; "Silly novels by lady novelists," 229; "uniformitarianism" of, 24–25, 221–24, 232–33, 235; *Westminster Review*, acting editor of, 224–27
enumerative induction. *See under* method, scientific
epic: evolutionary, 88–89, 100; geology and, 14–15, 88–89, 100, 111–12, 117, 123–24, 274; Charles Lapworth and, 88–89; Lord Byron and, 19, 127–30; Lyell, Charles, and, 102–3, 112–13, 17, 126–27; mock-epic, 121–25, 127–30; Niebuhr and, 119–21, 126–27; novel and, 18–19; promise of, in geology, 100–102, 117; *Quarterly Review* and, 117–21
Euclid, 42–43, 45, 49
evolution: Charles Darwin and, 24, 68, 180–81, 195, 242, 248; cosmology and, 15, 64, 65; evolutionary epic and, 88–89, 100; graptolites and, 92; illegal, unstamped, radical press, and, 66; inappropriate for understanding of geological writing, 24–25, 180–81, 195–97, 204–7, 248, 263–64, 271–73; Lyell's response to, 64, 110–11; plot or storytelling

and, 17, 20, 23–25, 64, 65, 66, 67–69, 107–8, 183–84, 195–97. *See also* Darwin, Charles; Lamarck

fieldwork. *See under* practice, geological
Fingal's Cave, 83–84, 86, 95–98
Fitton, William, 57–58
Fleming, John, 115–17
Fluvialists, 115
Forbes, Edward, "The future of geology," 224–27
form: breakdown and confusion of, 7–9, 26–27, 49–55, 59, 107–8, 110–13, 133–34, 137–38, 161–62, 181–84, 190–94, 215–19, 224–28, 231–33, 234–35, 239–48, 264, 275; geological, and ambivalence about plot, 24–26, 43–46, 55, 56–58, 61–68, 100–103, 110–13, 131; geological, in Kingsley, 181–83, 190–94, 198–99, 208–20; geological pageants, processions and retrospects, 21, 204–7; literary, as mode of scientific reasoning, 36–39, 42–43, 46, 48, 61–68; literary, of Charles Lyell's earth, 95, 108–17, 120–30; maps and, 131, 133, 161; mathematical reasoning and, 42–44, 48, 50, 62–64; metonymic, of maps and columns, 136, 138–40, 149, 148–50, 151–52, 154–60, 166–67; novel and evolutionary, linked, 24, 65–66, 67–68; as ontological category, 17, 227; outdated habits of thought and, 123–25, 224–28; philological, related to geological processes, 155–56, 156–58, 165; realist novel and, ambivalent about plot, 19, 26, 49–55, 56, 69; seductive, in geology and fiction, 18–19, 43–46, 48–49, 214–15, 235–37; spectacular, in works of Charles Dickens and Richard Owen, 254, 261, 67; strati-

368 INDEX

graphic, of history, 163, 166, 167, 170–72, 175, 208–20; stratigraphic, related to literary, 17–18, 21, 43–46, 61–62, 67–68, 87, 92–93, 100–103, 110–13, 134, 138–40, 154–60, 166–67, 170–72, 183–84, 208–20, 228. *See also* epic; narrative; novel; romance

Fortunes of Nigel. See Scott, Walter

fossils: ammonites, 14; dating of strata using, 9, 87, 89, 161, 163, 209; debate about progress in earth history and, 101–2, 107–8, 110–11, 117; geological retrospects and, 205; graptolites, 92, 265; incompleteness of fossil record, 106, 110–11, 117, 130, 224–28; Palaeozoic, 31, 59–60, 74, 161, 226–28, 242; reconstruction of fossil saurians, 23, 251, 272–73; tertiary, 111–12; trilobites, 1, 2; *Two Years Ago* and, 215; *Wooers and Winners* and, 173. *See also* primeval lizards

Fourier, Jean Baptiste, 101

Fraser's Magazine, 65

Friday Club, 40–41

Gaskell, Elizabeth, 147–50

Geikie, Archibald, 72, 74, 75, 86, 93, 274

geognosy, 4–5

Geological Society of London, 4–7, 16, 19, 54–55, 61, 63, 65, 78, 97, 113, 225, 273, 274; *Geological Inquiries*, 5–7, 9; gentlemanly deportment of, 9–10, 12, 31–34, 54, 55–56, 61–62, 72, 115–17, 132, 179–80, 184–88; inductive methods of, 5–10, 61–64, 85, 104–6, 114, 182, 184–88; Charles Kingsley and, 179–80; literary publication and, 56–57, 60, 78, 88, 99, 156–57; mapmaking and, 131–32, 154; meetings of, and papers given at, 31–34, 41, 61, 72–73, 85, 108, 114, 115–17, 156–57, 169; nomenclature of, 155, 166–67; *Principles of Geology* and, 113–17, 222–23; rejection of cosmology, cosmogony by, 16–17, 19, 25–26, 43, 54–55, 56–57, 61–67, 71, 85, 101, 106, 115, 131, 182, 196, 222–23, 225, 274; Romantic idealism of, 9; Walter Scott and, 56–57, 66; stratigraphy by members of, 7, 9–10, 17, 59, 61–62, 70–71, 85, 105, 107, 131–32, 154, 156–57, 166–67, 182, 224

Geological Survey, 132–33, 160, 208, 224

geology: definition of, 4–7, 57, 104; golden or heroic age of, 1, 9, 74, 179, 184–88; as historical science, 20; popular participation in, 11–13. *See also* columns, geological; cosmogony; cosmology; geognosy; Geological Society of London; Geological Survey; maps; strata

Goldsmith, Oliver, 47–48, 54

Glaucus. See under Kingsley, Charles

granite, 1, 10, 34–36, 45, 77–78, 83, 123, 211–12, 312n84

graptolites, 92, 265. *See also* fossils

Gray, Asa, 180

Greenough, George Bellas, 115–16, 132–33

Guide to Geology. See Phillips, John

Guy Mannering. See Scott, Walter

high jinks, 31–33, 41, 49, 54, 55, 56

Hardy, Thomas, 24

Hawkins, Benjamin Waterhouse, 251, 254

History of the Inductive Sciences. See Whewell, William

Homer, 15, 100, 121–22, 126, 149. *See also* epic; mock-epic

Household Words. See under Dickens, Charles

Horace, 42–43, 46, 48, 109

Horner, Leonard, 57
Howitt, William, 144–45
Humboldt, Alexander von, 57, 134
Hunt, Robert, 254, 269, 271
Hunterian Museum, 251, 254, 265, 271–72
Hutton, James: cosmology of, 55, 56, 225, 227; fieldwork and, 45; Huttonians, 41; hypothetico-deductive methods of, 45, 274; literary forms of his theory of the earth, 34–40, 43, 45–46; on granite, 34–36, 77–79; *Two Years Ago* and, 216
Huxley, Thomas, 180

Illustrated London News, 11, 253, 257
Illustrations of the Geology of Yorkshire. See Phillips, John
induction, enumerative: catastrophes and, 107–8; 184–88, 190–2, 196–97, 199–201, 207, 223; Geological Society and, 4–7, 57, 63–64, 72, 90, 102, 104–5, 108, 187–88, 194, 274; imagination and, 90, 118–21, 226; Charles Kingsley on, 182–83, 187–92, 194, 196, 197–201, 207; Charles Lapworth and, 90, 93; problematic, 44–45, 105–6, 110–12, 189–93, 197–201, 207, 230–31; storytelling and, 45–46, 70, 72, 104–5, 196–97, 222–23; theory and, 45, 196–97. *See also* method, scientific
Ivanhoe. See Scott, Walter

Jameson, Robert, 39, 41, 77–78, 97
Jeffrey, Francis, 40, 41, 48, 54, 69
Johnston, J. F. W., 135–36

Keighley Mechanics' Institute, 145
Kenilworth. See Scott, Walter
Kingsley, Charles, 22, 89, 179–220, 247, 274; *Alton Locke*, 26, 181, 195–97, 199–207; breakdown and evasion of plot in, 183–84, 194, 201–7; as geologist, 26, 179–80, 182–83; fieldwork and, 179, 184–88, 190–94, 197–203; *Glaucus*, 180, 186, 187, 196; *Madame How and Lady Why*, 180, 183; geology vs. storytelling and, 181–83, 194, 207; induction and, 182–83, 187–92, 194, 196, 197–201, 207; on perspective, scale, 211–20; strata and, 182, 208, 208–20; response to Darwin, 180–81; response to *Vestiges of the Natural History of Creation*, 181, 188, 195–97, 204; *Town Geology*; 180, 186, 196, 198–99; *Two Years Ago*, 26, 181, 182, 196, 208–20; *Water-Babies, The*, 180; *Westward Ho!*, 212; *Yeast*, 26, 181, 182, 183–84, 188–94, 196

"Lady of the Lake." See Scott, Walter
Lagrange, 62. *See also* algebra
Lamarck, 65, 66, 117. *See also* evolution
Landseer, Edwin, 73–74, 77
Laplace, 62. *See also* algebra
Lapworth, Charles, 88
literature and science, relations of, and between, 13–19, 22, 23–27, 41–42, 70–72, 134, 140–41
Lockhart, John Gibson, 41, 56–57, 70, 114
Lyell, Charles, 87, 157, 159, 179, 180, 197, 224, 225, 232, 274; William Buckland and, 98–192; Lord Byron and, 19, 95–98, 127–30; deductive writer, not inductive scientist, 65; Dante and, 103, 117–19, 123–24, 125–26; *Elements of Geology*, 7–9; epic and mock-epic and, 19, 102–3, 117–26; evolution and, 24–27, 64, 66–67, 110–11, Geological Society of London and, 113–17; Homer and, 121–22; "Lines on Staffa,"

95–98, 102, 103, 105, 123–24, 281–82;
Niebuhr and, 119–21, 122, 126–27;
Quarterly Review and, 57, 102–3,
114–15, 117–19, 122; realism and, 24,
25, 222; undergraduate at Oxford,
95–103; "uniformitarianism" of, 19,
24, 25, 105–13; William Whewell
and, 222–23
Lyell, Charles, *Principles of Geology*,
24, 97, 98–99, 104–30, 156, 235,
264; anthropocentrism of, 108, 224;
Byronic form of, 127–30; "catastrophism" and "catastrophists" and,
106–10, 115–16; cautious style of,
115–17; geologist as antiquary and,
112–13; geologist as census collector
and, 112; "historical sketch of the
progress of geology," 16–17, 103–4,
106, 113–14, 115; incompleteness of
geological record and, 106, 110–11;
as mock-epic, 19, 122–26; narrative
(or anti-narrative) form of, 26, 105–
13, 117, 127–30; positive response
of colleagues to, 108, 114; progress
and, 111; rejection of cosmology, cosmogony, romance, 16–17, 18, 103–4,
106, 122–23, 125; temple of Jupiter
Serapis in, 110; temple of Vesta in,
109–10; tertiary strata in, 111–12;
vera causa methodology and, 104–5,
274; Werner and, 123–24

Macbeth Moir, David, 90
MacCulloch, John, 78–87, 93, 98
Mackenzie, George Steuart, 40, 54
Macpherson, James, 156
Madame How and Lady Why. *See*
Kingsley, Charles
Mantell, Gideon, 60, 204, 251
maps: geological, 2, 5, 19, 21, 26, 131,
164; Arran, Isle of, 87, 366; of
Brontë country, 139–40, 141–42,
146, 150; correlation and, 9, 87, 134,
161, 209; of Devon, Wales, and the
Eifel, correlated, 208–11, 216–20;
of England and Wales, 132, 154,
163; ethnographic, 137, 163, 164;
Geological Atlas of Great Britain,
134; *Geological Map of Europe*, 133;
Geological Society and, 4–6, 7, 35,
131–32, 224; of Geological Survey,
132–33, 208; inferior to description,
79–80; Charles Lapworth and, 89;
of metonym and, 135–40, 159–60,
164, 166; mineralogical, 7, 131;
John Phillips and, 161–62, 167; as
predictors of national destiny, 137;
plots and, 131–33, 135, 136, 137–38,
222, 274; realism of, in *Wooers and
Winners*, 173–74; "romance" and,
especially in illustration, 79–83,
85, 88; scale of, 80, 89, 161, 213–20;
stratigraphical boundaries and,
161–62, 168; surface boundaries in,
133–34, 138, 140, 150, 152–54, 155,
161–67, 168, 170–72, 173–74; travel
and tourism and, 133; Yorkshire,
134–39, 154, 157, 158–60, 161–67
"Marmion." *See* Scott, Walter
masculinity, alleged importance of, to
geology, 10, 69, 70–72, 75, 184–94,
198–200, 202–7, 225, 236–37,
238–39
Maurice, F. D., 180
*Memorial by the Trustees of Cowgill
Chapel*. *See under* Sedgwick, Adam
metamorphosis, 9, 106, 155–56, 208
method, scientific: of analysis, 43–45,
46, 62–64; of correlation, 9, 87,
134, 161, 209; of dating rocks
(lithological), 87, 134, 159; of dating
rocks (palaeontological), 87, 110–13,
161, 163; of deduction, 4, 42, 45,
63–64, 274; geologist as antiquary

method, scientific (cont.)
and, 112–13; geologist as census
collector and, 112; imagination and,
14, 18, 37–39, 43–44, 62, 66–68, 75,
79–80, 88, 89–90, 118–21, 124–25,
127, 162, 225–28; literary style and,
63–64, 79–83; of Lyell, Charles,
as opposed to Geological Society
of London, 108, 110–11, 113, 222,
228; of Niebuhr's history, 120–21,
127; of synthesis, 42, 45, 46, 63, 64;
unobservability of the geological
past and, 104, 105–6, 110–13, 120–21,
122–25; *vera causa*, 105–6, 112–13,
274. *See also* induction, enumerative
Miall, L. C., 159
Miller, Hugh, 93, 204, 207, 257, 287n73
Mill on the Floss. See under Eliot,
George
Milton, 15, 69, 100, 102, 120, 125, 126.
See also epic
mineralogy, 4, 14, 77, 78, 97
"Minstrelsy of the Scottish Border." *See*
Scott, Walter
mock-epic, 19, 121–30. *See also* epic
Montgomery, James, 225
Mozley, Thomas, 99–100
Murchison, Roderick: admiration of
Walter Scott, 58, 73–76; on Arran,
Isle of, 77–78, 85–88; Cambrian-
Silurian dispute and, 58–61, 77, 88–
89, 92–93, 186; criticised for use of
"Palaeozoic," 226, 227–28; Devonian
controversy and, 208–10; Charles
Dickens and, 249; field practice of,
and Walter Scott, 77–88; *Geological
Map of Europe*, 133; Geological
Society of London and, 115–16;
Charles Kingsley and, 181, 186, 189,
190, 197; Edwin Landseer and,
73–74; on prestige of the novel, 58;
romantic geological style of, 73–76;
93, 274; *Silurian System*, 59, 61, 73,
166–67, 181, 189, 199, 249; social and
literary status of, 59–60, 186

narrative: antinarrative in Charles Lyell,
95–97, 107–13, 117, 122–30; anti-
epic in Niebuhr, 119–21; cautious,
circumscribed, or broken, in fiction,
16, 18–19, 26–27, 49–55, 69, 71, 72,
85, 148, 151–52, 174–77, 181–83,
183–84, 190–94, 195–97, 204–7,
221–24, 228–46, 275–76; cautious,
circumscribed, or broken, in geology, 16, 18–21, 22, 26–27, 57–59,
61–62, 100–102, 103–6, 111, 125,
156–58, 204–7, 225–28; geological
maps and, 136–37, 140–41, 149–52,
170–72, 176; importance of, for
geological writers, 14–15; *Origin
of Species* and, 68; plot and, 20–21;
"popular" geological writing and,
17, 22; of progress, 17, 100, 103, 106,
111, 204–7, 216, 219–20; romance
of fieldwork and, 72, 76–77, 93–94,
184–89; stratigraphic form and,
264; *Vestiges of the Natural History
of Creation* and, 64–65, 67–69,
188, 195–97, 204; visual forms of
Dickens's fiction and, 250–51, 262,
263–65. *See also* plot
natural history societies, 11
Newton, Isaac, 37, 43, 62, 63
Niebuhr, Barthold Georg, 119–21, 126,
127
Noah's Flood, 102, 106, 107, 115, 222,
235, 238, 243, 263, 266, 270
novel: evolutionary cosmology and,
65, 66–69, 195–97; as feminine,
48–49; geological breakdown and,
26–27, 49–54, 181–84, 190–94, 195,
208–20, 231–33, 240–46, 274–77;
geological "plots" and, 14, 23–26,

183–84, 194, 195–97, 204–7, 221–24, 232–33, 235–37, 245–46, 248, 263–64; geologists' reading of, 59, 69–70, 73, 86; as irreconcilable with geological methods of induction, 181–83, 191–94; print culture and, 65–66, 83–85; as prosaic, lacking in romance, 76–77, 83–85, 169, 173–76; as pursuit of the weak, idle, feminine, 48–49, 51, 54, 67, 69–71; as rival to science, 58, 60; regionalism and, 139–43, 147–53, 173–76; reinvented by Walter Scott, 19, 46–55, 56–57, 58; romance plots and, 16–19, 49–55, 88, 228–29, 237, 238–40, 244; spectacle and, 250–51, 261–62, 264–65, 266, 268, 271, 273; synthetic sweep of, 70. *See also* Banks, Isabella, *Wooers and Winners*; Brontë, Emily; Dickens, Charles; Eliot, George; Kingsley, Charles; Scott, Walter

observation. *See under* practice, geological
Origin of Species. *See* Darwin, Charles
Oxford Movement, 99–100
Our Mutual Friend. See under Dickens, Charles
Owen, Richard, 22, 31, 33, 251–54, 266, 271–73

panorama, 154, 248–49, 255–57, 259, 260, 261, 265–66, 264–65, 272
Pater, Walter, 23
Phillips, John, 160, 175; Brigantean Yorkshire, 162–63, 165–67; *Guide to Geology*, 132, 140; *Illustrations of the Geology of Yorkshire*, 152–54, 161; map of Yorkshire, 161–62, 167; *Rivers, Mountains and Sea-coast of Yorkshire, The*, 152, 154, 161, 162–63, 165–67, 168, 172, 371

Pictures from Italy. See under Dickens, Charles
Playfair, John, 36–46, 50, 54, 55, 57, 61, 62, 66, 67, 69, 72, 77–78, 274
plot: avoidance of, in geological pageants and retrospects, 21, 204–7; catastrophism and, 24–26, 107–10, 245–56; cosmology, cosmogony, and, 15, 19–21, 43, 54, 61, 64–69, 88, 89, 100–102, 106, 127–30; evolutionary, 17, 64–69, 183–84, 195–97; fragmentation and breakdown of, in geology and fiction, 26–27, 49–54, 100–102, 105–6, 110–13, 181–84, 190–94, 195, 231–33, 240–46, 274–77; geological maps and, 131, 135, 136, 222, 274; induction, collection, facts, and, 26, 50–53, 93, 104–6, 110–13, 181–83, 190–94, 196, 199–201, 204–7, 230–31; mock-epic and, 123–30; narrative as distinct from, 20–22; perceived lack of, in late novels of Charles Dickens, 250–51; problematic, for geology, 22, 26, 274; romance and, 16–17, 18, 20, 238; seductive, 48–49, 54, 234–38; Walter Scott and, 49–54; self-determining nature of, 15–17, 18, 20, 184, 224–28, 228–29, 235–37, 239; spectacle and, 263–65, 273; strata and, 27, 85, 100–102, 208–20, 224–28, 275; uniformitarianism and, 24–26, 106–13, 222–24. *See also* narrative
Pope, Alexander, 124–25
poststructuralism, 22, 23
practice, geological: classification, 4, 13, 31, 35–36, 58–61, 63, 74, 77–79, 88–90, 92–93, 110–13, 182–83, 186–87, 208–10, 219, 221, 224–28, 229, 233; collecting, 2, 4, 6, 10, 11–12, 13, 18, 22, 26, 50, 85, 89, 92, 93, 97,

practice, geological (cont.)
110, 173, 174, 179, 180, 187, 192, 196;
description, 4, 6–7 12, 13, 14, 26,
44–45, 52–53, 76, 78–83, 85, 93, 94,
128, 140, 143, 149, 152–54, 156–60,
162, 163, 174, 175–76, 182, 207,
208–20, 229–30, 251, 275; fieldwork, 4, 17, 70–72, 89, 179, 184–88,
197–203; literature as a mode of,
2–4, 13–19 24, 26, 42–47, 61–64,
67–68, 70–72, 74, 79–83, 85–88,
89–94, 100–103, 110–13, 114–17,
119–21, 122–30, 181–84, 186–92,
194, 196, 197–201, 207, 224–28, 249,
274–76; observation, 4–7, 22, 57, 63,
64, 196, 199, 214–15, 229–30, 244,
251. *See also* maps
Pre-Raphaelites, 212, 214–15
Principles of Geology. *See* Lyell, Charles,
Principles of Geology
Primary rocks, 1, 10, 34–36, 76, 77–78,
83, 111, 123, 163, 164–65, 166–67, 193,
211–12, 228, 312n84
primeval lizards: 1, 14, 18; ichthyosaurus, 261; iguanodon, 15, 251;
megalosaurus, 251, 262–63, 266, 272;
mylodon, 15. *See also* fossils

Quarterly Review, 49, 52, 56–57, 65, 67,
72, 102–3, 114–15, 117–21, 122

realism, 23, 26; breakdown of plot and,
181–83, 190–94, 195, 203, 228–29,
231, 237, 239, 240, 243–47; "catastrophism" and, 221–24, 264; induction
and, in Charles Kingsley and George
Eliot, 181–83, 190–94, 197–207,
230; romance, epic, and, 18–19, 25,
27; uniformitarianism and, 21–26,
221–24, 232–33, 264; Walter Scott
and, 49, 66; *Wooers and Winners*
and, 140–41, 173–76

regionalism: boundaries and, 134, 140,
149–50; place and time in, 150–51,
152, 166; in *Wuthering Heights*,
139–43, 147–53, 176; strata and, 166;
Wooers and Winners and, 173–76;
metonymic relations between
people and place in, 176, 135–36,
138–40, 147–52, 154–60, 166–67,
176, 231–32; stratigraphic imagination and, 175–76
Reynolds, J. M. W., 133
Riehl, W. H., 229–33
*Rivers, Mountains and Sea-Coast of
Yorkshire, The*. *See* Phillips, John
Rob Roy. *See* Scott, Walter
romance: "catastrophism" and, 24–25,
221–24, 263–65; enchantment of,
84–85, 86–87, 89, 90, 93–94; as
deceitful, too imaginative, 16, 19,
49, 67–68, 86, 113, 119, 188, 223,
244; of fieldwork, 71–72, 74–77,
179, 184–88, 197–203; as immersion
in violent, heroic past, contrasted
to the present, 72, 74–77, 79–88,
85, 86–87, 89, 90, 93, 119, 173–74;
medieval and chivalric, 14–15, 49,
58, 71–72, 93–94, 119, 184–88, 189,
185–86, 189–90, 197–203; plot and,
16–17, 18, 19, 20, 119, 223, 228–29,
230, 238, 239; promoting science
with, 18, 57–58, 87; realism and, 19,
25, 27; as wildness, immensity, and
contrast in nature, 15, 75, 79–81, 90,
93–94, 139–40, 148–50, 153–54, 159,
173–74; wonder of science and, 24/
Römische Geschichte. *See* Niebuhr,
Barthold Georg
Rossetti, Dante Gabriel, 23, 212. *See also*
epic
Rossetti Gabriele, 117–19, 120, 121, 122
Royal Society of Edinburgh, 39, 46–47,
54, 56

Royal Society of London, 16, 57, 62
Ruskin, John, 95, 212, 213

saurians. *See* primeval lizards
science wars, 22
Scott, Walter, 19, 31, 34, 39–40, 46–55, 58, 69, 70, 93–94, 114, 156, 247, 274; *Antiquary, The*, 52–53, 72; Charles Darwin and, 56; elitism and, 66; *Fortunes of Nigel*, 51; *Guy Mannering*, 31–34; "high jinks," 31–32; *Ivanhoe*, 72; *Kenilworth*, 65, 72; "Lady of the Lake," 75, 121; Edwin Landseer and, 73–74; Charles Lapworth and, 90–93; "Lord of the Isles," 77, 91; Charles Lyell and, 121; John MacCulloch and, 83–85; "Marmion," 91; "Minstrelsy of the Scottish Border," 91; Roderick Murchison and, 73, 76–77; paratexts (prefaces, epistles, introductions to Magnum Opus), 51–54, 244; as president of Royal Society of Edinburgh, 46–47; *Rob Roy*, 31, 84, 85, 86; 54; romance and, 72; Scott monument, 56; Adam Sedgwick and, 70–71, 76–77, 115–16, 119, 140; "St. Ronan's Well," 46; tourism and, 76–77; *Vestiges of the Natural History of Creation*, 65; Walter Scott Way, 91; *Waverley*, 49–50, 51, 72
Scenes of Clerical Life. *See under* Eliot, George
Secondary rocks, 35, 111, 163; chalk, 239, 241–42; of Devon, 208; of Yorkshire, 134–35, 139, 153, 154, 158, 161, 163–64
sections, columnar, 131–32; longitudinal, 154, 159, 169; by John Phillips, 161; transverse, 131, 154, 156–57, 159, 169; traverse, 131, 132, 167, 365, 367

Sedgwick, Adam: anti-slavery of, 143–44; Cambrian-Silurian dispute and, 58–61, 77, 88–89, 92–93, 186; "catastrophist," 108; Devonian controversy and, 208–10; on dialect terms in geology, 154–56, 157–58, 163; *Discourse on the Studies of the University of Cambridge*, 63, 196; on literary style and science, 59–61, 67–69, 154–58, 163; oral performances of, 59, 71–73; *Memorial by the trustees of Cowgill Chapel*, 146–47, 155, 173; mountain geometry, 62, 63–64; on John Phillips, 161; plot of *Wuthering Heights* and, 143–45; response to *Principles of Geology*, 108; review of *Vestiges of the Natural History of Creation*, 67–69, 225; stratigraphic geologist, 59, 105, 154, 156–58; "Three letters upon the geology of the Lake District," 87–88, 145, 158, 162, 168; Walter Scott and, 69–72, 76–77, 77–88; West Riding geology and, 154, 156–60, 163–64, 167, 169; *Wooers and Winners* and, 168–69; as Yorkshireman, 58, 145, 152, 154–55, 157–58
Settle Cave Exploration Committee, 169
Shakespeare, William, 70, 72
Silurian System, The. *See under* Murchison, Roderick
Smith, Sydney, 57, 67
Smith, William, 132, 152, 154, 160, 163, 369, 370–71
Society of Antiquaries, 169
Society of Natural Science (Chester), 180
Spasmodic poetry, 196, 214–15
Spencer, Herbert, 227–28, 242
Sprat, Thomas, 16. *See also* Royal Society of London

Stokes, George, 32
Stowe, Harriet Beecher, 70, 143, 143–44
strata: arbitrariness in classifications of, 161, 226, 227; architecture and, 3, 101–2, 110, 131, 132–33, 136, 154, 198; Cambrian, 59, 88–89, 92–93, 163, 166, 167, 208–10, 219; Cambrian-Silurian dispute, 58–61, 75, 77, 88–89, 92–93, 186; catastrophe and, 107–8; cave geology and, 170–71; as central research program of the Geological Society, 7, 9–10, 17, 57, 59, 61–62, 70–71, 85, 105, 107, 131–32, 154, 156–57, 166–67, 224; as chapters of earth history, 1, 3, 67, 106, 111, 182–83; Charles Kingsley and, 27, 182, 219–20; correlation of, 9, 87, 134, 161, 209; of Devon, 208–9; Devonian, 163, 167, 216–20; Devonian controversy, 208–10; of Dobbs Linn, 92; fieldwork and, 89, 187–88; Geological Survey, 132–33 and; of granite, 78–79; strata-hunting as middle-class leisure activity, 11, 12, 133, 266; incompleteness, confusion, violence of, 7–9, 27, 38, 78–79, 80, 92–93, 106, 107, 110–12, 121, 133, 134, 139, 153–54, 156–59, 162, 187, 208–10, 225, 228; of the Isle of Arran, 36, 45, 77–88, 365; lithology and, 87, 134, 159; narrative and, 57, 100–102, 106, 110–12, 264; maps and, 131–32, 133–34, 135, 138, 140, 150, 152–54, 155, 161–67, 168, 170–72, 173–74; as metaphor for history of language, archaeology, history, 100, 170–71, 211; novel writing opposed to, 182–83; Ordovician, 88–93; palaeontology and, 87, 110–13, 161, 163; Palaeozoic, 31, 59–60, 74, 161, 226–28, 242; place and, 138–40, 154–60, 166–67, 170–72; representations of the, 5, 6, 61–62, 87, 90, 92–93, 106, 111–12; Sedgwick's mountain geometry and, 62, 63–64; Silurian, 59, 74, 88–89, 92–93, 161, 163, 166, 167, 186, 208–10, 226, 227–28; and stratigraphy as chivalric, heroic, romantic, 93–94, 182–83, 186; tertiary, 111–12, 163, 180, 241; Abraham Gottlob Werner and interpretation of, 37, 39, 44–45, 78, 134; in Yorkshire, 134–38, 139, 153, 154, 156–57, 158–59, 161–67, 168. *See also* maps; columns, geological
St. Ronan's Well. See Scott, Walter
style: and scientific reasoning, 16, 42–46, 60–61; and William Buckland, 98–100; and Geological Society of London, 31–33, 71–73, 115–17; and James Hutton, 36; and Lamarck, 66; and Charles Lyell, 95–97, 98, 113–17, 122–25, 127–30; and John MacCulloch, 79–85; and Roderick Murchison, 60–61, 73–74, 75–76, 85–86, 88, 93, 297–98n83; and John Phillips, 152–54, 161–62, 166–67; and John Playfair, 36–40, 42–46; and quarterly reviews, 56–58, 69; and scientific method, 63–64; and Walter Scott, 46–48, 49–55, 56, 85; and Adam Sedgwick, 59–64, 85–86, 87–88, 71–73, 154–8, 163, 297–98n83; of *Vestiges of the Natural History of Creation*, 64–68, 196–97
Sunday League, 254
Swift, Jonathan, 60, 70
Swinburne, Algernon Charles, 23
synthesis. *See* method, scientific

Taylor, Isaac, 155, 163–64, 166–67
tertiary strata, 111–12, 163, 180, 241
Thackeray, William Makepeace, 58, 65, 69

Theatre Royal Edinburgh, 40
theories of the earth. *See* cosmogony; cosmology
"Three letters upon the geology of the Lake District." *See under* Sedgwick, Adam
toponyms, 155–56, 163–64, 209
Times, 60
transition rocks, 78
transmutation. *See* evolution; Lamarck; *Vestiges of the Natural History of Creation*
trilobites, 1, 2. *See also* fossils

uniformitarianism, 19, 104, 105, 106; "catastrophism" and "catastrophists" and, 24–26, 106–13, 115–16, 221; Charles Dickens and, 263–64; George Eliot and, 24–25, 221–24, 232–33, 235, 238, 246; misunderstood as plot of earth history, 24–26, 106–13, 222–24; realism and, 24–26, 221–23, 246; *vera causa* methodology and, 104–5, 274; William Whewell and, 104–5, 222–23
Ure, Andrew, 61

vera causa. *See under* method, scientific
Vestiges of the Natural History of Creation, 60, 64–68, 181, 188, 195–97, 204, 225, 237, 249, 263–64
Vicar of Wakefield, 47–48, 54
Virgil, 15, 100, 125. *See also* epic

Waverley. *See under* Scott, Walter
Wellington College, 186
Werner, Abraham Gottlob, 34–35, 37, 39, 44–45, 56, 78, 97, 123, 134, 225, 227; Wernerians, 40, 41, 42

Wernerian Natural History Society, 39, 41, 97
Westminster Review, 65, 138–40, 224–27, 228, 229–31
Whewell, William, 62–63, 64, 104–5, 113–14, 119, 222–23, 225, 232
Wooers and Winners. *See* Banks, Isabella, *Wooers and Winners*
Wordsworth, William, 145, 158, 162
Wuthering Heights. *See* Brontë, Emily
Wyld's Globe, 265

Yorkshire: Brigantian, 162–63, 165–67; center of industrial significance, 134–35, 139–40, 148, 151, 152, 166; decline of yeomen in, 146–48, 152; geological boundaries of, 141, 152–54, 158–59, 161–62, 168; metonym and, 135–36, 138–39, 147–52, 154–60; John Phillips and, 160; "romance" of, 139–40, 153–54; as rough, rugged, unhewn, 139–40, 149–51, 152, 156, 157–58; secondary strata of, 134–35, 139, 153, 154, 158, 163–64, 168; Adam Sedgwick and, 58, 140, 143–45, 146–47, 152, 154–60; site of Brontë novels, 135, 140–52; slaveholding economy of, 143–45; William Smith, and, 132; strata of, 8–9, 134–38, 139, 153, 154, 156–57, 158–59, 161–67, 168, 208; as symbolic heart of Britain, 134–35, 138–39, 140, 149, 150, 152; and *Wooers and Winners*, 167–69, 172–76
Yorkshire Geological and Polytechnic Society, 159, 169, 174
Yorkshire Philosophical Society, 174
Yorkshire Union of Mechanics' Institutes, 169